T0250226

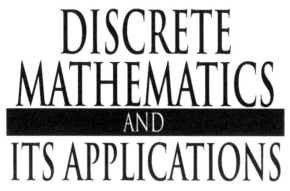

DISCRETE
MATHEMATICS
AND
ITS APPLICATIONS

Series Editor

Kenneth H. Rosen, Ph.D.

AT&T Laboratories
Middletown, New Jersey

Continued Titles

Network Reliability: Experiments with a Symbolic Algebra Environment,
Daryl D. Harms, Miroslav Kraetzl, Charles J. Colbourn, and John S. Devitt

RSA and Public-Key Cryptography
Richard A. Mollin

Quadratics, *Richard A. Mollin*

Verificaton of Computer Codes in Computational Science and Engineering,
Patrick Knupp and Kambiz Salari

DISCRETE MATHEMATICS AND ITS APPLICATIONS
Series Editor KENNETH H. ROSEN

FUNDAMENTALS of INFORMATION THEORY and CODING DESIGN

Roberto Togneri
Christopher J.S. deSilva

CHAPMAN & HALL/CRC

A CRC Press Company
Boca Raton   London   New York   Washington, D.C.

Published in 2003 by
Chapman & Hall/CRC
Taylor & Francis Group
6000 Broken Sound Parkway NW, Suite 300
Boca Raton, FL 33487-2742

No claim to original U.S. Government works
Printed in the United States of America on acid-free paper
10 9 8 7 6 5 4 3 2

International Standard Book Number-10: 1-58488-310-3(Hardcover)
International Standard Book Number-13: 978-1-58488-310-4 (Hardcover)
Library of Congress catalog number: 2002191160

Library of Congress Cataloging-in-Publication Data

Catalog record is available from the Library of Congress

Taylor & Francis Group
is the Academic Division of Informa plc.

Visit the Taylor & Francis Web site at
http://www.taylorandfrancis.com

and the CRC Press Web site at
http://www.crcpress.com

Preface

What is information? How do we quantify or measure the amount of information that is present in a file of data, or a string of text? How do we encode the information so that it can be stored efficiently, or transmitted reliably?

The main concepts and principles of information theory were developed by Claude E. Shannon in the 1940s. Yet only now, and thanks to the emergence of the information age and digital communication, are the ideas of information theory being looked at again in a new light. Because of information theory and the results arising from coding theory we now know how to quantify information, how we can efficiently encode it and how reliably we can transmit it.

This book introduces the main concepts behind how we model information sources and channels, how we code sources for efficient storage and transmission, and the fundamentals of coding theory and applications to state-of-the-art error correcting and error detecting codes.

This textbook has been written for upper level undergraduate students and graduate students in mathematics, engineering and computer science. Most of the material presented in this text was developed over many years at The University of Western Australia in the unit Information Theory and Coding 314, which was a core unit for students majoring in Communications and Electrical and Electronic Engineering, and was a unit offered to students enrolled in the Master of Engineering by Coursework and Dissertation in the Intelligent Information Processing Systems course.

The number of books on the market dealing with information theory and coding has been on the rise over the past five years. However, very few, if any, of these books have been able to cover the fundamentals of the theory without losing the reader in the complex mathematical abstractions. And fewer books are able to provide the important theoretical framework when discussing the algorithms and implementation details of modern coding systems. This book does not abandon the theoretical foundations of information and coding theory and presents working algorithms and implementations which can be used to fabricate and design real systems. The main emphasis is on the underlying concepts that govern information theory and the necessary mathematical background that describe modern coding systems. One of the strengths of the book are the many worked examples that appear throughout the book that allow the reader to immediately understand the concept being explained, or the algorithm being described. These are backed up by fairly comprehensive exercise sets at the end of each chapter (including exercises identified by an * which are more advanced or challenging).

The material in the book has been selected for completeness and to present a balanced coverage. There is discussion of cascading of information channels and additivity of information which is rarely found in modern texts. Arithmetic coding is fully explained with both worked examples for encoding and decoding. The connection between coding of extensions and Markov modelling is clearly established (this is usually not apparent in other textbooks). Three complete chapters are devoted to block codes for error detection and correction. A large part of these chapters deals with an exposition of the concepts from abstract algebra that underpin the design of these codes. We decided that this material should form part of the main text (rather than be relegated to an appendix) to emphasise the importance of understanding the mathematics of these and other advanced coding strategies.

Chapter 1 introduces the concepts of entropy and information sources and explains how information sources are modelled. In Chapter 2 this analysis is extended to information channels where the concept of mutual information is introduced and channel capacity is discussed. Chapter 3 covers source coding for efficient storage and transmission with an introduction to the theory and main concepts, a discussion of Shannon's Noiseless Coding Theorem and details of the Huffman and arithmetic coding algorithms. Chapter 4 provides the basic principles behind the various compression algorithms including run-length coding and dictionary coders. Chapter 5 introduces the fundamental principles of channel coding, the importance of the Hamming distance in the analysis and design of codes and a statement of what Shannon's Fundamental Coding Theorem tells us we can do with channel codes. Chapter 6 introduces the algebraic concepts of groups, rings, fields and linear spaces over the binary field and introduces binary block codes. Chapter 7 provides the details of the theory of rings of polynomials and cyclic codes and describes how to analyse and design various linear cyclic codes including Hamming codes, Cyclic Redundancy Codes and Reed-Muller codes. Chapter 8 deals with burst-correcting codes and describes the design of Fire codes, BCH codes and Reed-Solomon codes. Chapter 9 completes the discussion on channel coding by describing the convolutional encoder, decoding of convolutional codes, trellis modulation and Turbo codes.

This book can be used as a textbook for a one semester undergraduate course in information theory and source coding (all of Chapters 1 to 4), a one semester graduate course in coding theory (all of Chapters 5 to 9) or as part of a one semester undergraduate course in communications systems covering information theory and coding (selected material from Chapters 1, 2, 3, 5, 6 and 7).

We would like to thank Sean Davey and Nishith Arora for their help with the LATEX formatting of the manuscript. We would also like to thank Ken Rosen for his review of our draft manuscript and his many helpful suggestions and Sunil Nair from CRC Press for encouraging us to write this book in the first place!

Our examples on arithmetic coding were greatly facilitated by the use of the conversion calculator (which is one of the few that can handle fractions!) made available by www.math.com.

The manuscript was written in LaTeX and we are indebted to the open source software community for developing such a powerful text processing environment. We are especially grateful to the developers of LyX (www.lyx.org) for making writing the document that much more enjoyable and to the makers of xfig (www.xfig.org) for providing such an easy-to-use drawing package.

Roberto Togneri
Chris deSilva

Contents

Chapter 1

Entropy and Information

1.1 Structure

Structure is a concept of which we all have an intuitive understanding. However, it is not easy to articulate that understanding and give a precise definition of what structure is. We might try to explain structure in terms of such things as regularity, predictability, symmetry and permanence. We might also try to describe what structure is not, using terms such as featureless, random, chaotic, transient and aleatory.

Part of the problem of trying to define structure is that there are many different kinds of behaviour and phenomena which might be described as structured, and finding a definition that covers all of them is very difficult.

Consider the distribution of the stars in the night sky. Overall, it would appear that this distribution is random, without any structure. Yet people have found patterns in the stars and imposed a structure on the distribution by naming constellations.

Again, consider what would happen if you took the pixels on the screen of your computer when it was showing a complicated and colourful scene and strung them out in a single row. The distribution of colours in this single row of pixels would appear to be quite arbitrary, yet the complicated pattern of the two-dimensional array of pixels would still be there.

These two examples illustrate the point that we must distinguish between the presence of structure and our perception of structure. In the case of the constellations, the structure is imposed by our brains. In the case of the picture on our computer screen, we can only see the pattern if the pixels are arranged in a certain way.

Structure relates to the way in which things are put together, the way in which the parts make up the whole. Yet there is a difference between the structure of, say, a bridge and that of a piece of music. The parts of the Golden Gate Bridge or the Sydney Harbour Bridge are solid and fixed in relation to one another. Seeing one part of the bridge gives you a good idea of what the rest of it looks like.

The structure of pieces of music is quite different. The notes of a melody can be arranged according to the whim or the genius of the composer. Having heard part of the melody you cannot be sure of what the next note is going to be, leave alone any other part of the melody. In fact, pieces of music often have a complicated, multi-layered structure, which is not obvious to the casual listener.

In this book, we are going to be concerned with things that have structure. The kinds of structure we will be concerned with will be like the structure of pieces of music. They will not be fixed and obvious.

1.2   Structure in Randomness

Structure may be present in phenomena that appear to be random. When it is present, it makes the phenomena more predictable. Nevertheless, the fact that randomness is present means that we have to talk about the phenomena in terms of probabilities.

Let us consider a very simple example of how structure can make a random phenomenon more predictable. Suppose we have a fair die. The probability of any face coming up when the die is thrown is 1/6. In this case, it is not possible to predict which face will come up more than one-sixth of the time, on average.

On the other hand, if we have a die that has been biased, this introduces some structure into the situation. Suppose that the biasing has the effect of making the probability of the face with six spots coming up 55/100, the probability of the face with one spot coming up 5/100 and the probability of any other face coming up 1/10. Then the prediction that the face with six spots will come up will be right more than half the time, on average.

Another example of structure in randomness that facilitates prediction arises from phenomena that are correlated. If we have information about one of the phenomena, we can make predictions about the other. For example, we know that the IQ of identical twins is highly correlated. In general, we cannot make any reliable prediction about the IQ of one of a pair of twins. But if we know the IQ of one twin, we can make a reliable prediction of the IQ of the other.

In order to talk about structure in randomness in quantitative terms, we need to use probability theory.

1.3   First Concepts of Probability Theory

To describe a phenomenon in terms of probability theory, we need to define a *set of outcomes*, which is called the *sample space*. For the present, we will restrict consideration to sample spaces which are finite sets.

DEFINITION 1.1 Probability Distribution   *A probability distribution on a sample space* $S = \{s_1, s_2, \ldots, s_N\}$ *is a function P that assigns a probability to each outcome in the sample space. P is a map from S to the unit interval, $P : S \to [0, 1]$, which must satisfy $\sum_{i=1}^{N} P(s_i) = 1$.*

DEFINITION 1.2 Events   *Events are subsets of the sample space.*

We can extend a probability distribution P from S to the set of all subsets of S, which we denote by $\mathcal{P}(S)$, by setting $P(E) = \sum_{s \in E} P(s)$ for any $E \in \mathcal{P}(S)$. Note that $P(\emptyset) = 0$.

An event whose probability is 0 is impossible and an event whose probability is 1 is certain to occur.

If E and F are events and $E \cap F = \emptyset$ then $P(E \cup F) = P(E) + P(F)$.

DEFINITION 1.3 Expected Value   *If $S = \{s_1, s_2, \ldots, s_N\}$ is a sample space with probability distribution P, and $f : S \to V$ is a function from the sample space to a vector space V, the expected value of f is $\bar{f} = \sum_{i=1}^{N} P(s_i) f(s_i)$.*

NOTE   We will often have equations that involve summation over the elements of a finite set. In the equations above, the set has been $S = \{s_1, s_2, \ldots, s_N\}$ and the summation has been denoted by $\sum_{i=1}^{N}$. In other places in the text we will denote such summations simply by $\sum_{s \in S}$.

1.4   Surprise and Entropy

In everyday life, events can surprise us. Usually, the more unlikely or unexpected an event is, the more surprising it is. We can quantify this idea using a probability distribution.

DEFINITION 1.4 Surprise   *If E is an event in a sample space S, we define the surprise of E to be $s(E) = -\log(P(E)) = \log(1/P(E))$.*

Events for which $P(E) = 1$, which are certain to occur, have zero surprise, as we would expect, and events that are impossible, that is, for which $P(E) = 0$, have infinite surprise.

Defining the surprise as the negative logarithm of the probability not only gives us the appropriate limiting values as the probability tends to 0 or 1, it also makes surprise additive. If several independent events occur in succession, the total surprise they generate is the sum of their individual surprises.

DEFINITION 1.5 Entropy   *We can restrict the surprise to the sample space and consider it to be a function from the sample space to the real numbers. The expected value of the surprise is the* entropy *of the probability distribution.*

If the sample space is $S = \{s_1, s_2, \ldots, s_N\}$, with probability distribution P, the entropy of the probability distribution is given by

$$H(P) = -\sum_{i=1}^{N} P(s_i) \log(P(s_i)). \tag{1.1}$$

The concept of entropy was introduced into thermodynamics in the nineteenth century. It was considered to be a measure of the extent to which a system was disordered. The tendency of systems to become more disordered over time is described by the *Second Law of Thermodynamics*, which states that the entropy of a system cannot spontaneously decrease. In the 1940's, Shannon [6] introduced the concept into communications theory and founded the subject of information theory. It was then realised that entropy is a property of any stochastic system and the concept is now used widely in many fields. Today, information theory (as described in books such as [1], [2], [3]) is still principally concerned with communications systems, but there are widespread applications in statistics, information processing and computing (see [2], [4], [5]).

Let us consider some examples of probability distributions and see how the entropy is related to predictability. First, let us note the form of the function $s(p) = -p \log(p)$ where $0 < p \leq 1$ and log denotes the logarithm to base 2. (The actual base does not matter, but we shall be using base 2 throughout the rest of this book, so we may as well start here.) The graph of this function is shown in Figure 1.1.

Note that $-p \log(p)$ approaches 0 as p tends to 0 and also as p tends to 1. This means that outcomes that are almost certain to occur and outcomes that are unlikely to occur both contribute little to the entropy. Outcomes whose probability is close to 0.4 make a comparatively large contribution to the entropy.

EXAMPLE 1.1

$S = \{s_1, s_2\}$ with $P(s_1) = 0.5 = P(s_2)$. The entropy is

$$H(P) = -(0.5)(-1) - (0.5)(-1) = 1.$$

In this case, s_1 and s_2 are equally likely to occur and the situation is as unpredictable as it can be.   □

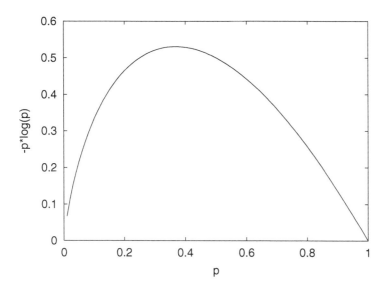

FIGURE 1.1
The graph of $-p\log(p)$.

EXAMPLE 1.2

$S = \{s_1, s_2\}$ with $P(s_1) = 0.96875$, and $P(s_2) = 0.03125$. The entropy is

$$H(P) = -(0.96875)(-0.0444) - (0.03125)(-5) \approx 0.20.$$

In this case, the situation is more predictable, with s_1 more than thirty times more likely to occur than s_2. The entropy is close to zero. □

EXAMPLE 1.3

$S = \{s_1, s_2\}$ with $P(s_1) = 1.0$, and $P(s_2) = 0.0$. Using the convention that $0 \log(0) = 0$, the entropy is 0. The situation is entirely predictable, as s_1 always occurs. □

EXAMPLE 1.4

$S = \{s_1, s_2, s_3, s_4, s_5, s_6\}$, with $P(s_i) = 1/6$ for $i = 1, 2, \ldots, 6$. The entropy is 2.585 and the situation is as unpredictable as it can be. □

EXAMPLE 1.5

$S = \{s_1, s_2, s_3, s_4, s_5, s_6\}$, with $P(s_1) = 0.995 \; P(s_i) = 0.001$ for $i = 2, 3, \ldots, 6$.

The entropy is 0.057 and the situation is fairly predictable as s_1 will occur far more frequently than any other outcome. ▯

EXAMPLE 1.6

$S = \{s_1, s_2, s_3, s_4, s_5, s_6\}$, with $P(s_1) = 0.498 = P(s_2)$ $P(s_i) = 0.001$ for $i = 3, 4, \ldots, 6$. The entropy is 1.042 and the situation is about as predictable as in Example 1.1 above, with outcomes s_1 and s_2 equally likely to occur and the others very unlikely to occur. ▯

Roughly speaking, a system whose entropy is E is about as unpredictable as a system with 2^E equally likely outcomes.

1.5   Units of Entropy

The units in which entropy is measured depend on the base of the logarithms used to calculate it. If we use logarithms to the base 2, then the unit is the *bit*. If we use natural logarithms (base e), the entropy is measured in *natural units*, sometimes referred to as *nits*. Converting between the different units is simple.

PROPOSITION 1.1

If H_e is the entropy of a probability distribution measured using natural logarithms, and H_r is the entropy of the same probability distribution measured using logarithms to the base r, then

$$H_r = \frac{H_e}{\ln(r)}. \qquad (1.2)$$

PROOF   Let the sample space be $S = \{s_1, s_2, \ldots, s_N\}$, with probability distribution P. For any positive number x,

$$\ln(x) = \ln(r) \log_r(x). \qquad (1.3)$$

It follows that

$$H_r(P) = -\sum_{i=1}^{N} P(s_i) \log_r(P(s_i))$$

$$= -\sum_{i=1}^{N} P(s_i) \frac{\ln(P(s_i))}{\ln(r)}$$

$$= \frac{-\sum_{i=1}^{N} P(s_i) \ln(P(s_i))}{\ln(r)}$$

$$= \frac{H_e(P)}{\ln(r)}. \tag{1.4}$$

⬜

1.6   The Minimum and Maximum Values of Entropy

If we have a sample space S with N elements, and probability distribution P on S, it is convenient to denote the probability of $s_i \in S$ by p_i. We can construct a vector in R^N consisting of the probabilities:

$$\mathbf{p} = \begin{bmatrix} p_1 \\ p_2 \\ \vdots \\ p_N \end{bmatrix}.$$

Because the probabilities have to add up to unity, the set of all probability distributions forms a *simplex* in R^N, namely

$$K = \left\{ \mathbf{p} \in R^N : \sum_{i=1}^{N} p_i = 1 \right\}.$$

We can consider the entropy to be a function defined on this simplex. Since it is a continuous function, extreme values will occur at the vertices of this simplex, at points where all except one of the probabilities are zero. If $\mathbf{p}_v$ is a vertex, then the entropy there will be

$$H(\mathbf{p}_v) = (N-1).0.\log(0) + 1.\log(1).$$

The logarithm of zero is not defined, but the limit of $x \log(x)$ as x tends to 0 exists and is equal to zero. If we take the limiting values, we see that at any vertex, $H(\mathbf{p}_v) = 0$, as $\log(1) = 0$. This is the minimum value of the entropy function.

The entropy function has a maximum value at an interior point of the simplex. To find it we can use *Lagrange multipliers*.

THEOREM 1.1
If we have a sample space with N elements, the maximum value of the entropy function is $\log(N)$.

PROOF   We want to find the maximum value of

$$H(\mathbf{p}) = -\sum_{i=1}^{N} p_i \log(p_i) \tag{1.5}$$

subject to the constraint

$$\sum_{i=1}^{N} p_i = 1. \tag{1.6}$$

We introduce the Lagrange multiplier λ, and put

$$G(\mathbf{p}) = H(\mathbf{p}) + \lambda \left(\sum_{i=1}^{N} p_i - 1 \right). \tag{1.7}$$

To find the maximum value we have to solve

$$\frac{\partial G}{\partial p_i} = 0 \tag{1.8}$$

for $i = 1, 2, \ldots, N$ and

$$\sum_{i=1}^{N} p_i = 1. \tag{1.9}$$

$$\frac{\partial G}{\partial p_i} = -\log(p_i) - 1 + \lambda \tag{1.10}$$

so

$$p_i = e^{\lambda - 1} \tag{1.11}$$

for each i. The remaining condition gives

$$N e^{\lambda - 1} = 1 \tag{1.12}$$

which can be solved for λ, or can be used directly to give

$$p_i = \frac{1}{N} \tag{1.13}$$

for all i. Using these values for the p_i, we get

$$H(\mathbf{p}) = -N \frac{1}{N} \log(1/N) = \log(N). \tag{1.14}$$

$\square$

1.7   A Useful Inequality

LEMMA 1.1

If $p_1, p_2, \ldots, p_N$ and $q_1, q_2, \ldots, q_N$ are all non-negative numbers that satisfy the conditions $\sum_{i=1}^{N} p_n = 1$ and $\sum_{i=1}^{N} q_n = 1$, then

$$- \sum_{i=1}^{N} p_i \log(p_i) \leq - \sum_{i=1}^{N} p_i \log(q_i) \tag{1.15}$$

with equality if and only if $p_i = q_i$ for all i.

PROOF   We prove the result for the natural logarithm; the result for any other base follows immediately from the identity

$$\ln(x) = \ln(r) \log_r(x). \tag{1.16}$$

It is a standard result about the logarithm function that

$$\ln x \leq x - 1 \tag{1.17}$$

for $x > 0$, with equality if and only if $x = 1$. Substituting $x = q_i/p_i$, we get

$$\ln(q_i/p_i) \leq q_i/p_i - 1 \tag{1.18}$$

with equality if and only if $p_i = q_i$. This holds for all $i = 1, 2, \ldots, N$, so if we multiply by p_i and sum over the i, we get

$$\sum_{i=1}^{N} p_i \ln(q_i/p_i) \leq \sum_{i=1}^{N} (q_i - p_i) = \sum_{i=1}^{N} q_i - \sum_{i=1}^{N} p_i = 1 - 1 = 0, \tag{1.19}$$

with equality if and only if $p_i = q_i$ for all i. So

$$\sum_{i=1}^{N} p_i \ln(q_i) - \sum_{i=1}^{N} p_i \ln(p_i) \leq 0, \tag{1.20}$$

which is the required result.   ☐

The inequality can also be written in the form

$$\sum_{i=1}^{N} p_i \log(q_i/p_i) \leq 0, \tag{1.21}$$

with equality if and only if $p_i = q_i$ for all i.

Note that putting $q_i = 1/N$ for all i in this inequality gives us an alternative proof that the maximum value of the entropy function is $\log(N)$.

1.8 Joint Probability Distribution Functions

There are many situations in which it is useful to consider sample spaces that are the Cartesian product of two or more sets.

DEFINITION 1.6 Cartesian Product   *Let $S = \{s_1, s_2, \ldots, s_M\}$ and $T = \{t_1, t_2, \ldots, t_N\}$ be two sets. The Cartesian product of S and T is the set $S \times T = \{(s_i, t_j) : 1 \leq i \leq M, 1 \leq j \leq N\}$.*

The extension to the Cartesian product of more than two sets is immediate.

DEFINITION 1.7 Joint Probability Distribution   *A joint probability distribution is a probability distribution on the Cartesian product of a number of sets.*

If we have S and T as above, then a joint probability distribution function assigns a probability to each pair (s_i, t_j). We can denote this probability by p_{ij}. Since these values form a probability distribution, we have

$$0 \leq p_{ij} \leq 1 \tag{1.22}$$

for $1 \leq i \leq M, 1 \leq j \leq N$, and

$$\sum_{i=1}^{M} \sum_{j=1}^{N} p_{ij} = 1. \tag{1.23}$$

If P is the joint probability distribution function on $S \times T$, the definition of entropy becomes

$$H(P) = -\sum_{i=1}^{M} \sum_{j=1}^{N} P(s_i, t_j) \log(P(s_i, t_j)) = -\sum_{i=1}^{M} \sum_{j=1}^{N} p_{ij} \log(p_{ij}). \tag{1.24}$$

If we want to emphasise the spaces S and T, we will denote the entropy of the joint probability distribution on $S \times T$ by $H(P_{S \times T})$ or simply by $H(S, T)$. This is known as the *joint entropy* of S and T.

If there are probability distributions P_S and P_T on S and T, respectively, and these are independent, the joint probability distribution on $S \times T$ is given by

$$p_{ij} = P_S(s_i) P_T(t_j) \tag{1.25}$$

for $1 \leq i \leq M, 1 \leq j \leq N$. If there are correlations between the s_i and t_j, then this formula does not apply.

DEFINITION 1.8 Marginal Distribution *If P is a joint probability distribution function on $S \times T$, the* marginal distribution *on S is $P_S : S \to [0,1]$ given by*

$$P_S(s_i) = \sum_{j=1}^{N} P(s_i, t_j), \qquad (1.26)$$

for $1 \leq i \leq N$ and the marginal distribution *on T is $P_T : T \to [0,1]$ given by*

$$P_T(t_j) = \sum_{i=1}^{M} P(s_i, t_j), \qquad (1.27)$$

for $1 \leq j \leq N$.

There is a simple relationship between the entropy of the joint probability distribution function and that of the marginal distribution functions.

THEOREM 1.2
If P is a joint probability distribution function on $S \times T$, and P_S and P_T are the marginal distributions on S and T, respectively, then

$$H(P) \leq H(P_S) + H(P_T), \qquad (1.28)$$

with equality if and only if the marginal distributions are independent.

PROOF

$$H(P_S) = -\sum_{i=1}^{M} P_S(s_i) \log(P_S(s_i))$$

$$= -\sum_{i=1}^{M} \sum_{j=1}^{N} P(s_i, t_j) \log(P_S(s_i)) \qquad (1.29)$$

and similarly

$$H(P_T) = -\sum_{i=1}^{M} \sum_{j=1}^{N} P(s_i, t_j) \log(P_T(t_j)). \qquad (1.30)$$

So

$$H(P_S) + H(P_T) = -\sum_{i=1}^{M} \sum_{j=1}^{N} P(s_i, t_j)[\log(P_S(s_i)) + \log(P_T(t_j))]$$

$$= -\sum_{i=1}^{M}\sum_{j=1}^{N} P(s_i, t_j) \log(P_S(s_i)P_T(t_j)). \tag{1.31}$$

Also,

$$H(P) = -\sum_{i=1}^{M}\sum_{j=1}^{N} P(s_i, t_j) \log(P(s_i, t_j)). \tag{1.32}$$

Since

$$\sum_{i=1}^{M}\sum_{j=1}^{N} P(s_i, t_j) = 1, \tag{1.33}$$

and

$$\sum_{i=1}^{M}\sum_{j=1}^{N} P_S(s_i)P_T(t_j) = \sum_{i=1}^{M} P_S(s_i) \sum_{j=1}^{N} P_T(t_j) = 1, \tag{1.34}$$

we can use the inequality of Lemma 1.1 to conclude that

$$H(P) \le H(P_S) + H(P_T) \tag{1.35}$$

with equality if and only if $P(s_i, t_j) = P_S(s_i)P_T(t_j)$ for all i and j, that is, if the two marginal distributions are independent. ▯

1.9 Conditional Probability and Bayes' Theorem

> **DEFINITION 1.9 Conditional Probability**   *If S is a sample space with a probability distribution function P, and E and F are events in S, the conditional probability of E given F is*
> $$P(E|F) = \frac{P(E \cap F)}{P(F)}. \tag{1.36}$$

It is obvious that

$$P(E|F)P(F) = P(E \cap F) = P(F|E)P(E). \tag{1.37}$$

Almost as obvious is one form of *Bayes' Theorem*:

> **THEOREM 1.3**
> *If S is a sample space with a probability distribution function P, and E and F are events in S, then*
> $$P(E|F) = \frac{P(F|E)P(E)}{P(F)}. \tag{1.38}$$

Bayes' Theorem is important because it enables us to derive probabilities of hypotheses from observations, as in the following example.

EXAMPLE 1.7

We have two jars, A and B. Jar A contains 8 green balls and 2 red balls. Jar B contains 3 green balls and 7 red balls. One jar is selected at random and a ball is drawn from it.

We have probabilities as follows. The set of jars forms one sample space, $S = \{A, B\}$, with

$$P(\{A\}) = 0.5 = P(\{B\})$$

as one jar is as likely to be chosen as the other.

The set of colours forms another sample space, $T = \{G, R\}$. The probability of drawing a green ball is

$$P(\{G\}) = 11/20 = 0.55,$$

as 11 of the 20 balls in the jars are green. Similarly,

$$P(\{R\}) = 9/20 = 0.45.$$

We have a joint probability distribution over the colours of the balls and the jars with the probability of selecting Jar A and drawing a green ball being given by

$$P(\{(G, A)\}) = 0.4.$$

Similarly, we have the probability of selecting Jar A and drawing a red ball

$$P(\{(R, A)\}) = 0.1,$$

the probability of selecting Jar B and drawing a green ball

$$P(\{(G, B)\}) = 0.15,$$

and the probability of selecting Jar B and drawing a red ball

$$P(\{(R, B)\}) = 0.35.$$

We have the conditional probabilities: given that Jar A was selected, the probability of drawing a green ball is

$$P(\{G\}|\{A\}) = 0.8,$$

and the probability of drawing a red ball is

$$P(\{R\}|\{A\}) = 0.2.$$

Given that Jar B was selected, the corresponding probabilities are:

$$P(\{G\}|\{B\}) = 0.3,$$

and

$$P(\{R\}|\{B\}) = 0.7.$$

We can now use Bayes' Theorem to work out the probability of having drawn from either jar, given the colour of the ball that was drawn. If a green ball was drawn, the probability that it was drawn from Jar A is

$$P(\{A\}|\{G\}) = \frac{P(\{G\}|\{A\})P(\{A\})}{P(\{G\})} = 0.8 \times 0.5/0.55 = 0.73,$$

while the probability that it was drawn from Jar B is

$$P(\{B\}|\{G\}) = \frac{P(\{G\}|\{B\})P(\{B\})}{P(\{G\})} = 0.3 \times 0.5/0.55 = 0.27.$$

If a red ball was drawn, the probability that it was drawn from Jar A is

$$P(\{A\}|\{R\}) = \frac{P(\{R\}|\{A\})P(\{A\})}{P(\{R\})} = 0.2 \times 0.5/0.45 = 0.22,$$

while the probability that it was drawn from Jar B is

$$P(\{B\}|\{R\}) = \frac{P(\{R\}|\{B\})P(\{B\})}{P(\{R\})} = 0.7 \times 0.5/0.45 = 0.78.$$

(In this case, we could have derived these conditional probabilities from the joint probability distribution, but we chose not to do so to illustrate how Bayes' Theorem allows us to go from the conditional probabilities of the colours given the jar selected to the conditional probabilities of the jars selected given the colours drawn.)   ☐

1.10   Conditional Probability Distributions and Conditional Entropy

In this section, we have a joint probability distribution P on a Cartesian product $S \times T$, where $S = \{s_1, s_2, \ldots, s_M\}$ and $T = \{t_1, t_2, \ldots, t_N\}$, with marginal distributions P_S and P_T.

DEFINITION 1.10 Conditional Probability of s_i given t_j   *For $s_i \in S$ and* $t_j \in T$, the conditional probability of s_i given t_j is

$$P(s_i|t_j) = \frac{P(s_i, t_j)}{P_T(t_j)} = \frac{P(s_i, t_j)}{\sum_{i=1}^{M} P(s_i, t_j)}. \tag{1.39}$$

DEFINITION 1.11 Conditional Probability Distribution given t_j *For a fixed* t_j, *the conditional probabilities* $P(s_i|t_j)$ *sum to 1 over* i, *so they form a probability distribution on* S, *the* conditional probability distribution given t_j. *We will denote this by* $P_{S|t_j}$.

DEFINITION 1.12 Conditional Entropy given t_j *The* conditional entropy given t_j *is the entropy of the conditional probability distribution on* S *given* t_j. *It will be denoted* $H(P_{S|t_j})$.

$$H(P_{S|t_j}) = -\sum_{i=1}^{M} P(s_i|t_j) \log(P(s_i|t_j)). \tag{1.40}$$

DEFINITION 1.13 Conditional Probability Distribution on S **given** T *The* conditional probability distribution on S given T *is the weighted average of the conditional probability distributions given* t_j *for all* j. *It will be denoted* $P_{S|T}$.

$$P_{S|T}(s_i) = \sum_{j=1}^{N} P_T(t_j) P_{S|t_j}(s_i). \tag{1.41}$$

DEFINITION 1.14 Conditional Entropy given T *The* conditional entropy given T *is the weighted average of the conditional entropies on* S *given* t_j *for all* $t_j \in T$. *It will be denoted* $H(P_{S|T})$.

$$H(P_{S|T}) = -\sum_{j=1}^{N} P_T(t_j) \sum_{i=1}^{M} P(s_i|t_j) \log(P(s_i|t_j)). \tag{1.42}$$

Since $P_T(t_j)P(s_i|t_j) = P(s_i, t_j)$, we can re-write this as

$$H(P_{S|T}) = -\sum_{i=1}^{M} \sum_{j=1}^{N} P(s_i, t_j) \log(P(s_i|t_j)). \tag{1.43}$$

We now prove two simple results about the conditional entropies.

THEOREM 1.4

$$H(P) = H(P_T) + H(P_{S|T}) = H(P_S) + H(P_{T|S}).$$

PROOF

$$H(P) = -\sum_{i=1}^{M}\sum_{j=1}^{N} P(s_i, t_j) \log(P(s_i, t_j))$$

$$= -\sum_{i=1}^{M}\sum_{j=1}^{N} P(s_i, t_j) \log(P_T(t_j)P(s_i|t_j))$$

$$= -\sum_{i=1}^{M}\sum_{j=1}^{N} P(s_i, t_j) \log(P_T(t_j)) - \sum_{i=1}^{M}\sum_{j=1}^{N} P(s_i, t_j) \log(P(s_i|t_j))$$

$$= -\sum_{j=1}^{N} P_T(t_j) \log(P_T(t_j)) - \sum_{i=1}^{M}\sum_{j=1}^{N} P(s_i, t_j) \log(P(s_i|t_j))$$

$$= H(P_T) + H(P_{S|T}). \qquad (1.44)$$

The proof of the other equality is similar. ☐

THEOREM 1.5

$H(P_{S|T}) \le H(P_S)$ *with equality if and only if P_S and P_T are independent.*

PROOF From the previous theorem, $H(P) = H(P_T) + H(P_{S|T})$.

From Theorem 1.2, $H(P) \le H(P_S) + H(P_T)$ with equality if and only if P_S and P_T are independent.

So $H(P_T) + H(P_{S|T}) \le H(P_S) + H(P_T)$.

Subtracting $H(P_T)$ from both sides we get $H(P_{S|T}) \le H(P_S)$, with equality if and only if P_S and P_T are independent.

☐

This result is obviously symmetric in S and T; so we also have $H(P_{T|S}) \le H(P_T)$ with equality if and only if P_S and P_T are independent. We can sum up this result by saying the *conditioning reduces entropy* or *conditioning reduces uncertainty*.

1.11 Information Sources

Most of this book will be concerned with random sequences. Depending on the context, such sequences may be called *time series, (discrete) stochastic processes* or *signals*. The first term is used by statisticians, the second by mathematicians and the third by engineers. This may reflect differences in the way these people approach the subject: statisticians are primarily interested in describing such sequences in terms of probability theory, mathematicians are interested in the behaviour of such series and the ways in which they may be generated and engineers are interested in ways of using such sequences and processing them to extract useful information from them.

A device or situation that produces such a sequence is called an *information source*. The elements of the sequence are usually drawn from a finite set, which may be referred to as the *alphabet*. The source can be considered to be emitting an element of the alphabet at each instant of a sequence of instants in time. The elements of the alphabet are referred to as *symbols*.

EXAMPLE 1.8

Tossing a coin repeatedly and recording the outcomes as heads (H) or tails (T) gives us a random sequence whose alphabet is $\{H, T\}$. 	⬜

EXAMPLE 1.9

Throwing a die repeatedly and recording the number of spots on the uppermost face gives us a random sequence whose alphabet is $\{1, 2, 3, 4, 5, 6\}$. 	⬜

EXAMPLE 1.10

Computers and telecommunications equipment generate sequences of bits which are random sequences whose alphabet is $\{0, 1\}$. 	⬜

EXAMPLE 1.11

A text in the English language is a random sequence whose alphabet is the set consisting of the letters of the alphabet, the digits and the punctuation marks. While we normally consider text to be meaningful rather than random, it is only possible to predict which letter will come next in the sequence in probabilistic terms, in general.

⬜

The last example above illustrates the point that a random sequence may not appear to be random at first sight. The difference between the earlier examples and the final example is that in the English language there are correlations between each letter in the sequence and those that precede it. In contrast, there are no such correlations in the cases of tossing a coin or throwing a die repeatedly. We will consider both kinds of information sources below.

An obvious question that is raised by the term "information source" is: What is the "information" that the source produces? A second question, perhaps less obvious, is: How can we measure the information produced by an information source?

An information source generates a sequence of symbols which has a certain degree of unpredictability. The more unpredictable the sequence is, the more information is conveyed by each symbol. The information source may impose structure on the sequence of symbols. This structure will increase the predictability of the sequence and reduce the information carried by each symbol.

The random behaviour of the sequence may be described by probability distributions over the alphabet. If the elements of the sequence are uncorrelated, a simple probability distribution over the alphabet may suffice. In other cases, conditional probability distributions may be required.

We have already seen that entropy is a measure of predictability. For an information source, the information content of the sequence that it generates is measured by the *entropy per symbol*. We can compute this if we make assumptions about the kinds of structure that the information source imposes upon its output sequences.

To describe an information source completely, we need to specify both the alphabet and the probability distribution that governs the generation of sequences. The entropy of the information source S with alphabet A and probability distribution P will be denoted by $H(S)$ in the following sections, even though it is actually the entropy of P. Later on, we will wish to concentrate on the alphabet and will use $H(A)$ to denote the entropy of the information source, on the assumption that the alphabet will have a probability distribution associated with it.

1.12 Memoryless Information Sources

For a *memoryless information source*, there are no correlations between the outputs of the source at different times. For each instant at which an output is emitted, there is a probability distribution over the alphabet that describes the probability of each symbol being emitted at that instant. If all the probability distributions are the same, the source is said to be *stationary*. If we know these probability distributions, we can calculate the information content of the sequence.

EXAMPLE 1.12

Tossing a fair coin gives us an example of a stationary memoryless information source. At any instant, the probability distribution is given by $P(H) = 0.5$, $P(T) = 0.50$. This probability distribution has an entropy of 1 bit; so the information content is 1 bit/symbol.   ▯

EXAMPLE 1.13

As an example of a non-stationary memoryless information source, suppose we have a fair coin and a die with H painted on four faces and T painted on two faces. Tossing the coin and throwing the die in alternation will create a memoryless information source with alphabet $\{H, T\}$. Every time the coin is tossed, the probability distribution of the outcomes is $P(H) = 0.5$, $P(T) = 0.5$, and every time the die is thrown, the probability distribution is $P(H) = 0.667$, $P(T) = 0.333$.

The probability distribution of the outcomes of tossing the coin has an entropy of 1 bit. The probability distribution of the outcomes of throwing the die has an entropy of 0.918 bits. The information content of the sequence is the average entropy per symbol, which is 0.959 bits/symbol. ⬚

Memoryless information sources are relatively simple. More realistic information sources have *memory*, which is the property that the emission of a symbol at any instant depends on one or more of the symbols that were generated before it.

1.13 Markov Sources and n-gram Models

Markov sources and *n-gram models* are descriptions of a class of information sources with memory.

> **DEFINITION 1.15 Markov Source**   *A Markov source consists of an alphabet A, a set of states Σ, a set of transitions between states, a set of labels for the transitions and two sets of probabilities. The first set of probabilities is the initial probability distribution on the set of states, which determines the probabilities of sequences starting with each symbol in the alphabet. The second set of probabilities is a set of transition probabilities. For each pair of states, σ_i and σ_j, the probability of a transition from σ_i to σ_j is $P(j|i)$. (Note that these probabilities are fixed and do not depend on time, so that there is an implicit assumption of stationarity.) The labels on the transitions are symbols from the alphabet.*

To generate a sequence, a state is selected on the basis of the initial probability distribution. A transition from this state to another state (or to the same state) is selected on the basis of the transition probabilities, and the label of this transition is output. This process is repeated to generate the sequence of output symbols.

It is convenient to represent Markov models diagrammatically in the form of a graph, with the states represented by vertices and the transitions by edges, as in the following example.

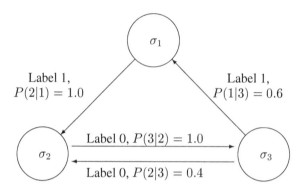

FIGURE 1.2
Diagrammatic representation of a Markov source.

EXAMPLE 1.14

Consider a Markov source with alphabet $\{0, 1\}$ and set of states $\Sigma = \{\sigma_1, \sigma_2, \sigma_3\}$. Suppose there are four transitions:

1. $\sigma_1 \rightarrow \sigma_2$, with label 1 and $P(2|1) = 1.0$;

2. $\sigma_2 \rightarrow \sigma_3$, with label 0 and $P(3|2) = 1.0$;

3. $\sigma_3 \rightarrow \sigma_1$, with label 1 and $P(1|3) = 0.6$;

4. $\sigma_3 \rightarrow \sigma_2$, with label 0 and $P(2|3) = 0.4$.

The initial probability distribution is $P(\sigma_1) = 1/3$, $P(\sigma_2) = 1/3$, $P(\sigma_3) = 1/3$.

The diagrammatic representation of this is shown in Figure 1.2.

The random sequences generated by this source all consist of subsequences of an even number of 0's separated by a pair of 1's, except at the very beginning of the sequence, where there may be a single 0 or 1.   □

It is possible to describe these sources without explicit reference to the states. In an *n-gram* model, the description is solely in terms of the probabilities of all the possible sequences of symbols of length n.

EXAMPLE 1.15

The following probabilities give us a 3-gram model on the language $\{0, 1\}$.

$$P(000) = 0.32, \quad P(001) = 0.08$$
$$P(010) = 0.15, \quad P(011) = 0.15$$
$$P(100) = 0.16, \quad P(101) = 0.04$$
$$P(110) = 0.06, \quad P(111) = 0.04.$$

☐

To describe the relationship between n-gram models and Markov sources, we need to look at special cases of Markov sources.

DEFINITION 1.16 *mth*-order Markov Source   *A Markov source whose states are sequences of m symbols from the alphabet is called an m*th-order Markov source.

When we have an mth-order Markov model, the transition probabilities are usually given in terms of the probabilities of single symbols being emitted when the source is in a given state. For example, in a second-order Markov model on $\{0, 1\}$, the transition probability from 01 to 10, which would be represented by $P(10|01)$, would be represented instead by the probability of emission of 0 when in the state 01, that is $P(0|01)$. Obviously, some transitions are impossible. For example, it is not possible to go from the state 01 to the state 00, as the state following 01 must have 1 as its first symbol.

We can construct a mth-order Markov model from an $(m+1)$-gram model and an m-gram model. The m-gram model gives us the probabilities of strings of length m, such as $P(s_1, s_2, \ldots, s_m)$. To find the emission probability of s from this state, we set

$$P(s|s_1, s_2, \ldots, s_m) = \frac{P(s_1, s_2, \ldots, s_m, s)}{P(s_1, s_2, \ldots, s_m)}, \tag{1.45}$$

where the probability $P(s_1, s_2, \ldots, s_m, s)$ is given by the $(m+1)$-gram model.

EXAMPLE 1.16

In the previous example 1.15 we had a 3-gram model on the language $\{0, 1\}$ given by

$$P(000) = 0.32, \quad P(001) = 0.08$$
$$P(010) = 0.15, \quad P(011) = 0.15$$
$$P(100) = 0.16, \quad P(101) = 0.04$$
$$P(110) = 0.06, \quad P(111) = 0.04.$$

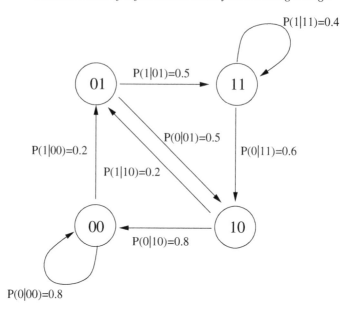

FIGURE 1.3
Diagrammatic representation of a Markov source equivalent to a 3-gram model.

If a 2-gram model for the same source is given by $P(00) = 0.4$, $P(01) = 0.3$, $P(10) = 0.2$ and $P(11) = 0.1$, then we can construct a second-order Markov source as follows:

$$P(0|00) = P(000)/P(00) = 0.32/0.4 = 0.8,$$
$$P(1|00) = P(001)/P(00) = 0.08/0.4 = 0.2,$$
$$P(0|01) = P(010)/P(01) = 0.15/0.3 = 0.5,$$
$$P(1|01) = P(011)/P(01) = 0.15/0.3 = 0.5,$$
$$P(0|10) = P(100)/P(10) = 0.16/0.2 = 0.8,$$
$$P(1|10) = P(101)/P(10) = 0.04/0.2 = 0.2,$$
$$P(0|11) = P(110)/P(11) = 0.06/0.1 = 0.6,$$
$$P(1|11) = P(111)/P(11) = 0.04/0.1 = 0.4.$$

Figure 1.3 shows this Markov source.

◻

To describe the behaviour of a Markov source mathematically, we use the *transition matrix* of probabilities. If the set of states is

$$\Sigma = \{\sigma_1, \sigma_2, \ldots, \sigma_N\},$$

the transition matrix is the $N \times N$ matrix

$$\Pi = \begin{bmatrix} P(1|1) & P(1|2) & \cdots & P(1|N) \\ P(2|1) & P(2|2) & \cdots & P(2|N) \\ \vdots & \vdots & \ddots & \vdots \\ P(N|1) & P(N|2) & \cdots & P(N|N) \end{bmatrix}. \tag{1.46}$$

The probability of the source being in a given state varies over time. Let w_i^t be the probability of the source being in state σ_i at time t, and set

$$W^t = \begin{bmatrix} w_1^t \\ w_2^t \\ \vdots \\ w_N^t \end{bmatrix}. \tag{1.47}$$

Then W^0 is the initial probability distribution and

$$W^{t+1} = \Pi W^t, \tag{1.48}$$

and so, by induction,

$$W^t = \Pi^t W^0. \tag{1.49}$$

Because they all represent probability distributions, each of the columns of Π must add up to 1, and all the w_i^t must add up to 1 for each t.

1.14 Stationary Distributions

The vectors W^t describe how the behaviour of the source changes over time. The asymptotic (long-term) behaviour of sources is of interest in some cases.

EXAMPLE 1.17

Consider a source with transition matrix

$$\Pi = \begin{bmatrix} 0.6 & 0.2 \\ 0.4 & 0.8 \end{bmatrix}.$$

Suppose the initial probability distribution is

$$W^0 = \begin{bmatrix} 0.5 \\ 0.5 \end{bmatrix}.$$

Then
$$W^1 = \Pi W^0 = \begin{bmatrix} 0.6 & 0.2 \\ 0.4 & 0.8 \end{bmatrix} \begin{bmatrix} 0.5 \\ 0.5 \end{bmatrix} = \begin{bmatrix} 0.4 \\ 0.6 \end{bmatrix}.$$

Similarly,
$$W^2 = \Pi W^1 = \begin{bmatrix} 0.36 \\ 0.64 \end{bmatrix},$$

$$W^3 = \Pi W^2 = \begin{bmatrix} 0.344 \\ 0.656 \end{bmatrix},$$

$$W^4 = \Pi W^3 = \begin{bmatrix} 0.3376 \\ 0.6624 \end{bmatrix},$$

and so on.

Suppose instead that
$$W_s^0 = \begin{bmatrix} 1/3 \\ 2/3 \end{bmatrix}.$$

Then
$$W_s^1 = \Pi W_s^0 = \begin{bmatrix} 0.6 & 0.2 \\ 0.4 & 0.8 \end{bmatrix} \begin{bmatrix} 1/3 \\ 2/3 \end{bmatrix} = \begin{bmatrix} 1/3 \\ 2/3 \end{bmatrix} = W_s^0,$$

so that
$$W_s^t = W_s^0$$

for all $t \geq 0$. This distribution will persist for all time.   ⬚

In the example above, the initial distribution W_s^0 has the property that
$$\Pi W_s^0 = W_s^0 \tag{1.50}$$

and persists for all time.

DEFINITION 1.17 Stationary Distribution   *A probability distribution W over the states of a Markov source with transition matrix Π that satisfies the equation $\Pi W = W$ is a stationary distribution.*

As shown in the example, if W^0 is a stationary distribution, it persists for all time, $W^t = W^0$ for all t. The defining equation shows that a stationary distribution W must be an eigenvector of Π with eigenvalue 1. To find a stationary distribution for Π, we must solve the equation $\Pi W = W$ together with the condition that $\sum_i w_i = 1$.

EXAMPLE 1.18

Suppose
$$\Pi = \begin{bmatrix} 0.25 & 0.50 & 0.00 \\ 0.50 & 0.00 & 0.25 \\ 0.25 & 0.50 & 0.75 \end{bmatrix}.$$

Then the equation $\Pi W = W$ gives

$$0.25w_1 + 0.50w_2 + 0.00w_3 = w_1$$
$$0.50w_1 + 0.00w_2 + 0.25w_3 = w_2$$
$$0.25w_1 + 0.50w_2 + 0.75w_3 = w_3.$$

The first equation gives us

$$0.75w_1 - 0.50w_2 = 0$$

and the other two give

$$0.50w_1 - 1.00w_2 + 0.25w_3 = 0$$
$$0.25w_1 + 0.50w_2 - 0.25w_3 = 0,$$

from which we get

$$-2.00w_2 + 0.75w_3 = 0.$$

So

$$w_1 = \frac{2}{3}w_2,$$

and

$$w_3 = \frac{8}{3}w_2.$$

Substituting these values in

$$w_1 + w_2 + w_3 = 1,$$

we get

$$\frac{2}{3}w_2 + w_2 + \frac{8}{3}w_2 = 1$$

which gives us

$$w_2 = \frac{3}{13}, w_1 = \frac{2}{13}, w_3 = \frac{8}{13}.$$

So the stationary distribution is

$$W = \begin{bmatrix} 2/13 \\ 3/13 \\ 8/13 \end{bmatrix}.$$

☐

In the examples above, the source has an unique stationary distribution. This is not always the case.

EXAMPLE 1.19

Consider the source with four states and probability transition matrix

$$\Pi = \begin{bmatrix} 1.0 \ 0.0 \ 0.5 \ 0.0 \\ 0.0 \ 0.0 \ 0.5 \ 0.0 \\ 0.0 \ 0.5 \ 0.0 \ 0.0 \\ 0.0 \ 0.5 \ 0.0 \ 1.0 \end{bmatrix}.$$

The diagrammatic representation of this source is shown in Figure 1.4.

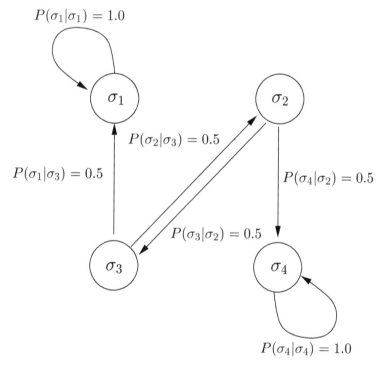

FIGURE 1.4
A source with two stationary distributions.

For this source, any distribution with $w_2 = 0.0$, $w_3 = 0.0$ and $w_1 + w_4 = 1.0$ satisfies the equation $\Pi W = W$. However, inspection of the transition matrix shows that once the source enters either the first state or the fourth state, it cannot leave it. The only stationary distributions that can occur are $w_1 = 1.0$, $w_2 = 0.0$, $w_3 = 0.0$, $w_4 = 0.0$ or $w_1 = 0.0$, $w_2 = 0.0$, $w_3 = 0.0$, $w_4 = 1.0$. □

Some Markov sources have the property that every sequence generated by the source has the same statistical properties. That is, the various frequencies of occurrence of symbols, pairs of symbols, and so on, obtained from any sequence generated by the source will, as the length of the sequence increases, approach some definite limit which is independent of the particular sequence. Sources that have this property are called *ergodic sources*.

The source of Example 1.19 is not an ergodic source. The sequences generated by that source fall into two classes, one of which is generated by sequences of states that end in the first state, the other of which is generated by sequences that end in the fourth state. The fact that there are two distinct stationary distributions shows that the source is not ergodic.

1.15   The Entropy of Markov Sources

There are various ways of defining the entropy of an information source. The following is a simple approach which applies to a restricted class of Markov sources.

DEFINITION 1.18 Entropy of the *i*th State of a Markov Source   *The entropy of the ith state of a Markov source is the entropy of the probability distribution on the set of transitions from that state.*

If we denote the probability distribution on the set of transitions from the *i*th state by P_i, then the entropy of the *i*th state is given by

$$H(P_i) = -\sum_{j=1}^{N} P(j|i) \log(P(j|i)). \tag{1.51}$$

DEFINITION 1.19 Unifilar Markov Source   *A unifilar Markov source is one with the property that the labels on the transitions from any given state are all distinct.*

We need this property in order to be able to define the entropy of a Markov source. We assume that the source has a stationary distribution.

DEFINITION 1.20 Entropy of a Unifilar Markov Source   *The* entropy
of a unifilar Markov source M, *whose stationary distribution is given by*
$w_1, w_2, \ldots, w_N$, *and whose transition probabilities are* $P(j|i)$ *for* $1 \le i \le N$,
$1 \le j \le N$, *is*

$$H(M) = \sum_{i=1}^{N} w_i H(P_i) = - \sum_{i=1}^{N} \sum_{j=1}^{N} w_i P(j|i) \log(P(j|i)). \qquad (1.52)$$

It can be shown that this definition is consistent with more general definitions of the
entropy of an information source.

EXAMPLE 1.20

For the Markov source of Example 1.14, there are three states, σ_1, σ_2 and σ_3. The
probability distribution on the set of transitions from σ_i is P_i for $i = 1, 2, 3$.

P_1 is given by

$$P_1(1) = P(1|1) = 0.0, P_1(2) = P(2|1) = 1.0, P_1(3) = P(3|1) = 0.0.$$

Its entropy is

$$H(P_1) = -(0.0)\log(0.0) - (1.0)(0.0) - (0.0)\log(0.0) = 0.0$$

using the usual convention that $0.0 \log(0.0) = 0.0$.

P_2 is given by

$$P_2(1) = P(1|2) = 0.0, P_2(2) = P(2|2) = 0.0, P_2(3) = P(3|2) = 1.0.$$

Its entropy is

$$H(P_2) = -(0.0)\log(0.0) - (0.0)\log(0.0) - (1.0)(0.0) = 0.0.$$

P_3 is given by

$$P_3(1) = P(1|3) = 0.6, P_3(2) = P(2|3) = 0.4, P_3(3) = P(3|3) = 0.0.$$

Its entropy is

$$H(P_3) = -(0.6)(-0.73697) - (0.4)(-1.32193) - (0.0)\log(0.0) = 0.97095.$$

The stationary distribution of the source is given by

$$w_1 = \frac{3}{13}, \quad w_2 = \frac{5}{13} \quad w_3 = \frac{5}{13}.$$

The entropy of the source is

$$H(M) = (0.0)(3/13) + (0.0)(5/13) + (0.97095)(5/13) = 0.37344.$$

☐

EXAMPLE 1.21

For the source of Example 1.16, the states are 00, 01, 10, 11.

P_{00} is given by $P_{00}(0) = P(0|00) = 0.8$, and $P_{00}(1) = P(1|00) = 0.2$. Its entropy is

$$H(P_{00}) = -(0.8)(-0.32193) - (0.2)(-2.32193) = 0.72193.$$

P_{01} is given by $P_{01}(0) = P(0|01) = 0.5$, and $P_{01}(1) = P(1|01) = 0.5$. Its entropy is

$$H(P_{01}) = -(0.5)(-1.0) - (0.5)(-1.0) = 1.0.$$

P_{10} is given by $P_{10}(0) = P(0|10) = 0.8$, and $P_{10}(1) = P(1|10) = 0.2$. Its entropy is

$$H(P_{10}) = -(0.8)(-0.32193) - (0.2)(-2.32193) = 0.72193.$$

P_{11} is given by $P_{11}(0) = P(0|11) = 0.6$, and $P_{11}(1) = P(1|11) = 0.4$. Its entropy is

$$H(P_{11}) = -(0.6)(-0.51083) - (0.4)(-1.32193) = 0.97095.$$

The stationary distribution of the source is given by

$$w_1 = \frac{120}{213}, w_2 = \frac{30}{213}, w_3 = \frac{35}{213}, w_4 = \frac{28}{213}.$$

The entropy of the source is

$$H(M) = \frac{120(0.72193) + 30(1.00000) + 35(0.72193) + 28(0.97095)}{213}$$

$$= 0.79383.$$

☐

1.16 Sequences of Symbols

It is possible to estimate the entropy of a Markov source using information about the probabilities of occurrence of sequences of symbols. The following results apply to ergodic Markov sources and are stated without proof. In a sense, they justify the use of the conditional probabilities of emission of symbols instead of transition probabilities between states in mth-order Markov models.

THEOREM 1.6

Given any $\epsilon > 0$ and any $\delta > 0$, we can find a positive integer N_0 such that all sequences of length $N \geq N_0$ fall into two classes: a set of sequences whose total probability is less than ϵ; and the remainder, for which the following inequality holds:

$$\left| \frac{\log(1/p)}{N} - H \right| < \delta, \tag{1.53}$$

where p is the probability of the sequence and H is the entropy of the source.

PROOF   See [6], Appendix 3.   ▯

THEOREM 1.7

Let M be a Markov source with alphabet $A = \{a_1, a_2, \ldots, a_n\}$, and entropy H. Let A^N denote the set of all sequences of symbols from A of length N. For $s \in A^N$, let $P(s)$ be the probability of the sequence s being emitted by the source. Define

$$G_N = -\frac{1}{N} \sum_{s \in A^N} P(s) \log(P(s)), \tag{1.54}$$

which is the entropy per symbol of the sequences of N symbols. Then G_N is a monotonic decreasing function of N and

$$\lim_{N \to \infty} G_N = H. \tag{1.55}$$

PROOF   See [6], Appendix 3.   ▯

THEOREM 1.8

Let M be a Markov source with alphabet $A = \{a_1, a_2, \ldots, a_n\}$, and entropy H. Let A^N denote the set of all sequences of symbols from A of length N. For $s \in A^{N-1}$, let $P(sa_i)$ be the probability of the source emitting the sequence s followed by the symbol a_i, and let $P(a_i|s)$ be the conditional probability of the symbol a_i being emitted immediately after the sequence s. Define

$$F_N = - \sum_{s \in A^{N-1}} \sum_{i=1}^{n} P(sa_i) \log(P(a_i|s)), \tag{1.56}$$

which is the conditional entropy of the next symbol when the $(N-1)$ preceding symbols are known. Then F_N is a monotonic decreasing function of N and

$$\lim_{N \to \infty} F_N = H. \tag{1.57}$$

PROOF   See [6], Appendix 3.   ⬜

THEOREM 1.9

If F_N and G_N are defined as in the previous theorems, then

$$F_N = NG_N - (N-1)G_{N-1}, \tag{1.58}$$

$$G_N = \frac{1}{N} \sum_{I=1}^{N} F_I, \tag{1.59}$$

and

$$F_N \leq G_N. \tag{1.60}$$

PROOF   See [6], Appendix 3.   ⬜

These results show that a series of approximations to the entropy of a source can be obtained by considering only the statistical behaviour of sequences of symbols of increasing length. The sequence of estimates F_N is a better approximation than the sequence G_N. If the dependencies in a source extend over no more than N_D symbols, so that the conditional probability of the next symbol knowing the preceding $(N_D - 1)$ symbols is the same as the conditional probability of the next symbol knowing the preceding $N \geq N_D$ symbols, then $F_{N_D} = H$.

1.17   The Adjoint Source of a Markov Source

It is possible to approximate the behaviour of a Markov source by a memoryless source.

> **DEFINITION 1.21 Adjoint Source of a Markov Source**   *The adjoint source of a Markov source is the memoryless source with the same alphabet which emits symbols independently of each other with the same probabilities as the Markov source.*

If we have an mth-order Markov source M with alphabet $A = \{a_1, a_2, \ldots, a_q\}$, the probabilities of emission of the symbols are

$$P(a_i) = \sum_j P(a_i | a_{j1} a_{j2} \ldots a_{jm}) P(a_{j1} a_{j2} \ldots a_{jm}), \qquad (1.61)$$

where $a_{j1} a_{j2} \ldots a_{jm}$ represents a sequence of m symbols from the alphabet of the source, $P(a_{j1} a_{j2} \ldots a_{jm})$ is the probability of this sequence in the stationary distribution of the Markov source and the summation over j indicates that all such sequences are included in the summation. The adjoint source of this Markov source, denoted $\bar{M}$, is the memoryless source that emits these symbols with the same probabilities.

EXAMPLE 1.22

For the 3-gram model of Example 1.15, we have transition probabilities

$$\begin{aligned}
P(0|00) &= 0.8, \quad P(1|00) = 0.2 \\
P(0|01) &= 0.5, \quad P(1|01) = 0.5 \\
P(0|10) &= 0.8, \quad P(1|10) = 0.2 \\
P(0|11) &= 0.6, \quad P(1|11) = 0.4,
\end{aligned}$$

which give us the transition matrix

$$\Pi = \begin{bmatrix} 0.8 & 0.0 & 0.8 & 0.0 \\ 0.2 & 0.0 & 0.2 & 0.0 \\ 0.0 & 0.5 & 0.0 & 0.6 \\ 0.0 & 0.5 & 0.0 & 0.4 \end{bmatrix}.$$

We need to find the stationary distribution of the source. The equation $\Pi W = W$ gives

$$\begin{aligned}
0.8w_1 + 0.0w_2 + 0.8w_3 + 0.0w_4 &= w_1 \\
0.2w_1 + 0.0w_2 + 0.2w_3 + 0.0w_4 &= w_2 \\
0.0w_1 + 0.5w_2 + 0.0w_3 + 0.6w_4 &= w_3 \\
0.0w_1 + 0.5w_2 + 0.0w_3 + 0.4w_4 &= w_4.
\end{aligned}$$

Solving these equations together with the constraint

$$w_1 + w_2 + w_3 + w_4 = 1.0,$$

we get the stationary distribution

$$P(00) = \frac{28}{46}, P(01) = \frac{6}{46}, P(10) = \frac{7}{46}, P(11) = \frac{5}{46}.$$

The probabilities for the adjoint source of the 3-gram models are

$$P(0) = P(0|00)P(00) + P(0|01)P(01) + P(0|10)P(10) + P(0|11)P(11)$$
$$= \frac{17}{23}$$

and

$$P(1) = P(1|00)P(00) + P(1|01)P(01) + P(1|10)P(10) + P(1|11)P(11)$$
$$= \frac{6}{23}.$$

Although the probabilities of emission of single symbols are the same for both the Markov source and its adjoint source, the probabilities of emission of sequences of symbols may not be the same. For example the probability of emission of the sequence 000 by the Markov source is $P(0|00) = 0.8$, while for the adjoint source it is $P(0)^3 = 0.4038$ (by the assumption of independence). $\quad\square$

Going from a Markov source to its adjoint reduces the number of constraints on the output sequence and hence increases the entropy. This is formalised by the following theorem.

THEOREM 1.10

If $\bar{M}$ is the adjoint of the Markov source M, their entropies are related by

$$H(M) \leq H(\bar{M}). \tag{1.62}$$

PROOF If M is an mth-order source with alphabet $\{a_1, a_2, \ldots, a_q\}$, we will denote the states, which are m-tuples of the a_i, by α_I, where $1 \leq I \leq q^m$. We assume that M has a stationary distribution.

The probabilities of emission of the symbols are

$$P(a_i) = \sum_I w_I P(a_i|\alpha_I), \tag{1.63}$$

where the summation is over all states and w_I is the probability of state α_I in the stationary distribution of the source.

The entropy of the adjoint is

$$H(\bar{M}) = -\sum_{i=1}^{q} P(a_i) \log(P(a_i))$$

$$= -\sum_{i=1}^{q}\sum_{I} w_I P(a_i|\alpha_I)\log(P(a_i))$$

$$= -\sum_{I} w_I \sum_{i=1}^{q} P(a_i|\alpha_I)\log(P(a_i)). \tag{1.64}$$

The entropy of the Ith state of M is

$$H(P_I) = -\sum_{i=1}^{q} P(a_i|\alpha_I)\log(P(a_i|\alpha_I)), \tag{1.65}$$

and the entropy of M is

$$H(M) = -\sum_{I}\sum_{i=1}^{q} w_I P(a_i|\alpha_I)\log(P(a_i|\alpha_I))$$

$$= -\sum_{I} w_I \sum_{i=1}^{q} P(a_i|\alpha_I)\log(P(a_i|\alpha_I)). \tag{1.66}$$

If we apply the inequality of Lemma 1.1 to each summation over i, the result follows.

$\square$

1.18   Extensions of Sources

In situations where codes of various types are being developed, it is often useful to consider sequences of symbols emitted by a source.

DEFINITION 1.22 Extension of a Stationary Memoryless Source   *The nth extension of a stationary memoryless source S is the stationary memoryless source whose alphabet consists of all sequences of n symbols from the alphabet of S, with the emission probabilities of the sequences being the same as the probabilities of occurrence of the sequences in the output of S.*

The nth extension of S will be denoted by S^n. Because the emission of successive symbols by S is statistically independent, the emission probabilities in S^n can be computed by multiplying the appropriate emission probabilities in S.

EXAMPLE 1.23

Consider the memoryless source S with alphabet $\{0, 1\}$ and emission probabilities $P(0) = 0.2, P(1) = 0.8$.

The second extension of S has alphabet $\{00, 01, 10, 11\}$ with emission probabilities

$$P(00) = P(0)P(0) = (0.2)(0.2) = 0.04,$$

$$P(01) = P(0)P(1) = (0.2)(0.8) = 0.16,$$

$$P(10) = P(1)P(0) = (0.8)(0.2) = 0.16,$$

$$P(11) = P(1)P(1) = (0.8)(0.8) = 0.64.$$

The third extension of S has alphabet $\{000, 001, 010, 011, 100, 101, 110, 111\}$ with emission probabilities

$$P(000) = P(0)P(0)P(0) = (0.2)(0.2)(0.2) = 0.008,$$

$$P(001) = P(0)P(0)P(1) = (0.2)(0.2)(0.8) = 0.032,$$

$$P(010) = P(0)P(1)P(0) = (0.2)(0.8)(0.2) = 0.032,$$

$$P(011) = P(0)P(1)P(1) = (0.2)(0.8)(0.8) = 0.128,$$

$$P(100) = P(1)P(0)P(0) = (0.8)(0.2)(0.2) = 0.032,$$

$$P(101) = P(1)P(0)P(1) = (0.8)(0.2)(0.8) = 0.128,$$

$$P(110) = P(1)P(1)P(0) = (0.8)(0.8)(0.2) = 0.128,$$

$$P(111) = P(1)P(1)P(1) = (0.8)(0.8)(0.8) = 0.512.$$

There is a simple relationship between the entropy of a stationary memoryless source and the entropies of its extensions.

THEOREM 1.11

If S^n is the nth extension of the stationary memoryless source S, their entropies are related by

$$H(S^n) = nH(S). \tag{1.67}$$

PROOF If the alphabet of S is $\{a_1, a_2, \ldots, a_q\}$, and the emission probabilities of the symbols are $P(a_i)$ for $i = 1, 2, \ldots, q$, the entropy of S is

$$H(S) = -\sum_{i=1}^{q} P(a_i) \log(P(a_i)). \tag{1.68}$$

The alphabet of S^n consists of all sequences $a_{i_1} a_{i_2} \ldots a_{i_n}$, where $i_j \in \{1, 2, \ldots, q\}$. The emission probability of $a_{i_1} a_{i_2} \ldots a_{i_n}$ is

$$P(a_{i_1} a_{i_2} \ldots a_{i_n}) = P(a_{i_1})P(a_{i_2}) \ldots P(a_{i_n}). \tag{1.69}$$

The entropy of S^n is

$$H(S^n) = -\sum_{i_1=1}^{q} \cdots \sum_{i_n=1}^{q} P(a_{i_1} a_{i_2} \ldots a_{i_n}) \log(P(a_{i_1} a_{i_2} \ldots a_{i_n}))$$

$$= -\sum_{i_1=1}^{q} \cdots \sum_{i_n=1}^{q} P(a_{i_1} a_{i_2} \ldots a_{i_n}) \sum_{j=1}^{n} \log(P(a_{i_j})). \qquad (1.70)$$

We can interchange the order of summation to get

$$H(S^n) = -\sum_{j=1}^{n} \sum_{i_1=1}^{q} \cdots \sum_{i_n=1}^{q} P(a_{i_1} a_{i_2} \ldots a_{i_n}) \log(P(a_{i_j})). \qquad (1.71)$$

Breaking $P(a_{i_1} a_{i_2} \ldots a_{i_n})$ into the product of probabilities, and rearranging, we get

$$H(S^n) = -\sum_{j=1}^{n} \sum_{i_1=1}^{q} P(a_{i_1}) \cdots \sum_{i_j=1}^{q} P(a_{i_j}) \log(P(a_{i_j})) \cdots \sum_{i_n=1}^{q} P(a_{i_n}). \quad (1.72)$$

Since

$$\sum_{i_k=1}^{q} P(a_{i_k}) = 1 \qquad (1.73)$$

for $k \neq j$, we are left with

$$H(S^n) = -\sum_{j=1}^{n} \sum_{i_j=1}^{q} P(a_{i_j}) \log(P(a_{i_j}))$$

$$= \sum_{j=1}^{n} H(S)$$

$$= nH(S). \qquad (1.74)$$

$\Box$

We also have extensions of Markov sources.

DEFINITION 1.23 Extension of an *mth*-order Markov Source   *Let m and n be positive integers, and let p be the smallest integer that is greater than or equal to m/n. The nth extension of the mth-order Markov source M is the pth-order Markov source whose alphabet consists of all sequences of n symbols from the alphabet of M and for which the transition probabilities between states are equal to the probabilities of the corresponding n-fold transitions of the mth-order source.*

We will use M^n to denote the nth extension of M.

EXAMPLE 1.24

Let M be the first-order Markov source with alphabet $\{0, 1\}$ and transition probabilities

$$P(0|0) = 0.3, \quad P(1|0) = 0.7, \quad P(0|1) = 0.4, \quad P(1|1) = 0.6.$$

The second extension of M has $m = 1$ and $n = 2$, so $p = 1$. It is a first-order source with alphabet $\{00, 01, 10, 11\}$. We can calculate the transition probabilities as follows.

$$
\begin{aligned}
P(00|00) &= P(0|0)P(0|0) = (0.3)(0.3) = 0.09 \\
P(01|00) &= P(0|0)P(1|0) = (0.3)(0.7) = 0.21 \\
P(10|00) &= P(1|0)P(0|1) = (0.7)(0.4) = 0.28 \\
P(11|00) &= P(1|0)P(1|1) = (0.7)(0.6) = 0.42 \\
P(00|11) &= P(0|1)P(0|0) = (0.4)(0.3) = 0.12 \\
P(01|11) &= P(0|1)P(1|0) = (0.4)(0.7) = 0.28 \\
P(10|11) &= P(1|1)P(0|1) = (0.6)(0.4) = 0.24 \\
P(11|11) &= P(1|1)P(1|1) = (0.6)(0.6) = 0.36
\end{aligned}
$$

□

EXAMPLE 1.25

Consider the second order Markov source with alphabet $\{0, 1\}$ and transition probabilities

$$
\begin{aligned}
P(0|00) &= 0.8, \quad P(1|00) = 0.2 \\
P(0|01) &= 0.5, \quad P(1|01) = 0.5 \\
P(0|10) &= 0.2, \quad P(1|10) = 0.8 \\
P(0|11) &= 0.6, \quad P(1|11) = 0.4
\end{aligned}
$$

The transition probabilities of the second extension are

$$P(00|00) = P(0|00)P(0|00) = (0.8)(0.8) = 0.64$$
$$P(01|00) = P(0|00)P(1|00) = (0.8)(0.2) = 0.16$$
$$P(10|00) = P(1|00)P(0|01) = (0.2)(0.5) = 0.10$$
$$P(11|00) = P(1|00)P(1|01) = (0.2)(0.5) = 0.10$$
$$P(00|01) = P(0|01)P(0|10) = (0.5)(0.2) = 0.10$$
$$P(01|01) = P(0|01)P(1|10) = (0.5)(0.8) = 0.40$$
$$P(10|01) = P(1|01)P(0|11) = (0.5)(0.6) = 0.30$$
$$P(11|01) = P(1|01)P(1|11) = (0.5)(0.4) = 0.20$$
$$P(00|10) = P(0|10)P(0|00) = (0.2)(0.8) = 0.16$$
$$P(01|10) = P(0|10)P(1|00) = (0.2)(0.2) = 0.04$$
$$P(10|10) = P(1|10)P(0|01) = (0.8)(0.5) = 0.40$$
$$P(11|10) = P(1|10)P(1|01) = (0.8)(0.5) = 0.40$$
$$P(00|11) = P(0|11)P(0|10) = (0.6)(0.2) = 0.12$$
$$P(01|11) = P(0|11)P(1|10) = (0.6)(0.8) = 0.48$$
$$P(10|11) = P(1|11)P(0|11) = (0.4)(0.6) = 0.24$$
$$P(11|11) = P(1|11)P(1|11) = (0.4)(0.4) = 0.16$$

If we denote the states of the second extension by $\alpha_1 = 00$, $\alpha_2 = 01$, $\alpha_3 = 10$ and $\alpha_4 = 11$, we have the transition probabilities of a first-order Markov source:

$$P(\alpha_1|\alpha_1) = 0.64, \quad P(\alpha_2|\alpha_1) = 0.16$$
$$P(\alpha_3|\alpha_1) = 0.10, \quad P(\alpha_4|\alpha_1) = 0.10$$
$$P(\alpha_1|\alpha_2) = 0.10, \quad P(\alpha_2|\alpha_2) = 0.40$$
$$P(\alpha_3|\alpha_2) = 0.30, \quad P(\alpha_4|\alpha_2) = 0.20$$
$$P(\alpha_1|\alpha_3) = 0.16, \quad P(\alpha_2|\alpha_3) = 0.04$$
$$P(\alpha_3|\alpha_3) = 0.40, \quad P(\alpha_4|\alpha_3) = 0.40$$
$$P(\alpha_1|\alpha_4) = 0.12, \quad P(\alpha_2|\alpha_4) = 0.48$$
$$P(\alpha_3|\alpha_4) = 0.24, \quad P(\alpha_4|\alpha_4) = 0.16$$

◻

EXAMPLE 1.26

Consider the fourth order Markov source with alphabet $\{0, 1\}$ and transition probabilities

$$
\begin{aligned}
&P(0|0000) = 0.9, \quad P(1|0000) = 0.1 \\
&P(0|0001) = 0.8, \quad P(1|0001) = 0.2 \\
&P(0|0010) = 0.7, \quad P(1|0010) = 0.3 \\
&P(0|0011) = 0.6, \quad P(1|0011) = 0.4 \\
&P(0|0100) = 0.5, \quad P(1|0100) = 0.5 \\
&P(0|0101) = 0.4, \quad P(1|0101) = 0.6 \\
&P(0|0110) = 0.3, \quad P(1|0110) = 0.7 \\
&P(0|0111) = 0.2, \quad P(1|0111) = 0.8 \\
&P(0|1000) = 0.1, \quad P(1|1000) = 0.9 \\
&P(0|1001) = 0.2, \quad P(1|1001) = 0.8 \\
&P(0|1010) = 0.3, \quad P(1|1010) = 0.7 \\
&P(0|1011) = 0.4, \quad P(1|1011) = 0.6 \\
&P(0|1100) = 0.5, \quad P(1|1100) = 0.5 \\
&P(0|1101) = 0.6, \quad P(1|1101) = 0.4 \\
&P(0|1110) = 0.7, \quad P(1|1110) = 0.3 \\
&P(0|1111) = 0.8, \quad P(1|1111) = 0.2
\end{aligned}
$$

We can use these probabilities to compute the transition probabilities of the second extension, for example,

$$
P(01|1010) = P(0|1010)P(1|0100) = (0.3)(0.5) = 0.15.
$$

If we denote the states of the second extension by $\alpha_1 = 00$, $\alpha_2 = 01$, $\alpha_3 = 10$ and $\alpha_4 = 11$, we have the transition probabilities of a second-order Markov source:

$$P(\alpha_1|\alpha_1\alpha_1) = 0.81, \quad P(\alpha_2|\alpha_1\alpha_1) = 0.09$$
$$P(\alpha_3|\alpha_1\alpha_1) = 0.08, \quad P(\alpha_4|\alpha_1\alpha_1) = 0.02$$
$$P(\alpha_1|\alpha_1\alpha_2) = 0.56, \quad P(\alpha_2|\alpha_1\alpha_2) = 0.24$$
$$P(\alpha_3|\alpha_1\alpha_2) = 0.12, \quad P(\alpha_4|\alpha_1\alpha_2) = 0.08$$
$$P(\alpha_1|\alpha_1\alpha_3) = 0.35, \quad P(\alpha_2|\alpha_1\alpha_3) = 0.35$$
$$P(\alpha_3|\alpha_1\alpha_3) = 0.12, \quad P(\alpha_4|\alpha_1\alpha_3) = 0.18$$
$$P(\alpha_1|\alpha_1\alpha_4) = 0.18, \quad P(\alpha_2|\alpha_1\alpha_4) = 0.42$$
$$P(\alpha_3|\alpha_1\alpha_4) = 0.08, \quad P(\alpha_4|\alpha_1\alpha_4) = 0.32$$
$$P(\alpha_1|\alpha_2\alpha_1) = 0.05, \quad P(\alpha_2|\alpha_2\alpha_1) = 0.45$$
$$P(\alpha_3|\alpha_2\alpha_1) = 0.10, \quad P(\alpha_4|\alpha_2\alpha_1) = 0.40$$
$$P(\alpha_1|\alpha_2\alpha_2) = 0.12, \quad P(\alpha_2|\alpha_2\alpha_2) = 0.28$$
$$P(\alpha_3|\alpha_2\alpha_2) = 0.24, \quad P(\alpha_4|\alpha_2\alpha_2) = 0.36$$
$$P(\alpha_1|\alpha_2\alpha_3) = 0.15, \quad P(\alpha_2|\alpha_2\alpha_3) = 0.15$$
$$P(\alpha_3|\alpha_2\alpha_3) = 0.42, \quad P(\alpha_4|\alpha_2\alpha_3) = 0.28$$
$$P(\alpha_1|\alpha_2\alpha_4) = 0.14, \quad P(\alpha_2|\alpha_2\alpha_4) = 0.06$$
$$P(\alpha_3|\alpha_2\alpha_4) = 0.64, \quad P(\alpha_4|\alpha_2\alpha_4) = 0.16$$
$$P(\alpha_1|\alpha_3\alpha_1) = 0.09, \quad P(\alpha_2|\alpha_3\alpha_1) = 0.01$$
$$P(\alpha_3|\alpha_3\alpha_1) = 0.72, \quad P(\alpha_4|\alpha_3\alpha_1) = 0.18$$
$$P(\alpha_1|\alpha_3\alpha_2) = 0.14, \quad P(\alpha_2|\alpha_3\alpha_2) = 0.06$$
$$P(\alpha_3|\alpha_3\alpha_2) = 0.48, \quad P(\alpha_4|\alpha_3\alpha_2) = 0.32$$
$$P(\alpha_1|\alpha_3\alpha_3) = 0.15, \quad P(\alpha_2|\alpha_3\alpha_3) = 0.15$$
$$P(\alpha_3|\alpha_3\alpha_3) = 0.28, \quad P(\alpha_4|\alpha_3\alpha_3) = 0.42$$
$$P(\alpha_1|\alpha_3\alpha_4) = 0.12, \quad P(\alpha_2|\alpha_3\alpha_4) = 0.28$$
$$P(\alpha_3|\alpha_3\alpha_4) = 0.12, \quad P(\alpha_4|\alpha_3\alpha_4) = 0.48$$
$$P(\alpha_1|\alpha_4\alpha_1) = 0.05, \quad P(\alpha_2|\alpha_4\alpha_1) = 0.45$$
$$P(\alpha_3|\alpha_4\alpha_1) = 0.10, \quad P(\alpha_4|\alpha_4\alpha_1) = 0.40$$
$$P(\alpha_1|\alpha_4\alpha_2) = 0.18, \quad P(\alpha_2|\alpha_4\alpha_2) = 0.42$$
$$P(\alpha_3|\alpha_4\alpha_2) = 0.16, \quad P(\alpha_4|\alpha_4\alpha_2) = 0.24$$
$$P(\alpha_1|\alpha_4\alpha_3) = 0.35, \quad P(\alpha_2|\alpha_4\alpha_3) = 0.35$$
$$P(\alpha_3|\alpha_4\alpha_3) = 0.18, \quad P(\alpha_4|\alpha_4\alpha_3) = 0.12$$
$$P(\alpha_1|\alpha_4\alpha_4) = 0.56, \quad P(\alpha_2|\alpha_4\alpha_4) = 0.24$$
$$P(\alpha_3|\alpha_4\alpha_4) = 0.16, \quad P(\alpha_4|\alpha_4\alpha_4) = 0.04$$

It is convenient to represent elements of the alphabet of M^n by single symbols as we have done in the examples above. If the alphabet of M is $\{a_1, a_2, \ldots, a_q\}$, then we will use α for a generic element of the alphabet of M^n, and $\alpha_{i_1 i_2 \ldots i_n}$ will stand for the sequence $a_{i_1} a_{i_2} \ldots a_{i_n}$. For further abbreviation, we will let I stand for $i_1 i_2 \ldots i_n$ and use α_I to denote $\alpha_{i_1 i_2 \ldots i_n}$, and so on.

The statistics of the extension M^n are given by the conditional probabilities $P(\alpha_J|\alpha_{I_1}\alpha_{I_2}$. In terms of the alphabet of M, we have

$$P(\alpha_J|\alpha_{I_1}\alpha_{I_2} \ldots \alpha_{I_p}) = P(a_{j_1} \ldots a_{j_n}|a_{i_{11}} \ldots a_{i_{1n}} \ldots a_{i_{pn}}). \tag{1.75}$$

This can also be written as

$$P(\alpha_J | \alpha_{I_1} \alpha_{I_2} \dots \alpha_{I_p}) = P(a_{j_1} | a_{i_{11}} \dots a_{i_{1n}} \dots a_{i_{pn}}) P(a_{j_2} | a_{i_{12}} \dots a_{i_{pn}} a_{j_1}) \dots$$
$$P(a_{j_n} | a_{i_{1n}} \dots a_{i_{pn}} a_{j_1} \dots a_{j_{n-1}}). \quad (1.76)$$

We can use this relationship to prove the following result.

THEOREM 1.12

If M^n is the nth extension of the Markov source M their entropies are related by

$$H(M^n) = nH(M). \quad (1.77)$$

The proof is similar to the proof of the corresponding result for memoryless sources. (Exercise 17 deals with a special case.)

Note that if $m < n$, then

$$H(M^m) = mH(M) < nH(M) = H(M^n). \quad (1.78)$$

Since an extension of a Markov source is a pth-order Markov source, we can consider its adjoint source. If M is an mth-order Markov source, M^n is an nth extension of M, and $\bar{M}^n$ the adjoint source of the extension, then we can combine the results of Theorem 1.10 and Theorem 1.12 to get

$$H(\bar{M}^n) \geq H(M^n) = nH(M). \quad (1.79)$$

1.19   Infinite Sample Spaces

The concept of entropy carries over to infinite sample spaces, but there are a number of technical issues that have to be considered.

If the sample space is countable, the entropy has to be defined in terms of the limit of a series, as in the following example.

EXAMPLE 1.27

Suppose the sample space is the set of *natural numbers*, $N = \{0, 1, 2, \dots\}$ and the probability distribution is given by

$$P(n) = 2^{-n-1} \quad (1.80)$$

for $n \in N$. The entropy of this distribution is

$$
\begin{aligned}
H(P) &= -\sum_{n=0}^{\infty} P(n) \log(P(n) \\
&= -\sum_{n=0}^{\infty} 2^{-n-1}(-n-1) \\
&= \sum_{n=0}^{\infty} \frac{n+1}{2^{n+1}}.
\end{aligned}
\tag{1.81}
$$

This infinite sum converges to 2, so $H(P) = 2$.   ⬜

If the sample space is a continuum, in particular, the real line, the summations become integrals. Instead of the probability distribution function, we use the probability density function f, which has the property that

$$
P([a, b]) = \int_a^b f(x)dx,
\tag{1.82}
$$

where $[a, b]$ is a closed interval and f is the probability density function.

The mean and variance of the probability density function are defined by

$$
\mu = \int_{-\infty}^{\infty} x f(x)dx,
\tag{1.83}
$$

and

$$
\sigma^2 = \int_{-\infty}^{\infty} (x - \mu)^2 f(x)dx.
\tag{1.84}
$$

The obvious generalisation of the definition of entropy for a probability density function f defined on the real line is

$$
H(f) = -\int_{-\infty}^{\infty} f(x) \log(f(x))dx,
\tag{1.85}
$$

provided this integral exists. This definition was proposed by Shannon, but has been the subject of debate because it is not invariant with respect to change of scale or change of co-ordinates in general. It is sometimes known as the *differential entropy*.

If we accept this definition of the entropy of a continuous distribution, it is easy to compute the entropy of a Gaussian distribution.

THEOREM 1.13
The entropy of a Gaussian distribution with mean μ and variance σ^2 is $\ln(\sqrt{2\pi e}\sigma)$ in natural units.

PROOF   The density function of the Gaussian distribution is

$$g(x) = \frac{1}{\sqrt{2\pi}\sigma} e^{(x-\mu)^2/2\sigma^2}. \tag{1.86}$$

Since this is a probability density function, we have

$$\int_{-\infty}^{\infty} g(x)dx = 1. \tag{1.87}$$

Taking the natural logarithm of g, we get

$$\ln(g(x)) = -\ln(\sqrt{2\pi}\sigma) - \frac{(x-\mu)^2}{2\sigma^2}. \tag{1.88}$$

By definition,

$$\sigma^2 = \int_{-\infty}^{\infty} (x-\mu)^2 g(x)dx; \tag{1.89}$$

this will be used below.

We now calculate the entropy:

$$\begin{aligned}
H(g) &= -\int_{-\infty}^{\infty} g(x)\ln(g(x))dx \\
&= -\int_{-\infty}^{\infty} g(x)\left(-\ln(\sqrt{2\pi}\sigma) - \frac{(x-\mu)^2}{2\sigma^2}\right)dx \\
&= \int_{-\infty}^{\infty} g(x)\ln(\sqrt{2\pi}\sigma)dx + \int_{-\infty}^{\infty} g(x)\frac{(x-\mu)^2}{2\sigma^2}dx \\
&= \ln(\sqrt{2\pi}\sigma)\int_{-\infty}^{\infty} g(x)dx + \frac{1}{2\sigma^2}\int_{-\infty}^{\infty} g(x)(x-\mu)^2 dx \\
&= \ln(\sqrt{2\pi}\sigma) + \frac{\sigma^2}{2\sigma^2} \\
&= \ln(\sqrt{2\pi}\sigma) + \frac{1}{2} \\
&= \ln(\sqrt{2\pi}\sigma) + \log(\sqrt{e}) \\
&= \ln(\sqrt{2\pi e}\sigma). \tag{1.90}
\end{aligned}$$

☐

If the probability density function is defined over the whole real line, it is not possible to find a specific probability density function whose entropy is greater than the entropy of all other probability density functions defined on the real line. However, if we restrict consideration to probability density functions with a given variance, it can be shown that the Gaussian distribution has the maximum entropy of all these distributions. (Exercises 20 and 21 outline the proof of this result.)

We have used $H(f)$ to denote the entropy of the probability distribution whose probability density function is f. If X is a random variable whose probability density function is f, we will denote its entropy by either $H(f)$ or $H(X)$.

1.20   Exercises

1. Let $S = \{s_1, s_2, s_3\}$ be a sample space with probability distribution P given by $P(s_1) = 0.2$, $P(s_2) = 0.3$, $P(s_3) = 0.5$. Let f be a function defined on S by $f(s_1) = 5$, $f(s_2) = -2$, $f(s_3) = 1$. What is the expected value of f?

2. Let $S = \{s_1, s_2\}$ be a sample space with probability distribution P given by $P(s_1) = 0.7$, $P(s_2) = 0.3$. Let f be the function from S to $\mathbb{R}^2$ given by

$$f(s_1) = \begin{pmatrix} 2.0 \\ 3.0 \end{pmatrix}$$

and

$$f(s_2) = \begin{pmatrix} 5.0 \\ -4.0 \end{pmatrix}.$$

What is the expected value of f?

3. Suppose that a fair die is tossed. What is the expected number of spots on the uppermost face of the die when it comes to rest? Will this number of spots ever be seen when the die is tossed?

4. Let $S = \{s_1, s_2, s_3, s_4\}$ be a sample space with probability distribution P given by $P(s_1) = 0.5$, $P(s_2) = 0.25$, $P(s_3) = 0.125$, $P(s_4) = 0.125$. There are sixteen possible events that can be formed from the elements of S. Compute the probability and surprise of these events.

5. Let $S = \{S_1, s_1, \ldots, s_N\}$ be a sample space for some N. Compute the entropy of each of the following probability distributions on S:

   (a) $N = 3$, $P(s_1) = 0.5$, $P(s_2) = 0.25$, $P(s_3) = 0.25$;

   (b) $N = 4$, $P(s_1) = 0.5$, $P(s_2) = 0.25$, $P(s_3) = 0.125$, $P(s_4) = 0.125$;

   (c) $N = 5$, $P(s_1) = 0.5$, $P(s_2) = 0.125$, $P(s_3) = 0.125$, $P(s_4) = 0.125$, $P(s_5) = 0.125$;

   (d) $N = 5$, $P(s_1) = 0.25$, $P(s_2) = 0.25$, $P(s_3) = 0.25$, $P(s_4) = 0.125$, $P(s_5) = 0.125$;

(e) $N = 5$, $P(s_1) = 0.0625$, $P(s_2) = 0.125$, $P(s_3) = 0.25$, $P(s_4) = 0.5$, $P(s_5) = 0.0625$.

6. Convert the following entropy values from bits to natural units: (a) $H(P) = 0.5$; (b) $H(P) = 1.0$; (c) $H(P) = 1.5$; (d) $H(P) = 2.0$; (e) $H(P) = 2.5$; (f) $H(P) = 3.0$.

7. Convert the following entropy values from natural units to bits: (a) $H(P) = 0.5$; (b) $H(P) = 1.0$; (c) $H(P) = 1.5$; (d) $H(P) = 2.0$; (e) $H(P) = 2.5$; (f) $H(P) = 3.0$.

8. Let $S = \{s_1, s_2\}$ and $T = \{t_1, t_2, t_3\}$. Let P be a joint probability distribution function on $S \times T$, given by $P(s_1, t_1) = 0.5$, $P(s_1, t_2) = 0.25$, $P(s_1, t_3) = 0.125$, $P(s_2, t_1) = 0.0625$, $P(s_2, t_2) = 0.03125$, $P(s_2, t_3) = 0.03125$. Compute the marginal distributions P_S and P_T and the entropies $H(P)$, $H(P_S)$ and $H(P_T)$. Also compute the conditional probability distributions $P_{S|T}$ and $P_{T|S}$ and their entropies $H(P_{S|T})$ and $H(P_{T|S})$.

9. Draw a diagram to represent the Markov source with alphabet $\{0, 1\}$ and set of states $\Sigma = \{\sigma_1, \sigma_2, \sigma_3\}$ with the following five transitions:

    (a) $\sigma_1 \to \sigma_2$, with label 1 and $P(2|1) = 0.4$;

    (b) $\sigma_1 \to \sigma_3$, with label 0 and $P(3|1) = 0.6$;

    (c) $\sigma_2 \to \sigma_1$, with label 0 and $P(1|2) = 0.8$;

    (d) $\sigma_2 \to \sigma_3$, with label 1 and $P(3|2) = 0.2$;

    (e) $\sigma_3 \to \sigma_1$, with label 1 and $P(1|3) = 1.0$.

    Write down the transition matrix for this source. Is it possible for this source to generate an output sequence that includes the subsequence 000? Is it possible for this source to generate an output sequence that includes the subsequence 111?

10. A 2-gram model on the language $\{A, B, C\}$ is given by the probabilities

$$P(AA) = 0.000, P(AB) = 0.200, P(AC) = 0.133,$$

$$P(BA) = 0.133, P(BB) = 0.000, P(BC) = 0.200,$$

$$P(CA) = 0.200, P(CB) = 0.133, P(CC) = 0.000.$$

The probabilities of the individual symbols are

$$P(A) = 1/3, P(B) = 1/3, P(C) = 1/3.$$

Construct a (first-order) Markov source from these models. Draw a diagram to represent the source and write down its transition matrix.

11. A 4-gram model on the language $\{0, 1\}$ is given by

$$P(1010) = 0.50, P(0101) = 0.50,$$

with all other probabilities 0. All the 3-gram probabilities are also 0, except for

$$P(010) = 0.50, P(101) = 0.50.$$

Construct a third-order Markov source from these models. Draw a diagram to represent the source and write down its transition matrix.

12. Consider a Markov source with transition matrix

$$\Pi = \begin{bmatrix} 0.3 & 0.8 \\ 0.7 & 0.2 \end{bmatrix}$$

and initial probability distribution

$$W^0 = \begin{bmatrix} 0.5 \\ 0.5 \end{bmatrix}.$$

Compute W^t for $t = 1, 2, 3, 4, 5$.

13. Find the stationary distribution of the Markov source whose transition matrix is

$$\Pi = \begin{bmatrix} 0.15 & 0.75 \\ 0.85 & 0.25 \end{bmatrix}.$$

*14. Prove that a Markov source with two states always has a stationary distribution, provided that none of the transition probabilities are 0.

15. Find the stationary distribution of the Markov source whose transition matrix is

$$\Pi = \begin{bmatrix} 0.4 & 0.3 & 0.2 \\ 0.5 & 0.2 & 0.6 \\ 0.1 & 0.5 & 0.2 \end{bmatrix}.$$

16. Compute the entropy of the Markov source in Exercise 9 above.

*17. Prove that if M^n is the nth extension of the first-order Markov source M, their entropies are related by $H(M^n) = nH(M)$.

18. Let $S = \{1, 2, 3, \ldots\}$ be the set of positive integers and let P be the probability distribution given by $P(k) = 2^{-k}$. Let f be the function on S defined by $f(k) = (-1)^{k+1}$. What is the expected value of f?

19. Let u be the probability density function of the uniform distribution over the closed and bounded interval $[a, b]$, so that $u(x) = (b - a)^{-1}$ if $a \leq x \leq b$ and $u(x) = 0$ otherwise. Compute the mean, variance and entropy (in natural units) of u.

The next two exercises prove the result that the Gaussian distribution has the maximum entropy of all distributions with a given variance.

20. Use the identity $\ln(x) \leq x - 1$ to show that if f and g are probability density functions defined on the real line then

$$\int_{-\infty}^{\infty} f(x) \ln(g(x)) dx \leq \int_{-\infty}^{\infty} f(x) \ln(f(x)) dx$$

provided both these integrals exist.

21. Show that if f is a probability density function with mean μ and variance σ^2 and g is the probability density function of a Gaussian distribution with the same mean and variance, then

$$-\int_{-\infty}^{\infty} f(x) \ln(g(x)) dx = \ln(\sqrt{2\pi e}\sigma).$$

Conclude that of all the probability density functions on the real line with variance σ^2, the Gaussian has the greatest entropy.

22. Compute the entropy of the probability density function of a uniform distribution on a closed interval whose variance is σ^2 and confirm that it is less than the entropy of a Gaussian distribution with variance σ^2. (Use the results of Exercise 19.)

The next four exercises are concerned with the *Kullback-Leibler Divergence*.

*23. Let P and Q be probability distributions on the sample space

$$S = \{s_1, s_2, \ldots, s_N\},$$

with $P(s_i) = p_i$ and $Q(s_i) = q_i$ for $1 \leq i \leq N$. The *Kullback-Leibler Divergence* of P and Q is defined by

$$D(P, Q) = \sum_{i=1}^{N} p_i \log(p_i/q_i).$$

Compute $D(P, Q)$ and $D(Q, P)$, when $N = 4$, $p_i = 0.25$ for $i = 1, 2, 3, 4$, and $q_1 = 0.125$, $q_2 = 0.25$, $q_3 = 0.125$, $q_4 = 0.5$. Is $D(P, Q) = D(Q, P)$?

*24. If P and Q are probability distributions on S, a finite sample space, show that $D(P, Q) \geq 0$ and $D(P, Q) = 0$ if and only if $P = Q$.

*25. If P, Q, and R are probability distributions on $S = \{s_1, s_2, \ldots, s_N\}$, with $P(s_i) = p_i$, $Q(s_i) = q_i$ and $R(s_i) = r_i$ for $1 \leq i \leq N$, show that

$$D(P, R) = D(P, Q) + D(Q, R).$$

if and only if

$$\sum_{i=1}^{N} (p_i - q_i) \log(q_i/r_i) = 0.$$

*26. If the sample space is the real line, it is easier to define the Kullback-Leibler Divergence in terms of probability density functions. If p and q are probability density functions on $\mathbb{R}$, with $p(x) > 0$ and $q(x) > 0$ for all $x \in \mathbb{R}$, then we define

$$D(p,q) = \int_{-\infty}^{\infty} p(x) \ln(p(x)/q(x)) dx.$$

Compute $D(p, q)$ when p is the probability density function of a Gaussian distribution with mean μ_1 and variance σ^2 and q is the probability density function of a Gaussian distribution with mean μ_2 and variance σ^2.

The next two exercises deal with the topic of *Maximum Entropy Estimation*.

*27. We have shown in Section 1.6 that the entropy of the uniform distribution on a sample space with N elements is $\log(N)$ and this is the maximum value of the entropy for any distribution defined on that sample space. In many situations, we need to find a probability distribution that satisfies certain constraints and has the maximum entropy of all the distributions that satisfy those constraints. One type of constraint that is common is to require that the mean of the distribution should have a certain value. This can be done using Lagrange multipliers. Find the probability distribution on $S = \{1, 2, 3, 4\}$ that has maximum entropy subject to the condition that the mean of the distribution is 2. To do this you have to find p_1, p_2, p_3 and p_4 that maximise the entropy

$$H(P) = \sum_{i=1}^{4} p_i \log(1/p_i)$$

subject to the two constraints

$$\sum_{i=1}^{4} p_i = 1$$

(so that the p_i form a probability distribution) and

$$\sum_{i=1}^{4} i p_i = 2.$$

*28. Find the probability distribution on $S = \{1, 2, 3, 4\}$ that has maximum entropy subject to the conditions that the mean of the distribution is 2 and the second moment of the distribution of the distribution is 5.2. In this case the constraints are

$$\sum_{i=1}^{4} p_i = 1,$$

$$\sum_{i=1}^{4} i p_i = 2,$$

and

$$\sum_{i=1}^{4} i^2 p_i = 5.2.$$

The next two exercises deal with the topic of *Mutual Information*.

*29. If P is a joint probability distribution on $S \times T$, the *Mutual Information of P*, denoted $I_M(P)$, is the Kullback-Leibler Divergence between P and the product of the marginals Q, given by

$$Q(s_i, t_j) = P_S(s_i) P_T(t_j).$$

Compute the Mutual Information of P when $S = \{s_1, s_2\}$, $T = \{t_1, t_2, t_3\}$, and P is given by $P(s_1, t_1) = 0.5$, $P(s_1, t_2) = 0.25$, $P(s_1, t_3) = 0.125$, $P(s_2, t_1) = 0.0625$, $P(s_2, t_2) = 0.03125$, $P(s_2, t_3) = 0.03125$.

*30. Show that if $S = \{s_1, s_2, \ldots, s_M\}$ and $T = \{t_1, t_2, \ldots, t_N\}$, an alternative expression for the Mutual Information of P, a joint probability distribution on $S \times T$, is given by

$$I_M(P) = \sum_{i=1}^{M} \sum_{j=1}^{N} P(s_i, t_j) \log(P(s_i|t_j)/P_S(s_i)).$$

1.21   References

[1] R. Ash, *Information Theory,* John Wiley & Sons, New York, 1965.

[2] T. M. Cover and J. A. Thomas, *Elements of Information Theory,* John Wiley & Sons, New York, 1991.

[3] D. S. Jones, *Elementary Information Theory,* Oxford University Press, Oxford, 1979.

[4] H. S. Leff and A. F. Rex, Eds., *Maxwell's Demon: Entropy, Information, Computing,* Adam Hilger, Bristol, 1990.

[5] R. D. Rosenkrantz, Ed., *E. T. Jaynes: Papers on Probability, Statistics and Statistical Physics,* D. Reidel, Dordrecht, 1983.

[6] C. E. Shannon and W. Weaver, *The Mathematical Theory Of Communication,* The University of Illinois Press, Urbana, IL, 1949.

Chapter 2

Information Channels

2.1 What Are Information Channels?

Information does not only have to be used or stored, it also has to be transmitted. In communication systems a transmitter converts, or encodes, a message to a form suitable for transmission through a communications medium, be it a fibre optic channel, satellite link or radio signal through space. The receiver then detects the transmitted signal and decodes it back to the original message. The encoding and decoding operations will be discussed in the following chapters. Although we will be dealing with digital data, physical transmission and storage has to deal with analog signals and media. Thus digital data has to be modulated in some way prior to transmission and storage, and detected by the receiver to reproduce the original digital sequence. Discussion of modern analog and digital communication systems is beyond the scope of this book and interested readers should refer to the standard textbooks in the area, for example [7].

As we will see, two forms of coding will be necessary. *Source coding* is required to efficiently store and transmit the information generated by a general-purpose source (e.g., ASCII text, images, audio, etc.) on the storage or transmission medium in use (e.g., computer disk or digital data networks, both requiring conversion to binary data). *Channel coding* is required to mitigate the effects of *noise* that may corrupt data during storage and transmission through physical systems. What is noise? Noise can be defined as any unwanted signal or effect in addition to the desired signal. As such there can be many causes of noise: interference from other signals, thermal and spurious effects generated by the electronic device being used to store or transmit the signal, environmental disturbances, etc.

In Chapter 1 we were concerned with defining the information content of a source. In this chapter we will be concerned with the equally important problem of defining or measuring the information carrying capacity of a channel. Intuitively a channel can carry no more information than is being pushed through the channel itself, that is, the entropy of the source or message being transmitted through the channel. But as we will see the presence of noise reduces the information carrying capacity of the channel and leads us to define another important quantity (alongside entropy) called *mutual information*. As we view the channel from an information theoretic point of view not surprisingly the concept of mutual information has uses beyond the information carrying capacity of a channel (as does entropy beyond measuring the information content of a source).

The following assumptions will apply in our modelling of an information channel:

Stationary The statistical nature of the channel and noise do not change with time

Memoryless The behaviour of the channel and the effect of the noise at time t will not depend on the behaviour of the channel or the effect of noise at any previous time

We now formally define a mathematical structure for an information channel.

DEFINITION 2.1 Information Channel   *An information channel is a triple* $\{A, B, P\}$, *where A is the* input alphabet, B *is the* output alphabet *and P is the* set of channel probabilities. $A = \{a_i : i = 1, 2, \ldots r\}$ *is a discrete set of $r = |A|$ symbols (where $|A|$ is the size of the input alphabet), and $B = \{b_j : j = 1, 2, \ldots s\}$ is a discrete set of $s = |B|$ symbols. The transmission behaviour of the channel is described by the probabilities in $P = \{P(b_j|a_i) : i = 1, 2, \ldots r; j = 1, 2, \ldots s\}$, where $P(b_j|a_i)$ is the probability that the output symbol b_j will be received if the input symbol a_i is transmitted.*

NOTE   The input alphabet represents the symbols transmitted into the channel and the output alphabet represents the symbols received from the channel. The definition of the channel implies that the input and output symbols may be different. In reality, one would expect that the received symbols are the same as those transmitted. However the effect of noise may introduce "new" symbols and thus we use different input and output alphabets to cater for such cases. For more general applications the channel models the matching of the input symbols to prescribed output symbols or classes which are usually different. In statistical applications the input and output symbols arise from two random variables and the channel models the joint relationship between the variables.

The conditional probabilities that describe an information channel can be represented conveniently using a matrix representation:

$$\mathbf{P} = \begin{bmatrix} P(b_1|a_1) & P(b_2|a_1) & \cdots & P(b_s|a_1) \\ P(b_1|a_2) & P(b_2|a_2) & \cdots & P(b_s|a_2) \\ \vdots & \vdots & \ddots & \vdots \\ P(b_1|a_r) & P(b_2|a_r) & \cdots & P(b_s|a_r) \end{bmatrix} \tag{2.1}$$

where $\mathbf{P}$ is the *channel matrix* and for notational convenience we may sometimes rewrite this as:

$$\mathbf{P} = \begin{bmatrix} P_{11} & P_{12} & \cdots & P_{1s} \\ P_{21} & P_{22} & \cdots & P_{2s} \\ \vdots & \vdots & \ddots & \vdots \\ P_{r1} & P_{r2} & \cdots & P_{rs} \end{bmatrix} \tag{2.2}$$

where we have defined $P_{ij} = P(b_j|a_i)$.

A graphical representation of an information channel is given in Figure 2.1.

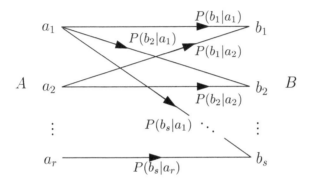

FIGURE 2.1
Graphical representation of an information channel.

The channel matrix exhibits the following properties and structure:

- Each row of $\mathbf{P}$ contains the probabilities of all possible outputs from the same input to the channel

- Each column of $\mathbf{P}$ contains the probabilities of all possible inputs to a particular output from the channel

- If we transmit the symbol a_i we must receive an output symbol with probability 1, that is:

$$\sum_{j=1}^{s} P(b_j|a_i) = 1 \quad \text{for } i = 1, 2, \ldots, r \tag{2.3}$$

that is, the probability terms in each row must sum to 1.

EXAMPLE 2.1

Consider a binary source and channel with input alphabet $\{0, 1\}$ and output alphabet $\{0, 1\}$.

Noiseless: If the channel is noiseless there will be no error in transmission, the channel matrix is given by $\mathbf{P} = \begin{bmatrix} 1 & 0 \\ 0 & 1 \end{bmatrix}$ and the channel is

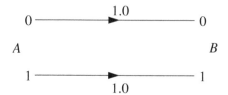

A typical input-output sequence from this channel could be:

$$\text{input:} \quad 0\ 1\ 1\ 0\ 0\ 1\ 0\ 1\ 1\ 0$$
$$\text{output:} \quad 0\ 1\ 1\ 0\ 0\ 1\ 0\ 1\ 1\ 0$$

Noisy: Say the channel is noisy and introduces a bit inversion 1% of the time, then the channel matrix is given by $\mathbf{P} = \begin{bmatrix} 0.99 & 0.01 \\ 0.01 & 0.99 \end{bmatrix}$ and the channel is

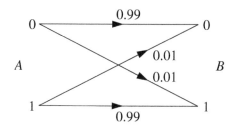

A typical input-output sequence from this channel could be:

$$\text{input:} \quad 0\ 1\ 1\ 0\ 0\ 1\ 0\ 1\ 1\ 0$$
$$\text{output:} \quad 0\ 1\ 1\ 0\ 0\ 1\ 1\ 1\ 1\ 0$$

2.2 BSC and BEC Channels

In digital communication systems the input to the channel will be the binary digits $\{0,1\}$ and this set will be the input alphabet and, ideally, also be the output alphabet. Furthermore, the effect of noise will not depend on the transmission pattern, that is, the channel is assumed *memoryless*. Two possible scenarios on the effect of noise are possible.

Ideally if there is no noise a transmitted 0 is detected by the receiver as a 0, and a transmitted 1 is detected by the receiver as a 1. However in the presence of noise the receiver may produce a different result.

The most common effect of noise is to force the detector to detect the wrong bit (bit inversion), that is, a 0 is detected as a 1, and a 1 is detected as a 0. In this case the information channel that arises is called a *binary symmetric channel* or *BSC* where $P(b = 1|a = 0) = P(b = 0|a = 1) = q$ is the probability of error (also called bit error probability, *bit error rate (BER)*, or "crossover" probability) and the output alphabet is also the set of binary digits $\{0, 1\}$. The parameter q fully defines the behaviour of the channel. The BSC is an important channel for digital communication systems as noise present in physical transmission media (fibre optic cable, copper wire, etc.) typically causes bit inversion errors in the receiver.

A BSC has channel matrix $\mathbf{P} = \begin{bmatrix} p & q \\ q & p \end{bmatrix}$ where $p = 1 - q$ and is depicted in Figure 2.2.

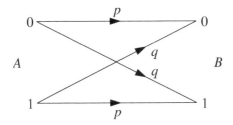

FIGURE 2.2
Binary symmetric channel.

Another effect that noise (or more usually, loss of signal) may have is to prevent the receiver from deciding whether the symbol was a 0 or a 1. In this case the output alphabet includes an additional symbol, ? , called the "erasure" symbol that denotes a bit that was not able to be detected. Thus for binary input $\{0, 1\}$, the output alphabet consists of the three symbols, $\{0, ?, 1\}$. This information channel is called a *binary erasure channel* or *BEC* where $P(b =?|a = 0) = P(b =?|a = 1) = q$ is the probability of error (also called the "erasure" probability). Strictly speaking a BEC does not model the effect of bit inversion; thus a transmitted bit is either received correctly (with probability $p = 1 - q$) or is received as an "erasure" (with probability q). A BEC is becoming an increasingly important model for wireless mobile and satellite communication channels, which suffer mainly from dropouts and loss of signal leading to the receiver failing to detect any signal.

A BEC has channel matrix $\mathbf{P} = \begin{bmatrix} p & q & 0 \\ 0 & q & p \end{bmatrix}$ and is depicted in Figure 2.3.

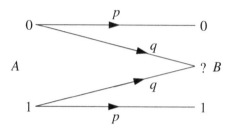

FIGURE 2.3
Binary erasure channel.

2.3 Mutual Information

To fully specify the behaviour of an information channel it is necessary to specify the characteristics of the input as well as the channel matrix. We will assume that the input characteristics are described by a probability distribution over the input alphabet, with $P(a_i)$ denoting the probability of symbol a_i being input to the channel. Then a channel will be fully specified if the input source probabilities, $P(A) = \{P(a_1), P(a_2), \ldots, P(a_r)\}$, and channel probabilities, $\mathbf{P} = [P(b_j|a_i)]_{j,i=1,1}^{s,r}$, are given.

If a channel is fully specified then the output probabilities, $P(B) = \{P(b_1), P(b_2), \ldots, P(b_s)\}$, can be calculated by:

$$P(b_j) = \sum_{i=1}^{r} P(b_j|a_i)P(a_i) \tag{2.4}$$

The probabilities $P(b_j|a_i)$ are termed the *forward probabilities* where forward indicates that the direction of channel use is with input symbol a_i being transmitted and output symbol b_j being received (i.e., a_i then b_j or b_j given a_i). We can similarly define the *backward probabilities* as $P(a_i|b_j)$ indicating the channel is running backwards: output symbol b_j occurs first followed by input symbol a_i (i.e., b_j then a_i, or a_i given b_j). The backward probabilities can be calculated by application of Bayes' Theorem as follows:

$$P(a_i|b_j) = \frac{P(a_i, b_j)}{P(b_j)} = \frac{P(b_j|a_i)P(a_i)}{P(b_j)} = \frac{P(b_j|a_i)P(a_i)}{\sum_{i=1}^{r} P(b_j|a_i)P(a_i)} \tag{2.5}$$

where $P(a_i, b_j)$ is the joint probability of a_i and b_j.

EXAMPLE 2.2

Consider the binary information channel fully specified by:

$$\mathbf{P} = \begin{bmatrix} 3/4 & 1/4 \\ 1/8 & 7/8 \end{bmatrix} \quad \text{and} \quad \begin{array}{l} P(a = 0) = 2/3 \\ P(a = 1) = 1/3 \end{array} \tag{2.6}$$

which is usually represented diagrammatically as:

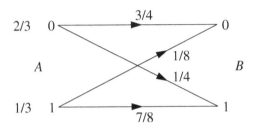

The output probabilities are calculated as follows:

$$
\begin{aligned}
P(b = 0) &= P(b = 0|a = 0)P(a = 0) + P(b = 0|a = 1)P(a = 1) \\
&= \frac{3}{4} \times \frac{2}{3} + \frac{1}{8} \times \frac{1}{3} = \frac{13}{24} \\
P(b = 1) &= 1 - P(b = 0) = \frac{11}{24}
\end{aligned}
$$

and the backward probabilities by:

$$
\begin{aligned}
P(a = 0|b = 0) &= \frac{P(b = 0|a = 0)P(a = 0)}{P(b = 0)} = \frac{\left(\frac{3}{4}\right)\left(\frac{2}{3}\right)}{\left(\frac{13}{24}\right)} = \frac{12}{13} \\
P(a = 1|b = 0) &= 1 - P(a = 0|b = 0) = \frac{1}{13} \\
P(a = 1|b = 1) &= \frac{P(b = 1|a = 1)P(a = 1)}{P(b = 1)} = \frac{\left(\frac{7}{8}\right)\left(\frac{1}{3}\right)}{\left(\frac{11}{24}\right)} = \frac{7}{11} \\
P(a = 0|b = 1) &= 1 - P(a = 1|b = 1) = \frac{4}{11}
\end{aligned}
$$

□

Conceptually we can characterise the probabilities as *a priori* if they provide the probability assignment *before* the channel is used (without any knowledge), and as *a posteriori* if the probability assignment is provided *after* the channel is used (given knowledge of the channel response). Specifically:

$P(b_j)$　a priori probability of output symbol b_j if we *do not know* which input symbol was sent

$P(b_j|a_i)$　a posteriori probability of output symbol b_j if we *know* that input symbol a_i was sent

$P(a_i)$　a priori probability of input symbol a_i if we *do not know* which output symbol was received

$P(a_i|b_j)$　a posteriori probability of input symbol a_i if we *know* that output symbol b_j was received

We can similarly refer to the *a priori entropy of A*:

$$H(A) = \sum_{a \in A} P(a) \log \frac{1}{P(a)} \tag{2.7}$$

as the average uncertainty we have about the input *before* the channel output is observed and the *a posteriori entropy of A given b_j*:

$$H(A|b_j) = \sum_{a \in A} P(a|b_j) \log \frac{1}{P(a|b_j)} \tag{2.8}$$

as the average uncertainty we have about the input *after* the channel output b_j is observed.

How does our average uncertainty about the input change after observing the output of the channel? Intuitively, we expect our uncertainty to be reduced as the channel output provides us with knowledge and knowledge reduces uncertainty. However, as we will see in the following example the output can sometimes increase our uncertainty (i.e., be more of a hindrance than a help!).

EXAMPLE 2.3

Consider the binary information channel from Example 2.2:

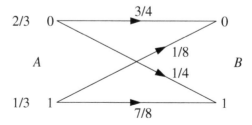

What is our uncertainty of the input that is transmitted through the channel before we observe an output from the channel? This is provided by the entropy based on the given a priori input probabilities, $P(a = 0) = \frac{2}{3}$ and $P(a = 1) = \frac{1}{3}$, yielding the a priori entropy of A, $H(A) = 0.918$.

What is our uncertainty of the input that is transmitted through the channel after we observe an output from the channel? Say we observe an output of $b = 0$, then the a posteriori input probabilities, $P(a = 0|b = 0) = \frac{12}{13}$ and $P(a = 1|b = 0) = \frac{1}{13}$, yield an a posteriori entropy of A, $H(A|b = 0) = 0.391$. Thus we reduce our uncertainty once we observe an output of $b = 0$. But what if we observe an output of $b = 1$? Then the a posteriori input probabilities become $P(a = 1|b = 1) = \frac{7}{11}$ and $P(a = 0|b = 1) = \frac{4}{11}$ and the a posteriori entropy of A, $H(A|b = 1) = 0.946$. Our uncertainty is in fact increased! The reason is that our high expectation of an input of 0, from the a priori probability $P(a = 0) = \frac{2}{3}$, is negated by receiving an output of 1. Thus $P(a = 0|b = 1) = \frac{4}{11}$ is closer to equi-probable than $P(a = 0) = \frac{2}{3}$ and this increases the a posteriori entropy.

Notwithstanding the fact that $H(A|b = 1) > H(A)$ even though $H(A|b = 0) < H(A)$ we can show that if we average across all possible outputs the channel will indeed reduce our uncertainty, that is:

$$
\begin{aligned}
H(A|B) &= \sum_{j=1}^{2} P(b_j) H(A|b_j) \\
&= P(b = 0)H(A|b = 0) + P(b = 1)H(A|b = 1) \\
&= 0.645
\end{aligned}
$$

and thus $H(A|B) < H(A)$. ◻

The average of the a posterior entropies of A, $H(A|B)$, calculated in Example 2.3 is sometimes referred to as the *equivocation of A with respect to B* where equivocation is used to refer to the fact that $H(A|B)$ measures the amount of uncertainty or equivocation we have about the input A when observing the output B. Together with the a priori entropy of A, $H(A)$, we can now establish a measure of how well a channel transmits information from the input to the output. To derive this quantity consider the following interpretations:

$H(A)$ average uncertainty (or surprise) of the input to the channel *before* observing the channel output;

$H(A|B)$ average uncertainty (or equivocation) of the input to the channel *after* observing the channel output;

$H(A) - H(A|B)$ reduction in the average uncertainty of the input to the channel *provided* or *resolved* by the channel.

> **DEFINITION 2.2 Mutual Information**   *For input alphabet A and output alpha-bet B the term*
>
> $$I(A;B) = H(A) - H(A|B) \tag{2.9}$$
>
> *is the* mutual information *between A and B*

The mutual information, $I(A;B)$, indicates the information about A, $H(A)$, that is provided by the channel minus the degradation from the equivocation or uncertainty, $H(A|B)$. The $H(A|B)$ can be construed as a measure of the "noise" in the channel since the noise directly contributes to the amount of uncertainty we have about the channel input, A, given the channel output B.

Consider the following cases:

noisefree If $H(A|B) = 0$ this implies $I(A;B) = H(A)$ which means the channel is able to provide all the information there is about the input, i.e., $H(A)$. This is the best the channel will ever be able to do.

noisy If $H(A|B) > 0$ but $H(A|B) < H(A)$ then the channel is noisy and the input information, $H(A)$, is reduced by the noise, $H(A|B)$, so that the channel is only able to provide $I(A;B) = H(A) - H(A|B)$ amount of information about the input.

ambiguous If $H(A|B) = H(A)$ the amount of noise totally masks the contribution of the channel and the channel provides $I(A;B) = 0$ information about the input. In other words the channel is useless and is no better than if the channel was not there at all and the outputs were produced independently of the inputs!

An alternative expression to Equation 2.9 for the mutual information can be derived as follows:

$$
\begin{aligned}
I(A;B) &= H(A) - H(A|B) \\
&= \sum_{a \in A} P(a) \log \frac{1}{P(a)} - \sum_{a \in A} \sum_{b \in B} P(a,b) \log \frac{1}{P(a|b)} \\
&= \sum_{a \in A} \sum_{b \in B} P(a,b) \log \frac{1}{P(a)} - \sum_{a \in A} \sum_{b \in B} P(a,b) \log \frac{1}{P(a|b)} \\
&= \sum_{a \in A} \sum_{b \in B} P(a,b) \log \frac{P(a|b)}{P(a)} \tag{2.10}
\end{aligned}
$$

Using Equation 2.5 the mutual information can be expressed more compactly:

> **RESULT 2.1 Alternative expressions for Mutual Information**
>
> $$I(A;B) = \sum_{a \in A} \sum_{b \in B} P(a,b) \log \frac{P(a,b)}{P(a)P(b)} = \sum_{a \in A} \sum_{b \in B} P(a)P(b|a) \log \frac{P(b|a)}{P(b)} \tag{2.11}$$

2.3.1 Importance of Mutual Information

The mutual information has been defined in the context of measuring the information carrying capacity of communication channels. However the concept of mutual information has had far-reaching effects on solving difficult estimation and data analysis problems in biomedical applications [8], image processing and signal processing. In these applications the key in using mutual information is that it provides a measure of the independence between two random variables or distributions.

In image processing [12] and speech recognition [9] the use of the *maximum mutual information* or *MMI* between the observed data and available models has yielded powerful strategies for training the models based on the data in a discriminative fashion. In signal processing for communication systems the idea of minimising the mutual information between the vector components for separating mutually interfering signals [3] has led to the creation of a new area of research for signal separation based on the idea of *independent component analysis* or *ICA* .

2.3.2 Properties of the Mutual Information

From Example 2.3 we saw that for specific values of the output alphabet, B, either $H(A) > H(A|b_j)$ or $H(A) < H(A|b_j)$ but when we averaged over the output alphabet then $H(A) > H(A|B)$. This implies that $I(A; B) > 0$ for this case. But what can we say about $I(A; B)$ for other cases? We restate the following result from Chapter 1:

$$\sum_{i=1}^{N} p_i \log \frac{q_i}{p_i} \leq 0 \tag{2.12}$$

for two sources of size N with symbol probabilities, p_i and q_i, respectively, and equality only if $p_i = q_i$ for $i = 1, 2, \ldots, N$. Let $p_i \equiv P(a, b)$ and $q_i \equiv P(a)P(b)$ and $N \equiv |A \times B|$. Then from Equations 2.11 and 2.12 we get:

$$-I(A; B) = \sum_{a \in A} \sum_{b \in B} P(a, b) \log \frac{P(a)P(b)}{P(a, b)}$$

$$\equiv \sum_{a \in A} \sum_{b \in B} p_i \log \frac{r_i}{p_i} \leq 0 \tag{2.13}$$

That is:

RESULT 2.2

The mutual information is a non-negative quantity:

$$I(A; B) \geq 0 \tag{2.14}$$

with $I(A; B) = 0$ if and only if $P(a, b) = P(a)P(b)$ $\forall a, b$, i.e., the input and output alphabets are statistically independent.

The expression for $I(A; B)$ provided by Equation 2.11 is symmetric in the variables a and b. Thus by exchanging A and B we get the following result:

$$I(A; B) = I(B; A) \qquad (2.15)$$

or

$$H(A) - H(A|B) = H(B) - H(B|A) \qquad (2.16)$$

NOTE   The information the channel provides about A upon observing B is the same as the information the channel provides about B upon noting A was sent.

The term $H(B|A)$ is sometimes referred to as the *equivocation of B with respect to A* and measures the uncertainty or equivocation we have about the output B given the input A.

RESULT 2.3
From the fact that $I(A; B) = I(B; A) \geq 0$ it can be stated that, in general and on average, uncertainty is decreased when we know something, *that is:*

$$H(A|B) \leq H(A) \quad \text{and} \quad H(B) \leq H(B|A) \qquad (2.17)$$

Other expressions and relationships between the entropies, $H(A)$, $H(B)$, the equivocation, $H(A|B)$, $H(B|A)$, the joint entropy, $H(A, B)$ and the mutual information, $I(A; B) = I(B; A)$, can be derived. All these relations can be summarised by the Venn diagram of Figure 2.4. From Figure 2.4 additional expressions involving the

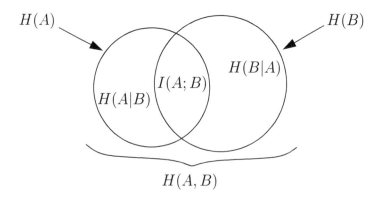

FIGURE 2.4
Relationship between all entropy and mutual information expressions.

joint entropy can be derived:

$$H(A, B) = H(A) + H(B) - I(A; B)$$
$$= H(A) + H(B|A)$$
$$= H(B) + H(A|B) \qquad (2.18)$$

The relation $H(A, B) = H(A) + H(B) - I(A; B)$ can be stated conceptually as:

"The total uncertainty in both A and B $[H(A, B)]$ is the sum of the uncertainties in A and B $[H(A) + H(B)]$ minus the information provided by the channel $[I(A; B)]$"

whereas the relation $H(A, B) = H(A) + H(B|A)$ becomes:

"The total uncertainty in both A and B $[H(A, B)]$ is the sum of the uncertainty in A plus the remaining uncertainty in B after we are given A."

REMARK 2.1   Given the input probabilities, $P(a)$, and channel matrix, **P**, only three quantities need to be calculated directly from the individual probabilities to completely determine the Venn diagram of Figure 2.4 and the remaining three can be derived from the existing entropy expressions.   ⬜

EXAMPLE 2.4

From Example 2.3 we calculated:

$$H(A) = 0.918$$
$$H(A|B) = 0.645$$

and from Example 2.2 we can calculate:

$$H(B) = \sum_{b \in B} P(b) \log \frac{1}{P(b)} = 0.995$$

The quantities $H(A)$, $H(B)$ and $H(A|B)$ completely determine the Venn diagram and the remaining quantities can be derived by:

$$I(A; B) = H(A) - H(A|B) = 0.273$$
$$H(B|A) = H(B) - I(A; B) = 0.722$$
$$H(A, B) = H(A) + H(B) - I(A; B) = 1.640$$

⬜

The mutual information of a BSC can be expressed algebraically by defining:

$$w = P(a = 0) \Rightarrow \text{probability that a 0 is transmitted}$$
$$q = P(b = 1|a = 0) \Rightarrow \text{bit error probability}$$

and hence:

$$\bar{w} = (1 - w) \Rightarrow \text{probability that a 1 is transmitted}$$
$$p = (1 - q) \Rightarrow \text{probability of no error}$$

Then the output probabilities are given by:

$$P(b = 0) = wp + \bar{w}q$$
$$P(b = 1) = \bar{w}p + wq$$

A sketch of the BSC showing both input and output probabilities is given in Figure 2.5.

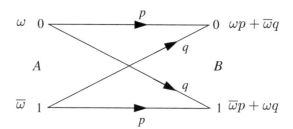

FIGURE 2.5
Mutual information of a BSC.

The expression for $H(B)$ is:

$$H(B) = (wp + \bar{w}q) \log \frac{1}{(wp + \bar{w}q)} + (\bar{w}p + wq) \log \frac{1}{(\bar{w}p + wq)} \qquad (2.19)$$

The simplified expression for $H(B|A)$ is:

$$H(B|A) = \sum_{a \in A} P(a) \sum_{b \in B} P(b|a) \log \frac{1}{P(b|a)}$$

$$= w \left(q \log \frac{1}{q} + p \log \frac{1}{p} \right) + \bar{w} \left(q \log \frac{1}{q} + p \log \frac{1}{p} \right)$$

$$= q \log \frac{1}{q} + p \log \frac{1}{p} \qquad (2.20)$$

and the mutual information is: $I(A; B) = H(B) - H(B|A)$.

EXAMPLE 2.5

What is the mutual information of a BSC for the following cases?

1. $p = q = 0.5$: The channel operates in an *ambiguous* manner since the errors are as likely as no errors, the output symbols are equally likely, $H(B) = 1$ no matter what is happening at the input (since $w + \overline{w} = 1$), the equivocation is $H(B|A) = 1$ and the mutual information is $I(A; B) = 0$.

2. $p = 1.0$: The channel is noisefree, $H(B) = H(A) = w \log \frac{1}{w} + \overline{w} \log \frac{1}{\overline{w}}$, $H(B|A) = H(A|B) = 0$ and the mutual information is $I(A; B) = H(B) = H(A)$.

3. $w = 0.5$: The source exhibits maximum entropy with $H(A) = 1$, the output entropy also exhibits maximum entropy with $H(B) = 1$ (since $p + q = 1$) and the mutual information is given by $I(A; B) = 1 - \left(q \log \frac{1}{q} + p \log \frac{1}{p} \right)$.

4. $w = 1.0$: The source contains no information since $H(A) = 0$, the output entropy, $H(B)$, is the same as the channel uncertainty, $H(B|A)$, and the mutual information is $I(A; B) = 0$ since there is no information to transmit.

☐

The mutual information of a BEC can be similarly expressed algebraically. Adopting the same notation used for the BSC the output probabilities of the BEC are given by:

$$P(b = 0) = wp$$
$$P(b = 1) = \overline{w}p$$
$$P(b =?) = wq + \overline{w}q = q$$

A sketch of the BEC showing both input and output probabilities is given in Figure 2.6.

The expression for $H(B)$ is:

$$H(B) = wp \log \frac{1}{wp} + \overline{w}p \log \frac{1}{\overline{w}p} + q \log \frac{1}{q} \tag{2.21}$$

The simplified expression for $H(B|A)$ reduces to the same expression as Equation 2.20:

$$H(B|A) = q \log \frac{1}{q} + p \log \frac{1}{p} \tag{2.22}$$

and the mutual information is: $I(A; B) = H(B) - H(B|A)$.

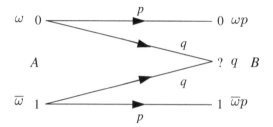

FIGURE 2.6
Mutual information of a BEC.

2.4   Noiseless and Deterministic Channels

We have already discussed the BSC and BEC structures as important channel models for describing modern digital communication systems. We have also loosely referred to channels as being *noisefree*, *noisy* and *ambiguous*. We now formally define *noiseless* channels as those channels that are not subject to the effects of noise (i.e., no uncertainty of the input that was transmitted given the output that was received). We also define a dual class of channels called *deterministic* channels for which we can determine the output that will be received given the input that was transmitted.

2.4.1   Noiseless Channels

A noiseless channel will have either the same or possibly more output symbols than input symbols and is such that there is no noise, ambiguity, or uncertainty of which input caused the output.

> **DEFINITION 2.3 Noiseless Channel**   *A channel in which there are at least as many output symbols as input symbols, but in which each of the output symbols can be produced by the occurrence only of a particular one of the input symbols is called a* noiseless channel. *The channel matrix of a noiseless channel has the property that there is one, and only one, non-zero element in each column.*

EXAMPLE 2.6

The following channel with 6 outputs and 3 inputs is noiseless because we know, with certainty 1, which input symbol, $\{a_1, a_2, a_3\}$, was transmitted given knowledge of the received output symbol, $\{b_1, b_2, b_3, b_4, b_5, b_6\}$.

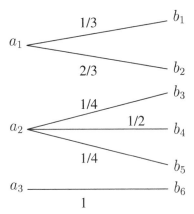

The corresponding channel matrix is given by:

$$P = \begin{bmatrix} 1/3 & 2/3 & 0 & 0 & 0 & 0 \\ 0 & 0 & 1/4 & 1/2 & 1/4 & 0 \\ 0 & 0 & 0 & 0 & 0 & 1 \end{bmatrix}$$

and we note that there is only one non-zero element in each column. ⬚

What can we say about the mutual information through a noiseless channel? Let b_j be the received output. Then, from the definition of a noiseless channel, we know which input, say a_{i*}, was transmitted and we know this with certainty 1. That is, $P(a_{i*}|b_j) = 1$ for a_{i*} and hence $P(a_i|b_j) = 0$ for all other a_i. The equivocation $H(A|B)$ becomes:

$$H(A|B) = \sum_{a \in A} \sum_{b \in B} P(a, b) \log \frac{1}{P(a|b)}$$

$$= \sum_{b \in B} P(b) \sum_{a \in A} P(a|b) \log \frac{1}{P(a|b)}$$

$$= 0$$

since:

$$\text{if } P(a|b) = 0 \text{ then } 0 \log \frac{1}{0} = 0$$

$$\text{if } P(a|b) = 1 \text{ then } 1 \log \frac{1}{1} = 0$$

and hence $\sum_{a \in A} P(a|b) \log \frac{1}{P(a|b)} = 0$.

The mutual information is then given by the following result.

RESULT 2.4

The mutual information for noiseless channels is given by:

$$I(A; B) = H(A) \qquad (2.23)$$

That is, the amount of information provided by the channel is the same as the information sent through the channel.

2.4.2 Deterministic Channels

A deterministic channel will have either the same or possibly more input symbols than output symbols and is such that we can determine which output symbol will be received when a particular input symbol is transmitted.

DEFINITION 2.4 Deterministic Channel   *A channel in which there are at least as many input symbols as output symbols, but in which each of the input symbols is capable of producing only one of the output symbols is called a* deterministic channel. *The channel matrix of a deterministic channel has the property that there is one, and only one, non-zero element in each row, and since the entries along each row must sum to 1, that non-zero element is equal to 1.*

EXAMPLE 2.7

The following channel with 3 outputs and 6 inputs is deterministic because we know, with certainty 1, which output symbol, $\{b_1, b_2, b_3\}$, will be received given knowledge of the transmitted input symbol, $\{a_1, a_2, a_3, a_4, a_5, a_6\}$.

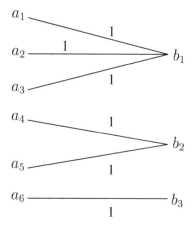

The corresponding channel matrix is given by:

$$P = \begin{bmatrix} 1 & 0 & 0 \\ 1 & 0 & 0 \\ 1 & 0 & 0 \\ 0 & 1 & 0 \\ 0 & 1 & 0 \\ 0 & 0 & 1 \end{bmatrix}$$

and we note that there is only one nonzero element in each row and that the element is 1. ⬚

What is the mutual information of a deterministic channel? Let a_i be the transmitted input symbol. Then, from the definition of a deterministic channel, we know that the received output will be, say, b_{j^*} and we know this with certainty 1. That is, $P(b_{j^*}|a_i) = 1$ for b_{j^*} and hence $P(b_j|a_i) = 0$ for all other b_j. The equivocation $H(B|A)$ then becomes:

$$H(B|A) = \sum_{a \in A} \sum_{b \in B} P(a,b) \log \frac{1}{P(b|a)}$$

$$= \sum_{a \in A} P(a) \sum_{b \in B} P(b|a) \log \frac{1}{P(b|a)}$$

$$= 0$$

since $\sum_{b \in B} P(b|a) \log \frac{1}{P(b|a)} = 0$.

The mutual information is given by the following result.

RESULT 2.5

The mutual information for deterministic channels is given by:

$$I(A; B) = H(B) \tag{2.24}$$

That is, the amount of information provided by the channel is the same as the information produced by the channel output.

2.5 Cascaded Channels

In most typical cases of information transmission and storage the data are passed through a cascade of different channels rather than through just the one channel.

One example of where this happens is in modern data communication systems where data can be sent through different physical transmission media links (e.g., copper wire, optical fibre) and wireless media links (e.g., satellite, radio) from transmitter to receiver. Each of these links can be modelled as an independent channel and the complete transmission path as a cascade of such channels. What happens to the information when passed through a cascade of channels as compared to a single channel only?

Intuitively we would expect additive loss of information arising from the cumulative effects of uncertainty (or equivocation) from each channel in the cascade. Only for the case of a noiseless channel would we expect no additional loss of information in passing data through that channel. We verify these and other results when, without loss of generality, we derive the mutual information of a cascade of two channels and compare this with the mutual information through the first channel.

Consider a pair of channels in cascade as shown in Figure 2.7. In Figure 2.7 the

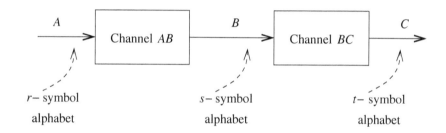

FIGURE 2.7
Cascade of two channels.

output of channel AB is connected to the input of channel BC. Thus the symbol alphabet B of size s is both the output from channel AB and the input to channel BC. Say the input symbol a_i is transmitted through channel AB and this produces b_j as the output from channel AB. Then b_j forms the input to channel BC which, in turn, produces c_k as the output from channel BC. The output c_k depends solely on b_j, not on a_i. Thus we can define a cascade of channels as occurring when the following condition holds:

$$P(c_k|a_i, b_j) = P(c_k|b_j) \quad \forall i, j, k \tag{2.25}$$

Similarly we can also state that the following will also be true for a cascade of channels:

$$P(a_i|b_j, c_k) = P(a_i|b_j) \quad \forall i, j, k \tag{2.26}$$

The problem of interest is comparing the mutual information through channel AB only, $I(A; B)$, with the mutual information through the cascade of channel AB with channel BC, that is, $I(A; C)$. To this end we first show that $H(A|C) - H(A|B) \geq 0$ as follows:

$$H(A|C) - H(A|B) = \sum_{a \in A} \sum_{c \in C} P(a, c) \log \frac{1}{P(a|c)}$$

$$- \sum_{a \in A} \sum_{b \in B} P(a, b) \log \frac{1}{P(a|b)}$$

$$= \sum_{a \in A} \sum_{b \in B} \sum_{c \in C} P(a, b, c) \log \frac{P(a|b)}{P(a|c)} \qquad (2.27)$$

Equation 2.26 gives $P(a|b) = P(a|b, c)$, noting that $P(a, b, c) = P(b, c)P(a|b, c)$, and given that $\ln \frac{1}{x} \geq 1 - x$ with equality when $x = 1$ we can now state:

$$H(A|C) - H(A|B) \geq \frac{1}{\ln 2} \sum_{a \in A} \sum_{b \in B} \sum_{c \in C} P(b, c)P(a|b, c) \left(1 - \frac{P(a|c)}{P(a|b, c)}\right)$$

$$\geq \frac{1}{\ln 2} \sum_{b \in B} \sum_{c \in C} P(b, c) \left(\sum_{a \in A} P(a|b, c) - \sum_{a \in A} P(a|c)\right)$$

$$\geq \frac{1}{\ln 2} \sum_{b \in B} \sum_{c \in C} P(b, c) (1 - 1) = 0 \qquad (2.28)$$

with equality when $P(a|b) = P(a|c)$ such that $P(b, c) \neq 0$.

Since $I(A; B) = H(A) - H(A|B)$ and $I(A; C) = H(A) - H(A|C)$ we have the following result.

RESULT 2.6
For the cascade of channel AB with channel BC it is true that:

$$I(A; B) \geq I(A; C) \quad \text{with equality iff } P(a|c) = P(a|b) \text{ when } P(b, c) \neq 0$$
$$(2.29)$$

That is, channels tend to leak information and the amount of information out of a cascade can be no greater (and is usually less) than the information from the first channel.

CLAIM 2.1

If channel BC is noiseless then $I(A; B) = I(A; C)$

PROOF For noiseless channel BC if $P(b, c) \neq 0$ then this implies that $P(b|c) = 1$. For the cascade of channel AB with BC the condition for $I(A; B) = I(A; C)$ when

$P(b, c) \neq 0$ is $P(a|c) = P(a|b)$. From Bayes' Theorem we can show that:

$$P(a|c) = \sum_{b \in B} P(a|b, c)P(b|c) \qquad (2.30)$$

For a cascade we know that $P(a|b, c) = P(a|b)$ and since $P(b|c) = 1$ when $P(b, c) \neq 0$, and $P(b|c) = 0$ otherwise, for noiseless channel BC this gives the required result that:

$$P(a|c) = P(a|b) \qquad (2.31)$$

and hence $I(A; B) = I(A; C)$.   □

The converse of Claim 2.1, however, is not true since, surprisingly, there may exist particular channel combinations and input distributions which give rise to $I(A; B) = I(A; C)$ even if channel BC is not noiseless. The following example illustrates this point.

EXAMPLE 2.8

Consider the cascade of channel AB:

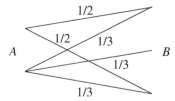

with channel BC:

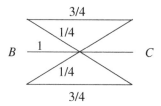

which produces the cascaded channel AC:

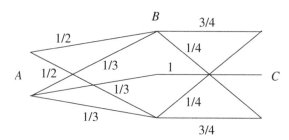

The corresponding channel matrices for AB and BC are:

$$\mathbf{P}_{AB} = \begin{bmatrix} 1/2 & 0 & 1/2 \\ 1/3 & 1/3 & 1/3 \end{bmatrix} \qquad \mathbf{P}_{BC} = \begin{bmatrix} 3/4 & 0 & 1/4 \\ 0 & 1 & 0 \\ 1/4 & 0 & 3/4 \end{bmatrix}$$

Obviously channel BC is <u>not</u> noiseless. Nevertheless it is true that:

$$I(A; B) = I(A; C)$$

by virtue of the fact that the channel matrix for the cascaded channel AC:

$$\mathbf{P}_{AC} = \mathbf{P}_{AB}\mathbf{P}_{BC} = \begin{bmatrix} 1/2 & 0 & 1/2 \\ 1/3 & 1/3 & 1/3 \end{bmatrix} \begin{bmatrix} 3/4 & 0 & 1/4 \\ 0 & 1 & 0 \\ 1/4 & 0 & 3/4 \end{bmatrix} = \begin{bmatrix} 1/2 & 0 & 1/2 \\ 1/3 & 1/3 & 1/3 \end{bmatrix}$$

is identical to the channel matrix for channel AB! ⬚

2.6 Additivity of Mutual Information

In the previous section it was shown that when information channels are cascaded there is a tendency for information loss, unless the channels are noiseless. Of particular interest to communication engineers is the problem of how to reduce information loss, especially when confronted with transmission through a noisy channel. The practical outcome of this is the development of channel codes for reliable transmission (channel codes are discussed in Chapters 5 to 9). From a purely information theoretic point of view channel coding represents a form of additivity of mutual information. Additivity is achieved when we consider the average information provided by the channel about a single input symbol upon observing a *succession of output symbols*. Such a multiplicity of outputs can occur spatially or temporally. Spatial multiplicity occurs when the same input is transmitted simultaneously through more than one channel, thereby producing as many outputs as there are channels. Temporal multiplicity occurs when the same input is repeatedly transmitted through the same channel, thereby producing as many outputs as repeat transmissions. Both cases minimise information loss by exploiting redundancy, either extra channel space or extra time to repeat the same input. We now develop an expression for the mutual information for the special case of two channel outputs (original plus repeat) and show that there is a gain of mutual information over using the single original channel output.

Consider the channel system shown in Figure 2.8 where A is the input alphabet of size $r = |A|$, B is the original output alphabet of size $s = |B|$ and C is the repeat output alphabet of size $t = |C|$. The output B can be construed as the output from

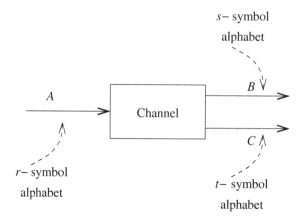

FIGURE 2.8
Channel system with two outputs.

the first channel AB (if there are two physical channels) or first transmission (if there is one physical channel) and output C can be construed as the output of the second channel AC or second (repeat) transmission.

To develop an expression for the mutual information we extend our notion of *a priori* and *a posteriori* probabilities and entropies as discussed in Section 2.3 to include the contribution of a second output as follows:

$P(a_i)$: a priori probability of input symbol a_i if we *do not know* which output symbol was received

$P(a_i|b_j)$: a posteriori probability of input symbol a_i if we *know* that output symbol b_j was received

$P(a_i|b_j, c_k)$: a posteriori probability of input symbol a_i if we *know* that *both* output symbols, b_j and c_k, were received.

Thus we can define the following a posteriori entropy:

$$H(A|b_j, c_k) = \sum_{a \in A} P(a|b_j, c_k) \log \frac{1}{P(a|b_j, c_k)} \qquad (2.32)$$

the equivocation of A with respect to B and C by:

$$H(A|B, C) = \sum_{b \in B} \sum_{c \in C} P(b, c) H(A|b, c) \qquad (2.33)$$

and the amount of information channels AB and AC provide about A, that is the mutual information of BC and A, by:

$$I(A; B, C) = H(A) - H(A|B, C) \qquad (2.34)$$

What can we say about $I(A; B) = H(A) - H(A|B)$, the amount of information channel AB provides about A, in comparison to $I(A; B, C)$, the amount of information both channels AB and AC provide about A? Expand Equation 2.34 as follows:

$$I(A; B, C) = \{H(A) - H(A|B)\} + \{H(A|B) - H(A|B, C)\} \qquad (2.35)$$
$$= I(A; B) + I(A; C|B)$$

where $I(A; C|B)$ is the amount of information channel AC additionally provides about A, after using channel AB. It can be shown that:

$$\mathrm{I}(A; C|B) \geq 0 \quad \text{with equality iff } H(A|B) = H(A|B, C) \qquad (2.36)$$

Thus we have the following result.

RESULT 2.7
For a channel with input A and dual outputs B and C it is true that:

$$I(A; B, C) \geq I(A; B) \quad \text{with equality iff } H(A|B) = H(A|B, C) \qquad (2.37)$$

That is, dual use of a channel provides more information than the single use of a channel.

EXAMPLE 2.9

Consider a BSC where the input symbol, a_i, is transmitted through the channel twice. Let b_j represent the output from the original transmission and let c_k be the output from the repeat transmission. Since the same channel is used:

$$\mathbf{P}_{AB} = \mathbf{P}_{AC} = \begin{bmatrix} p & q \\ q & p \end{bmatrix}$$

For simplicity we assume that $P(a = 0) = P(a = 1) = 0.5$. From Example 2.5 we have that for $\omega = 0.5$:

$$I(A; B) = 1 - \left(q \log \frac{1}{q} + p \log \frac{1}{p} \right)$$

What is the expression for $I(A; B, C)$ and how does it compare to $I(A; B)$? A direct extension of Equation 2.11 yields the following expression for $I(A; B, C)$:

$$I(A; B, C) = \sum_{a \in A} \sum_{b \in B} \sum_{c \in C} P(a, b, c) \log \frac{P(a, b, c)}{P(a)P(b, c)} \qquad (2.38)$$

from which we can state that:

$$P(a) = 0.5$$

$$P(b, c) = \sum_{a \in A} P(a, b, c)$$

$$P(a, b, c) = P(a)P(b|a)P(c|b, a) = \frac{1}{2}P(b|a)P(c|a)$$

where we note that $P(c|b, a) = P(c|a)$ since the repeat output does not depend on the original output. We list the contribution of each term in the expression of Equation 2.38 as shown by Table 2.1.

Table 2.1 Individual terms of $I(A; B, C)$ expression

a bc	$P(a, b, c)$	$P(b, c)$	Type
0 00	$\frac{1}{2}p^2$	$\frac{1}{2}(p^2 + q^2)$	X
1 11	$\frac{1}{2}p^2$		X
0 01	$\frac{1}{2}pq$	pq	Z
1 01	$\frac{1}{2}pq$		Z
0 10	$\frac{1}{2}pq$	pq	Z
1 10	$\frac{1}{2}pq$		Z
0 11	$\frac{1}{2}q^2$	$\frac{1}{2}(p^2 + q^2)$	Y
1 00	$\frac{1}{2}q^2$		Y

The **Type** column assigns a class label to each term. The type **X** terms represent the case of no error in both the original or repeat outputs and these positively reinforce the information provided by the dual use of the channel. Collecting the type **X** terms we get:

$$2 \cdot \left(\frac{1}{2}p^2 \log \frac{\frac{1}{2}p^2}{\frac{1}{2} \cdot \frac{1}{2}(p^2 + q^2)} \right) = p^2 \log \frac{2p^2}{p^2 + q^2}$$

The type **Y** terms represent the case of complete error in both the original and repeat outputs and these negatively reinforce the information provided by the dual use of the channel. Collecting the type **Y** terms we get:

$$2 \cdot \left(\frac{1}{2}q^2 \log \frac{\frac{1}{2}q^2}{\frac{1}{2} \cdot \frac{1}{2}(p^2 + q^2)} \right) = q^2 \log \frac{2q^2}{p^2 + q^2}$$

The type **Z** terms, however, indicate contradictory original and repeat outputs which effectively negate any information that the dual use of the channel may provide. Collecting the type **Z** terms we see that these make no contribution to the mutual information:

$$4. \left(\frac{1}{2}pq \log \frac{\frac{1}{2}pq}{\frac{1}{2} \cdot pq} \right) = 0$$

Putting it all together we get the final expression for $I(A; B, C)$:

$$I(A; B, C) = p^2 \log \frac{2p^2}{p^2 + q^2} + q^2 \log \frac{2q^2}{p^2 + q^2}$$

By plotting $I(A; B)$ and $I(A; B, C)$ for different values of q we see from Figure 2.9 that $I(A; B, C) \geq I(A; B)$. It is interesting to note the conditions for equality,

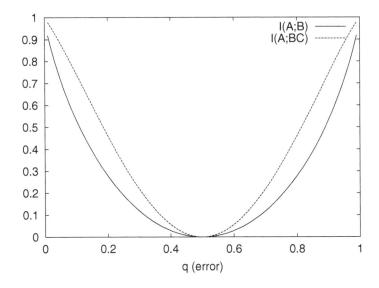

FIGURE 2.9
Comparison of $I(A; B)$ with $I(A; B, C)$.

$I(A; B) = I(A; B, C)$. This happens when:

- $q = 0$, there is 100% no error, so $I(A; B) = I(A; B, C) = 1$
- $q = 1$, there is 100% bit reversal, so $I(A; B) = I(A; B, C) = 1$
- $q = 0.5$, there is total ambiguity, so $I(A; B) = I(A; B, C) = 0$

2.7   Channel Capacity: Maximum Mutual Information

Consider an information channel with input alphabet A, output alphabet B and channel matrix $\mathbf{P}_{AB}$ with conditional channel probabilities $P(b_j|a_i)$. The mutual information:

$$I(A; B) = \sum_{a \in A} \sum_{b \in B} P(a, b) \log \frac{P(a, b)}{P(a)P(b)} \tag{2.39}$$

which, if we now assume the logarithm is base 2, indicates the amount of information the channel is able to carry in bits per symbol transmitted. The maximum amount of information carrying capacity for the channel is $H(A)$, the amount of information that is being transmitted through the channel. But this is reduced by $H(A|B)$, which is an indication of the amount of "noise" present in the channel.

The expression for mutual information depends not only on the channel probabilities, $P(b_j|a_i)$, which uniquely identify a channel, but also on how the channel is used, the input or source probability assignment, $P(a_i)$. As such $I(A; B)$ cannot be used to provide a unique and comparative measure of the information carrying capacity of a channel since it depends on how the channel is used. One solution is to ensure that the same probability assignment (or input distribution) is used in calculating the mutual information for different channels. The questions then is: which probability assignment should be used? Obviously we can't use an input distribution with $H(A) = 0$ since that means $I(A; B) = 0$ for whatever channel we use! Intuitively an input distribution with maximum information content (i.e., maximum $H(A)$) makes sense. Although this allows us to compare different channels the comparison will not be fair since another input distribution may permit certain channels to exhibit a higher value of mutual information.

The usual solution is to allow the input distribution to vary for each channel and to determine the input distribution that produces the maximum mutual information for that channel. That is we attempt to calculate the maximum amount of information a channel can carry in any single use (or source assignment) of that channel, and we refer to this measure as the capacity of the channel.

DEFINITION 2.5 Channel Capacity   *The maximum average mutual information, $I(A; B)$, in any single use of a channel defines the* channel capacity. *Mathematically, the channel capacity, C, is defined as:*

$$C = \max_{\{P_A(a)\}} I(A; B) \tag{2.40}$$

that is, the maximum mutual information over all possible input probability assignments, $P_A(a)$.

The channel capacity has the following properties:

1. $C \geq 0$ since $I(A; B) \geq 0$

2. $C \leq \min\{\log|A|, \log|B|\}$ since $C = \max I(A; B)$ and $\max I(A; B) \leq \max H(A) = \log|A|$ when considering the expression $I(A; B) = H(A) - H(A|B)$, and $\max I(A; B) \leq \max H(B) = \log|B|$ when considering the expression $I(A; B) = H(B) - H(B|A)$

The calculation of C involves maximisation of $I(A; B)$ over $r = \log|A|$ independent variables (the input probabilities, $\{P(a_i) : i = 1, 2, \ldots r\}$) subject to the two constraints:

1. $P(a_i) \geq 0 \quad \forall a_i$

2. $\sum_{i=1}^{r} P(a_i) = 1$

This constrained maximisation of a non-linear function is not a trivial task. Methods that can be used include:

- standard constrained maximisation techniques like the method of Lagrangian multipliers

- gradient search algorithms

- derivation for special cases (e.g., weakly symmetric channels)

- the iterative algorithms developed by Arimoto [1] and Blahut [2]

2.7.1 Channel Capacity of a BSC

For a BSC, $I(A; B) = H(B) - H(B|A)$ where $H(B)$ is given by Equation 2.19 and $H(B|A)$ is given by Equation 2.20. We want to examine how the mutual information varies with different uses of the same channel. For the same channel the channel probabilities, p and q, remain constant. However, for different uses the input probability, $\omega = P(a = 0)$, varies from 0 to 1. From Figure 2.10 the maximum mutual information occurs at $P(a = 0) = \frac{1}{2}$.

For $\omega = P(a = 0) = \frac{1}{2}$ the mutual information expression simplifies to:

$$I(A; B) = 1 - \left(p \log \frac{1}{p} + q \log \frac{1}{q}\right) \tag{2.41}$$

Since this represents the maximum possible mutual information we can then state:

$$C_{BSC} = 1 - \left(p \log \frac{1}{p} + q \log \frac{1}{q}\right) \tag{2.42}$$

How does the channel capacity vary for different error probabilities, q? From Figure 2.11 the following observations can be made:

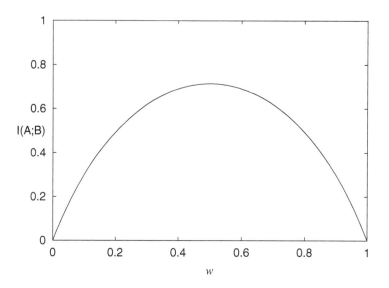

FIGURE 2.10
Variation of $I(A; B)$ with different source assignments for a typical BSC.

- When $q = 0$ or 1, that is, no error or 100% bit inversion, the BSC channel will provide its maximum capacity of 1 bit

- When $q = 0.5$ the BSC channel is totally ambiguous or useless and has a capacity of 0 bits

2.7.2   Channel Capacity of a BEC

For a BEC, $I(A; B) = H(B) - H(B|A)$ where $H(B)$ is given by Equation 2.21 and $H(B|A)$ is given by Equation 2.22. The expression can be further simplified to:

$$I(A; B) = p \left(\omega \log \frac{1}{\omega} + \overline{\omega} \log \frac{1}{\overline{\omega}} \right) = (1 - q) H(A) \qquad (2.43)$$

from which we can immediately see that:

$$C_{BEC} = \max_{\{P_A(a)\}} I(A; B) = \max_{\{P_A(a)\}} (1 - q) H(A) = 1 - q \qquad (2.44)$$

since we know that for the binary input alphabet, A, that $\max_{\{P_A(a)\}} H(A) = \log |A| = \log 2 = 1$ occurs when $P(a = 0) = \frac{1}{2}$. The following observations can be made:

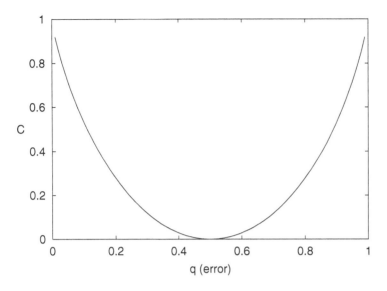

FIGURE 2.11
Variation of channel capacity of a BSC with q.

- when $q = 0$, that is, no erasures, the BEC channel will provide its maximum capacity of 1 bit

- when $q = 1$, the BEC channel will be producing only erasures and have a capacity of 0 bits

2.7.3 Channel Capacity of Weakly Symmetric Channels

The calculation of the channel capacity, C, is generally quite involved since it represents a problem in constrained optimisation of a non-linear function. However there is a special class of channels where the derivation of the channel capacity can be stated explicitly. Both *symmetric* and *weakly symmetric* channels are examples of this special class. We now provide a formal definition of symmetric and weakly symmetric channels.

DEFINITION 2.6 (Weakly) Symmetric Channels   *A channel is said to be* symmetric *if the rows and columns of the channel matrix are permutations of each other. A channel is said to be* weakly symmetric *if the rows of the channel matrix are permutations of each other and the column sums, $c_j = \sum_{a \in A} P(b_j|a_i)$, are equal. Obviously a symmetric channel will also have equal column sums since columns are a permutation of each other.*

Consider the weakly symmetric channel AB with mutual information:

$$I(A; B) = H(B) - H(B|A)$$

$$= H(B) - \sum_{a \in A} P(a_i) \sum_{b \in B} P(b|a_i) \log \frac{1}{P(b|a_i)}$$

$$\leq \log |B| - \sum_{a \in A} P(a_i) \{H(B|a_i)\} \tag{2.45}$$

where $H(B|a_i) = \sum_{b \in B} P(b|a_i) \log \frac{1}{P(b|a_i)}$ and the summation involves terms in the ith row of the channel matrix. Since each row is a permutation of, say, the first row, then $H(B|a_1) = H(B|a_2) = \ldots = H(B|a_r) \equiv H(B|a)$ where $r = |A|$. This means that $\sum_{a \in A} P(a_i) \{H(B|a_i)\} = \sum_{a \in A} P(a_i) \{H(B|a)\} = H(B|a)$ since $\sum_{a \in A} P(a_i) = 1$. Since $I(A; B)$ is bounded above, then the upper bound of $\log |B| - H(B|a)$ would be the channel capacity *if it can be achieved by an appropriate input distribution*. Let $P(a_i) = \frac{1}{r}$ then $P(b_j) = \sum_{a \in A} P(b_j|a_i) P(a_i) = \frac{1}{r} \sum_{a \in A} P(b_j|a_i) = \frac{c_j}{r}$. Since the channel is weakly symmetric then the column sums, c_j, are all equal, $c_j \equiv c$, and we have that $P(b_j) = \frac{c}{r}$, that is, the *output probabilities are equal*. Since we know that maximum entropy is achieved with equal symbol probabilities then it follows that $H(B) = \log |B|$ when $P(a_i) = \frac{1}{r}$. We have established that:

THEOREM 2.1

For a symmetric or weakly symmetric channel AB, the channel capacity can be stated explicitly as:

$$C = \log |B| - H(B|a) \tag{2.46}$$

where $H(B|a) = \sum_{b \in B} P(b|a_i) \log \frac{1}{P(b|a_i)}$ can be calculated for any row i. Channel capacity is achieved when the inputs are uniformly distributed, that is, $P(a) = \frac{1}{|A|}$.

EXAMPLE 2.10

Consider the BSC channel matrix:

$$\mathbf{P} = \begin{bmatrix} p & q \\ q & p \end{bmatrix}$$

The BSC is a symmetric channel and we can use Theorem 2.1 to explicitly derive the channel capacity as follows:

$$\log |B| = \log 2 = 1$$

$$H(B|a) = p \log \frac{1}{p} + q \log \frac{1}{q}$$

$$\Rightarrow C = 1 - \left(p \log \frac{1}{p} + q \log \frac{1}{q} \right)$$

when $P(a) = \frac{1}{2}$, which is the same expression that was derived in Section 2.7.1.

Now consider a channel with the following channel matrix:

$$\mathbf{P} = \begin{bmatrix} 0.5 & 0.3 & 0.4 \\ 0.3 & 0.5 & 0.4 \end{bmatrix}$$

Obviously the channel is not symmetric, but since the second row is a permutation of the first row and since the column sums are equal then the channel is *weakly symmetric* and the channel capacity can be explicitly stated as:

$$\log |B| = \log 3 = 1.5850$$
$$H(B|a) = 0.5 \log \frac{1}{0.5} + 0.3 \log \frac{1}{0.3} + 0.4 \log \frac{1}{0.4} = 1.5499$$
$$C = 1.5850 - 1.5499 = 0.0351$$

when $P(a) = \frac{1}{2}$. ⬚

2.8   Continuous Channels and Gaussian Channels

We extend our analysis of information channels to the case of continuous valued input and output alphabets and to the most important class of continuous channel, the Gaussian channel. In digital communication systems noise analysis at the most basic level requires consideration of continuous valued random variables rather than discrete quantities. Thus the Gaussian channel represents the most fundamental form of all types of communication channel systems and is used to provide meaningful insights and theoretical results on the information carrying capacity of channels. The BSC and BEC models, on the other hand, can be considered as high-level descriptions of the practical implementations and operations observed in most digital communication systems.

When considering the continuous case our discrete-valued symbols and discrete probability assignments are replaced by continuous-valued random variables, X, with associated probability density functions, $f_X(x)$.

DEFINITION 2.7 Mutual Information of two random variables   *Let X and Y be two random variables with joint density $f_{XY}(x,y)$ and marginal densities $f_X(x)$ and $f_Y(y)$. Then the mutual information between X and Y is defined as:*

$$I(X;Y) = \int f_{XY}(x,y) \log \frac{f_{XY}(x,y)}{f_X(x)f_Y(y)} \, dx\, dy \qquad (2.47)$$

$$= H(Y) - H(Y|X)$$

where $H(Y)$ is the differential entropy*:*

$$H(Y) = -\int f_Y(y) \log f_Y(y) \, dy \qquad (2.48)$$

and $H(Y|X)$ is the conditional differential entropy*:*

$$H(Y|X) = -\int f_{XY}(x,y) \log f_Y(y|x) \, dy\, dx \qquad (2.49)$$

The mutual information, $I(X;Y)$, provides a measure of the amount of information that can be carried by the continuous channel XY. Let us now consider the Gaussian Channel shown in Figure 2.12 where $N_i \sim \mathcal{N}(0, \sigma_N^2)$ is a zero-mean Gaussian random variable with variance σ_N^2. Let X_i be the value of the input to the channel at time i, and let Y_i be the value of the output from the channel at time i.

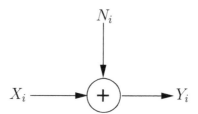

FIGURE 2.12
The Gaussian channel.

The output of the channel is given by:

$$Y_i = X_i + N_i \qquad (2.50)$$

That is, the channel output is perturbed by *additive white Gaussian noise* (AWGN).

We assume that the channel is *band-limited* to W Hz. Two immediate implications of this are that:

1. Assume the AWGN has a power spectral density of $N_o/2$. Then the noise variance, or average power, is band-limited and given by $\sigma_N^2 = E[N_i^2] = \int_{-W}^{W} \frac{N_o}{2} df = N_o W$

2. To faithfully reproduce any signals transmitted through the channel the signals must be transmitted at a rate not exceeding the Nyquist rate of $2W$ samples per second

We further assume that the channel is *power-limited* to P. That is:

$$E[X_i^2] = \sigma_X^2 \leq P \tag{2.51}$$

This *band-limited, power-limited Gaussian channel* just described is not only of theoretical importance in the field of information theory but of practical importance to communication engineers since it provides a fundamental model for many modern communication channels, including wireless radio, satellite and fibre optic links.

2.9 Information Capacity Theorem

In Section 2.7 we defined the channel capacity as the maximum of the mutual information over all possible input distributions. Of importance to communication engineers is the channel capacity of a band-limited, power-limited Gaussian channel. This is given by the following maximisation problem:

$$\begin{aligned} C &= \max_{\{f_X(x):E[X^2]\leq P\}} I(X;Y) \\ &= \max_{\{f_X(x):E[X^2]\leq P\}} \{H(Y) - H(Y|X)\} \end{aligned} \tag{2.52}$$

We now provide the details of deriving an important and well-known closed-form expression for C for Gaussian channels. The result is the *Information Capacity Theorem* which gives the capacity of a Gaussian communication channel in terms of the two main parameters that confront communication engineers when designing such systems: the signal-to-noise ratio and the available bandwidth of the system.

CLAIM 2.2

If $Y = X + N$ and X is uncorrelated with N then:

$$H(Y|X) = H(N) \tag{2.53}$$

PROOF   We note from Bayes' Theorem that $f_{XY}(x,y) = f_Y(y|x)f_X(x)$. Also since $Y = X + N$ and since X and N are uncorrelated we have that $f_Y(y|x) = f_N(y-x)$. Using these in the expression for $H(Y|X)$ gives:

$$H(Y|X) = -\int_{XY} f_{XY}(x,y) \log f_Y(y|x) \, dy \, dx$$

$$= -\int_X f_X(x) \left\{ \int_Y f_Y(y|x) \log f_Y(y|x) \, dy \right\} dx$$

$$= -\int_X f_X(x) \left\{ \int_N f_N(n) \log f_N(n) \, dn \right\} dx$$

$$= H(N) \int_X f_X(x) \, dx = H(N) \qquad (2.54)$$

☐

From Chapter 1 we stated that for a Gaussian random variable the differential entropy attains the maximum value of $\log\left(\sqrt{2\pi e}\sigma\right)$; so for the Gaussian random variable, N, we know that:

$$H(N) = \log\left(\sqrt{2\pi e}\sigma\right) = \frac{1}{2}\log\left(2\pi e\sigma_N^2\right) \qquad (2.55)$$

For random variable $Y = X + N$ where X and N are uncorrelated we have that:

$$H(Y) \le \frac{1}{2}\log\left(2\pi e\sigma_Y^2\right) = \frac{1}{2}\log\left[2\pi e\left(\sigma_X^2 + \sigma_N^2\right)\right] \qquad (2.56)$$

with the maximum achieved when Y is a Gaussian random variable. Thus:

$$I(X;Y) = H(Y) - H(Y|X)$$
$$= H(Y) - H(N)$$
$$\le \frac{1}{2}\log\left[2\pi e\left(\sigma_X^2 + \sigma_N^2\right)\right] - \frac{1}{2}\log\left(2\pi e\sigma_N^2\right)$$
$$\le \frac{1}{2}\log\left(\frac{\sigma_X^2 + \sigma_N^2}{\sigma_N^2}\right) = \frac{1}{2}\log\left(1 + \frac{\sigma_X^2}{\sigma_N^2}\right) \qquad (2.57)$$

If X is chosen to be a Gaussian random variable with $\sigma_X^2 = P$ then Y will also be a Gaussian random variable and the maximum mutual information or channel capacity will be achieved:

$$C = \frac{1}{2}\log\left(1 + \frac{P}{\sigma_N^2}\right) \quad \text{bits per transmission} \qquad (2.58)$$

Since the channel is also band-limited to W Hz then there can be no more than $2W$ symbols transmitted per second and $\sigma_N^2 = N_oW$. This provides the final form of the channel capacity, stated as Shannon's most famous *Information Capacity Theorem* [10], which is also known as the *Shannon-Hartley Law* in recognition of the early work by Hartley [6].

RESULT 2.8 Information Capacity Theorem

The information capacity of a continuous channel of bandwidth W Hz, perturbed by AWGN of power spectral density $N_o/2$ and bandlimited also to W Hz, is given by:

$$C = W \log \left(1 + \frac{P}{N_o W}\right) \quad \text{bits per second} \qquad (2.59)$$

where P is the average transmitted power and $P/N_o W$ is the signal-to-noise ratio or SNR.

Equation 2.59 provides the theoretical capacity or upper bound on the bits per second that can be transmitted for error-free transmission through a channel for a given transmitted power, P, and channel bandwidth, W, in the presence of AWGN noise with power spectral density, $N_o/2$. Thus the information capacity theorem defines a fundamental limit that confronts communication engineers on the rate for error-free transmission through a power-limited, band-limited Gaussian channel.

EXAMPLE 2.11

What is the minimum signal-to-noise ratio that is needed to support a 56k modem?

A 56k modem requires a channel capacity of 56,000 bits per second. We assume a telephone bandwidth of $W = 3600$ Hz. From Equation 2.59 we have:

$$56,000 = 3600 \log (1 + \text{SNR})$$
$$\Rightarrow \text{SNR} = 48159$$

or:

$$\text{SNR}_{dB} = 10 \log_{10} (48159)$$
$$\text{SNR}_{dB} = 47 \, dB$$

Thus a SNR of at least 48 dB is required to support running a 56k modem. In real telephone channels other factors such as crosstalk, co-channel interference, and echoes also need to be taken into account. ⬜

2.10   Rate Distortion Theory

Consider a discrete, memoryless source with alphabet $X = \{x_i : i = 1, 2, \ldots q\}$ and associated probabilities $\{P(x_i) : i = 1, 2, \ldots q\}$. Source coding is usually required to efficiently transmit or store messages from the source in the appropriate representation for the storage or communication medium. For example, with digital communication channels and storage, source symbols need to be encoded as a binary representation. Furthermore, source symbols transmitted through a communication channel may be output in a different symbol representation. Thus the source symbols will appear as symbols from a *representation or code word alphabet* $Y = \{y_j : j = 1, 2, \ldots r\}$. If there are no losses or distortions in the coding or transmission there is perfect representation and x_i can be recovered fully from its representation y_j. But in the following situations:

- lossiness in the source coding where the code alphabet and permitted code words do not allow exact representation of the source symbols and the decoding is subject to errors,

- insufficient redundancy in the channel code such that the rate of information is greater than the channel capacity,

the representation is not perfect and there are unavoidable errors or distortions in the representation of the source symbol x_i by the representation symbol y_j.

Rate distortion theory, first developed by Shannon [11], deals with the minimum mutual information (equivalent to the information or code rate) that the channel must possess, for the given source symbol probability distributions, to ensure that the average distortion is guaranteed not to exceed a specified threshold D. To derive this value we first define what we mean by an information or code rate (this is for the general case; in Chapter 5 we define the code rate for the specific case that applies with channel codes).

DEFINITION 2.8 Code Rate (General Case)   *Assume one of M possible source messages is represented as a code word of length n. Let $H(M)$ be the average number of bits transmitted with each source message, then the code rate R is defined as:*

$$R = \frac{H(M)}{n} \tag{2.60}$$

For the case of equally likely source messages, $H(M) = \log M$. For the general case, $H(M)$ is the entropy of the source message.

DEFINITION 2.9 Distortion Measures   *The* single-letter distortion measure, $d(x_i, y_j)$, *is defined as the measure of the cost incurred in representing the source symbol x_i by the representation symbol y_j. The* average distortion, $\overline{d}$, *is defined as the average of $d(x_i, y_j)$ over all possible source symbol and representation symbol combinations:*

$$\overline{d} = \sum_{i=1}^{q} \sum_{j=1}^{r} P(x_i) P(y_j | x_i) d(x_i, y_j) \tag{2.61}$$

where the $P(y_j | x_i)$ are the channel or transition *probabilities.*

Consider all possible conditional probability assignments for $P(y_j | x_i)$ for the given source and representation alphabets. An assignment is deemed to be D-admissible if and only if the average distortion, $\overline{d}$, is less than or equal to some acceptable or specified threshold, D. The set of all D-admissible conditional probability assignments is denoted by:

$$P_D = \left\{ P(y_j | x_i) : \overline{d} \leq D \right\} \tag{2.62}$$

For each D-admissible conditional probability assignment we have an associated mutual information, or information rate, given by:

$$I(X; Y) = \sum_{i=1}^{q} \sum_{j=1}^{r} P(x_i) P(y_j | x_i) \log \frac{P(y_j | x_i)}{P(y_j)} \tag{2.63}$$

where $P(y_j) = \sum_{i=1}^{q} P(y_j | x_i) P(x_i)$ are the representation alphabet probabilities.

DEFINITION 2.10 Rate Distortion Function   *For given D and fixed source probability distribution $\{P(x_i) : i = 1, 2, \ldots q\}$ the minimum information rate we require for the given average distortion D is given by the rate distortion function:*

$$R(D) = \min_{P(y_j | x_i) \in P_D} I(X; Y) \tag{2.64}$$

which is derived by finding the D-admissible conditional probability assignment that minimises the mutual information subject to the constraints:

$$\sum_{j=1}^{r} P(y_j | x_i) = 1 \quad \text{for } i = 1, 2, \ldots r \tag{2.65}$$

Intuitively, we expect that tolerating a larger distortion permits the use of lower information rates (as is the case of achieving low bit rate transmission by lossy compression) and conversely that by increasing the information rate lower distortion is possible.

The calculation of $R(D)$ from Equation 2.64 involves a minimisation of a non-linear function over an unknown but constrained set of probabilities, which is a very difficult problem analytically.

EXAMPLE 2.12

Consider a binary source with symbols $\{x_1 = 0, x_2 = 1\}$ that are detected at the receiver as the ternary representation alphabet with symbols $\{y_1 = 0, y_2 =?, y_3 = 1\}$. This representation covers the following three cases:

- $x_1 = 0$ is detected as $y_1 = 0$ and $x_2 = 1$ is detected as $y_3 = 1$, that is, there is no error in representation.

- $x_1 = 0$ is detected as $y_2 =?$ and $x_2 = 1$ is detected as $y_2 =?$, that is, the receiver fails to detect the symbol and there is an *erasure*.

- $x_1 = 0$ is detected as $y_3 = 1$ and $x_2 = 1$ is detected as $y_1 = 0$, that is, the receiver detects the symbol incorrectly and there is a bit inversion.

The corresponding channel is shown in Figure 2.13 where $P_{ij} = P(y_j|x_i)$ are the transition probabilities and we are given that $P(x_1) = P(x_2) = 0.5$.

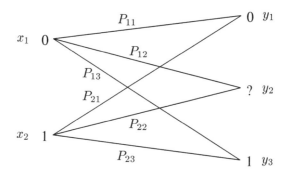

FIGURE 2.13
Graphical representation of information channel.

We define a distortion measure for this channel as follows:

- $d(x_1, y_1) = d(x_2, y_3) = 0$ to indicate there is no distortion (or distortion of 0) when there is no error in the representation.

- $d(x_1, y_2) = d(x_2, y_2) = 1$ to indicate there is a distortion or cost of 1 when there is an *erasure*.

- $d(x_1, y_3) = d(x_2, y_1) = 3$ to indicate there is a distortion of 3 where there is a bit inversion.

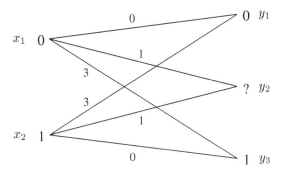

FIGURE 2.14
Graphical representation of distortion measure.

Note that a higher distortion of 3 is attributed to bit inversions compared to a lower distortion of 1 when there is an erasure. The set of distortion measures are represented graphically in Figure 2.14.

Assume the following conditional probability assignment for the channel:

$$\mathbf{P} = \begin{bmatrix} 0.7 & 0.2 & 0.1 \\ 0.1 & 0.2 & 0.7 \end{bmatrix} \qquad (2.66)$$

We calculate the representation alphabet probabilities: $P(y_1) = 0.4$, $P(y_2) = 0.2$, $P(y_3) = 0.4$, and the mutual information (from Equation 2.63) is $I(X;Y) = 0.365$.

The average distortion is then:

$$\bar{d} = \sum_{i=1}^{2} \sum_{j=1}^{3} P(x_i)P(y_j|x_i)d(x_i, y_j)$$
$$= 2 \cdot \frac{1}{2} \{(0.7)(0) + (0.2)(1) + (0.1)(3)\}$$
$$= 0.5$$

Consider the derivation of $R(0.5)$, that is, the rate distortion function for $D = 0.5$. The conditional probability assignment of Equation 2.66 is D-admissible and provides an information rate of $I(X;Y) = 0.365$. Is there another probability assignment with $\bar{d} \leq 0.5$ and information rate $I(X;Y) \leq 0.365$? Intuitively one expects the minimum information rate to occur at the maximum allowable distortion of $\bar{d} = 0.5$. To investigate this we consider the derivation of the corresponding BSC and BEC equivalents of Figure 2.13 such that $\bar{d} = 0.5$. The following conditional probability assignments:

$$\mathbf{P}_{BSC} = \begin{bmatrix} \frac{5}{6} & 0 & \frac{1}{6} \\ \frac{1}{6} & 0 & \frac{5}{6} \end{bmatrix} \qquad \mathbf{P}_{BEC} = \begin{bmatrix} \frac{1}{2} & \frac{1}{2} & 0 \\ 0 & \frac{1}{2} & \frac{1}{2} \end{bmatrix}$$

both yield the same average distortion of $\overline{d} = 0.5$. For the BSC equivalent the mutual information is calculated as $I_{BSC}(X;Y) = 0.350$ and for the BEC equivalent we have $I_{BEC}(X;Y) = 0.5$. Thus the BSC equivalent is the channel with lowest information rate for the same level of average distortion. To find $R(0.5)$ we need to consider all D-admissible conditional probability assignments and find the one providing the minimum value for the mutual information.
⬚

EXAMPLE 2.13

An important example of rate distortion analysis is for the case of analog to digital conversion when the analog source signal has to be represented as a digital source signal. An important component of this conversion is the *quantisation* of the continuous-valued analog signal to a discrete-valued digital sample.

Consider source symbols generated from a discrete-time, memoryless Gaussian source with zero mean and variance σ^2. Let x denote the value of the source symbol or sample generated by this source. Although the source symbol is discrete in time it is continuous-valued and a discrete-valued (quantised) representation of x is needed for storage and transmission through digital media. Let y be a symbol from the discrete-valued representation alphabet Y that is used to represent the continuous-valued x. For example Y can be the set of non-negative integers and $y = \text{round}(x)$ is the value of x rounded to the nearest integer. It should be noted that, strictly speaking, a finite representation is needed since digital data are stored in a finite number of bits. This is usually achieved in practice by assuming a limited dynamic range for x and limiting the representation alphabet Y to a finite set of non-negative integers. Thus we refer to y as the quantised version of x and the mapping of x to y as the quantisation operation.

The most intuitive distortion measure between x and y is a measure of the error in representing x by y and the most widely used choice is the *squared error distortion*:

$$d(x, y) = (x - y)^2 \tag{2.67}$$

It can be shown with the appropriate derivation (see [4, 5]) that the rate distortion function for the quantisation of a Gaussian source with variance σ^2 with a squared error distortion is given by:

$$R(D) = \begin{cases} \frac{1}{2}\log\left(\frac{\sigma^2}{D}\right), & 0 \leq D \leq \sigma^2 \\ 0, & D > \sigma^2 \end{cases} \tag{2.68}$$

⬚

2.10.1 Properties of $R(D)$

By considering the minimum and maximum permissible values for $R(D)$ and the corresponding distortion threshold, D, a more intuitive understanding of the behaviour of the rate distortion function, $R(D)$, is possible. Obviously the minimum value of $R(D)$ is 0, implying that there is no minimum rate of information, but what does this mean and what does the corresponding distortion threshold imply? Intuitively the maximum value of $R(D)$ should be $H(X)$ and this should happen when $D = 0$ (perfect reconstruction), but is this really the case? We answer these questions by stating and proving the following result.

RESULT 2.9

The rate distortion function is a monotonically decreasing function of D, limited in range by:

$$0 \le R(D) \le H(X) \tag{2.69}$$

where $R(D_{min}) = H(X)$ indicates that the upper bound occurs for the minimum possible value of the distortion, D_{min}, and $R(D_{max}) = 0$ indicates that there is a maximum permissible value D_{max} such that $R(D) = 0$ for $D \ge D_{max}$.

A typical plot of $R(D)$ as a function of D for the case when $D_{min} = 0$ is shown in Figure 2.15.

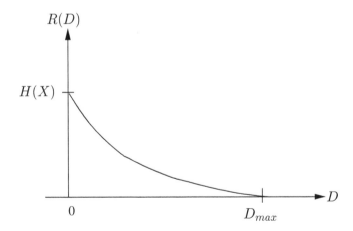

FIGURE 2.15
Sketch of typical function $R(D)$.

We can prove that $R(D_{min}) = H(X)$ by considering D_{min} the minimum permissible value for the distortion D. Since $\bar{d} \le D$ this is equivalent to finding the smallest possible $\bar{d}$. That is what we want to find the conditional probability assignment that minimises:

$$\bar{d} = \sum_{i=1}^{q}\sum_{j=1}^{r} P(x_i)P(y_j|x_i)d(x_i, y_j) \qquad (2.70)$$

The minimum $\bar{d}$ occurs by considering, for each x_i, the value of y_j that minimises $d(x_i, y_j)$ and setting $P(y_j|x_i) = 1$ for these values, with all other conditional probabilities set to zero. For each x_i define $d(x_i) = \min_j d(x_i, y_j)$ as the minimum distortion for x_i and $y_{j(x_i)}$, where $j(x_i) = \arg\min_j d(x_i, y_j)$, as the representation symbol that yields the minimum distortion. We set $P(y_{j(x_i)}|x_i) = 1$ with all other conditional probabilities $P(y_j|x_i)$ for $y_j \neq y_{j(x_i)}$ being set to zero. Hence:

$$D_{min} = \sum_{i=1}^{q} P(x_i)d(x_i) \qquad (2.71)$$

In typical applications $y_{j(x_i)}$ is the representation alphabet symbol that uniquely identifies the source symbol x_i and in such cases the above conditional probability assignment implies that $H(Y|X) = 0$ (perfect reconstruction since there is no uncertainty) and hence $I(X;Y) = H(X)$. Thus we have that $R(D_{min}) = H(X)$. Furthermore we typically assign a distortion of 0 for the pair $(x_i, y_{j(x_i)})$ and this means that $D_{min} = 0$; thus $R(0) = H(X)$.

Derivation of the maximum permissible D_{max} relies on the observation that this condition occurs when $R(D_{max}) = 0$. That is, $I(X;Y) = 0$ and thus $P(y_j|x_i) = P(y_j)$, the representation symbols are independent of the source symbols and there is no information conveyed by the channel. The average distortion measure for a channel with $I(X;Y) = 0$ is given by:

$$\bar{d} = \sum_{j=1}^{r} P(y_j)\left\{\sum_{i=1}^{q} P(x_i)d(x_i, y_j)\right\} \qquad (2.72)$$

Then since we are looking for D_{max} such that $R(D) = 0$ for $D \geq D_{max}$, then D_{max} is given by the minimum value of Equation 2.72 over all possible probability assignments for $\{P(y_j)\}$. Define $j^* = \arg\min_j \{\sum_{i=1}^{q} P(x_i)d(x_i, y_j)\}$. The minimum of Equation 2.72 occurs when $P(y_{j^*}) = 1$ for that y_{j^*} such that the expression $\{\sum_{i=1}^{q} P(x_i)d(x_i, y_j)\}$ is smallest, and $P(y_j) = 0$ for all other $y_j \neq y_{j^*}$. This gives:

$$D_{max} = \min_{j=1}^{r}\left\{\sum_{i=1}^{q} P(x_i)d(x_i, y_j)\right\} \qquad (2.73)$$

Equation 2.73 implies that if we are happy to tolerate an average distortion that is as much as D_{max} then there is a choice of representation y_j that is independent of x_i such that $\bar{d} \leq D_{max}$.

EXAMPLE 2.14

Consider the channel from Example 2.12. The following conditional probability assignment:

$$P = \begin{bmatrix} \frac{1}{3} & \frac{1}{3} & \frac{1}{3} \\ \frac{1}{3} & \frac{1}{3} & \frac{1}{3} \end{bmatrix}$$

obviously implies that $I(X;Y) = 0$. The resulting average distortion is:

$$\overline{d} = \left(\frac{1}{2}\right)\left(\frac{1}{3}\right)\{0+1+3+3+1+0\} = 1\frac{1}{3}$$

However this may not be D_{max} since there may be other conditional probability assignments for which $I(X;Y) = 0$ that provide a lower distortion. Indeed using Equation 2.73 gives:

$$D_{max} = \frac{1}{2}\min_{j=1}^{3}\{d(x_1, y_j) + d(x_2, y_j)\}$$
$$= \frac{1}{2}\min\{(0+3),(1+1),(3+0)\} = 1$$

Thus $D_{max} = 1$ and this occurs with the following conditional probability assignment:

$$P = \begin{bmatrix} 0 & 1 & 0 \\ 0 & 1 & 0 \end{bmatrix}$$

Interestingly this represents the condition where, no matter what input is transmitted through the channel, the output will always be $y_2 = ?$ with an average cost of 1. Thus if we can tolerate an average distortion of 1 or more we might as well not bother! □

EXAMPLE 2.15

Consider the rate distortion function given by Equation 2.68 in Example 2.13 for the quantisation of a Gaussian source with squared error distortion. Since the source x is continuous-valued and y is discrete-valued then for $D = D_{min} = 0$ the rate distortion function must be infinite since no amount of information will ever be able to reconstruct x from y with no errors, and the quantisation process, no matter what discrete-valued representation is used, will always involve a loss of information. That is, from Equation 2.68:

$$R(D) \to \infty \text{ as } D \to 0$$

It should be noted that this does not contradict Result 2.9 since that result implicitly assumed that both x and y were discrete-valued.

For the case of zero rate distortion Equation 2.68 gives:

$$R(D) = 0 \text{ for } D = \sigma^2$$

This result can be confirmed intuitively by observing that if no information is provided (i.e., the receiver or decoder does not have access to y) the best estimate for x is its mean value from which the average squared error will, of course, be the variance, σ^2.

$\square$

2.11   Exercises

1. A binary channel correctly transmits a 0 (as a 0) twice as many times as transmitting it incorrectly (as a 1) and correctly transmits a 1 (as a 1) three times more often then transmitting it incorrectly (as a 0). The input to the channel can be assumed equiprobable.

   (a) What is the channel matrix $\mathbf{P}$? Sketch the channel.

   (b) Calculate the output probabilities, $P(b)$.

   (c) Calculate the backward channel probabilities, $P(a|b)$.

2. Secret agent 101 communicates with her source of information by phone, unfortunately in a foreign language over a noisy connection. Agent 101 asks questions requiring only *yes* and *no* answers from the source. Due to the noise and language barrier, Agent 101 hears and interprets the answer correctly only 75% of the time, she fails to understand the answer 10% of the time and she misinterprets the answer 15% of the time. Before asking the question, Agent 101 expects the answer *yes* 80% of the time.

   (a) Sketch the communication channel that exists between Agent 101 and her source.

   (b) Before hearing the answer to the question what is Agent 101's average uncertainty about the answer?

   (c) Agent 101 interprets the answer over the phone as *no*. What is her average uncertainty about the answer? Is she is more uncertain or less uncertain about the answer given her interpretation of what she heard? Explain!

   (d) Agent 101 now interprets the answer over the phone as *yes*. What is her average uncertainty about the answer? Is she is more uncertain or less uncertain about the answer given her interpretation of what she heard? Explain and compare this with the previous case.

3. Calculate the equivocation of A with respect to B, $H(A|B)$, for the communication channel of Qu. 2. Now calculate the mutual information of the channel as $I(A; B) = H(A) - H(A|B)$. What can you say about the channel given your answers for Qu. 2(c) and 2(d)? Now verify that $I(A; B) = H(B) - H(B|A)$ by calculating $H(B)$ and $H(B|A)$.

*4. A friend of yours has just seen your exam results and has telephoned to tell you whether you have passed or failed. Alas the telephone connection is so bad that if your friend says "pass" you mistake that for "fail" 3 out of 10 times and if your friend says "fail" you mistake that for "pass" 1 out of 10 times. Before talking to your friend you were 60% confident that you had passed the exam. How confident are you of having passed the exam if you heard your friend say you have passed?

5. Which is better "erasure" or "crossover"?

   (a) Consider a fibre-optic communication channel with crossover probability of 10^{-2} and a wireless mobile channel with erasure probability of 10^{-2}. Calculate the mutual information assuming equiprobable inputs for both types of communication channels. Which system provides more information for the same bit error rate?

   *(b) Let us examine the problem another way. Consider a fibre-optic communication channel with crossover probability of q_{BSC} and a wireless mobile channel with erasure probability of q_{BEC}. Assume equiprobable inputs and express the mutual information $I(A; B)$ for the fibre optic channel as a function of q_{BSC} and for the wireless mobile channel as a function of q_{BEC}. Calculate q_{BSC} and q_{BEC} for the same mutual information of 0.95. Which channel can get away with a higher bit error rate and still provide the same amount of mutual information?

6. Given $H(B|A) = 0.83$, $H(A) = 0.92$ and $H(A|B) = 0.73$ find $H(B)$, $H(A, B)$ and $I(A; B)$.

7. Consider the following binary channel:

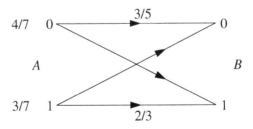

Calculate $I(A; B)$, $H(A)$, $H(B)$, $H(B|A)$, $H(A|B)$ and $H(A, B)$ as painlessly as possible.

8. Here are some quick questions:

    (a) Prove that if a BSC is shorted (i.e., all the outputs are grounded to 0) then the channel provides no information about the input.

    (b) Consider the statement: "Surely if you know the information of the source, $H(A)$, and the information provided by the channel, $I(A; B)$, you will know the information of the output, $H(B)$?" Prove whether this statement is true or not. If not, under what conditions, if any, would it be true?

    (c) Conceptually state (in plain English) what $H(A|B) = H(A) - I(A; B)$ means.

9. Sketch a sample channel with r inputs and s outputs, find the expression for the mutual information, the channel capacity and the input probability distribution to achieve capacity, for the following cases:

    (a) a noiseless, non-deterministic channel

    (b) a deterministic, non-noiseless channel

    (c) a noiseless, deterministic channel

*10. It was established that the minimum value of $I(A; B)$ is 0, that is $I(A; B) \geq 0$. What is the maximum value of $I(A; B)$?

11. Show that the mutual information expression for a BEC can be simplified to:

$$I(A; B) = p \left(\omega \log \frac{1}{\omega} + \bar{\omega} \log \frac{1}{\bar{\omega}} \right)$$

12. Consider the following *errors-and-erasure* channel:

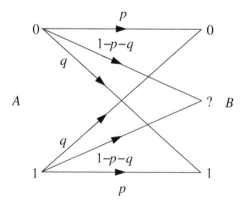

Find all the values of p and q for which the above channel is:

(a) totally ambiguous (i.e., $I(A; B) = 0$),

(b) noiseless,

(c) deterministic.

13. Channel AB has channel matrix:

$$\mathbf{P}_{AB} = \begin{bmatrix} 1 & 0 \\ \frac{2}{3} & \frac{1}{3} \\ 0 & 1 \end{bmatrix}$$

and is connected to channel BC with matrix:

$$\mathbf{P}_{BC} = \begin{bmatrix} \frac{1}{4} & \frac{3}{4} & 0 \\ 0 & 0 & 1 \end{bmatrix}$$

The ternary input A to the channel system has the following statistics: $P(a_1) = 1/8$, $P(a_2) = 3/8$, and $P(a_3) = 1/2$. The output of the channel AB is B and the output of channel BC is C.

(a) Calculate $H(A)$, $H(B)$ and $H(C)$.

(b) Calculate $I(A; B)$, $I(B; C)$ and $I(A; C)$.

(c) What can you say about channel BC? Explain!

14. The input to a BEC is repeated as shown below:

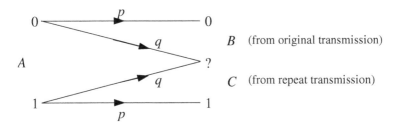

Given equiprobable binary inputs derive the expression for $I(A; B, C)$ and show that $I(A; B, C) \geq I(A; B)$ for:

(a) $p = q = 1/2$

(b) $p = 1/3$

(c) $p = 2/3$

15. Agent 01 contacts two of his sources by email for some straight "yes" and "no" answers. The first source he contacts is known to be unreliable and to give wrong answers about 30% of the time. Hence Agent 01 contacts his second source to ask for the same information. Unfortunately, the second source insists on using a non-standard email encoding and Agent 01 finds that only 60% of the answers are intelligible.

(a) What is the average uncertainty Agent 01 has about the input given the answers he receives from his first source? Hence what is the mutual information of the first source?

(b) What is the average uncertainty Agent 01 has about the input given the answers he receives from both the first and second source? Hence what is the mutual information from both sources?

*16. In order to improve utilisation of a BSC, special input and output electronics are designed so that the input to the BSC is sent twice and the output is interpreted as follows:

- if two 0's are received the channel outputs a single 0
- if two 1's are received the channel outputs a single 1
- if either 01 or 10 is received the channel outputs a single ?

Derive the expression for the mutual information through this augmented BSC assuming equiprobable inputs and compare this with the mutual information of a standard BSC. Either analytically or numerically show whether this augmented BSC is superior to the standard BSC. What price is paid for this "improved" performance?

17. The mutual information between two random variables, X and Y, is defined by Equation 2.47. Explain how $I(X;Y)$ provides a measure of the amount of independence between X and Y.

*18. A digital transmission channel consists of a terrestrial fibre-optic link with a measured cross-over (bit error) probability of 0.1 followed by a satellite link with an erasure probability of 0.2. No prior statistics regarding the source of information being transmitted through the channel are available. What is the average amount of information that can be resolved by the channel? In order to improve the reliability of the channel it is proposed that the same data being transmitted through the channel also be transmitted through a cheaper and less reliable terrestrial/marine copper channel with cross-over probability of 0.3. What is the average amount of information that can now be transmitted through the combined channel system? Compare with your previous answer. What is the cost of this improvement?

19. Consider the following *errors-and-erasure* channel:

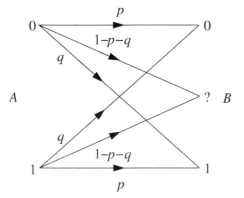

Under what conditions will the channel be *weakly symmetric*? What is the expression for the channel capacity when the channel is *weakly symmetric*?

*20. Consider the following *errors-and-erasure* channel:

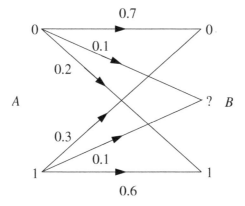

Express $I(A; B)$ as a function of $\omega = P(a = 0)$. Hence derive the channel capacity by finding the value of ω that maximises $I(A; B)$. You can do this numerically, graphically or analytically.

*21. Consider the following channel:

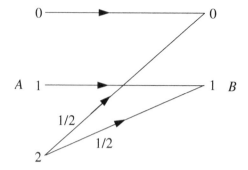

Let $P(a = 0) = w_1$ and $P(a = 1) = w_2$; thus $P(a = 2) = 1 - w_1 - w_2$. Calculate the mutual information for the following cases:

(a) $w_1 + w_2 = 1$

(b) $w_1 = \frac{1}{3}, w_2 = \frac{1}{3}$

Now derive an algebraic expression for the mutual information as a function of w_1 and w_2 and graphically, numerically or otherwise try to find the condition for channel capacity. Explain your findings!

22. A proposed monochrome television picture standard consists of 3×10^6 pixels per frame, each occupying one of 16 grayscale levels with equal probability. Calculate the minimum bandwidth required to support the transmission of 40 frames per second when the signal-to-noise ratio is 25 dB. Given that for efficient spectrum usage the bandwidth should not exceed 10 MHz what do you think happened to this standard?

23. You are asked to consider the design of a cable modem utilising a broadband communications network. One of the requirements is the ability to support full duplex 10 Mbps data communications over a standard television channel bandwidth of 5.5 MHz. What is the minimum signal-to-noise ratio that is required to support this facility?

24. Consider a BSC with channel matrix:

$$\mathbf{P}_{BSC} = \begin{bmatrix} p & q \\ q & p \end{bmatrix}$$

with input $\{a_1 = 0, a_2 = 1\}$ and output $\{b_1 = 0, b_2 = 1\}$. Define the following single-letter distortion measure:

$$d(a_i, b_j) = \begin{cases} 0 \text{ if } i = j \\ 1 \text{ if } i \neq j \end{cases}$$

Assuming equiprobable inputs derive the expression for the average distortion, $\bar{d}$. Hence what is the expression for the rate-distortion function $R(D)$ as a function of D?

25. For Qu. 24, what is D_{min} and $R(D_{min})$?

26. For Qu. 24, what is D_{max} and what is a value of q that yields $\bar{d} = D_{max}$?

*27. Consider the following *errors-and-erasure* channel:

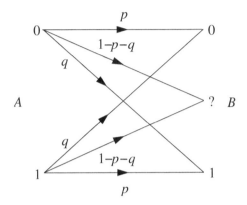

with input $\{a_1 = 0,\ a_2 = 1\}$ and output $\{b_1 = 0,\ b_2 = 1,\ b_3 =?\}$. Define the following single-letter distortion measure:

$$d(a_i, b_j) = \begin{cases} 0 \text{ if } i = j \\ 1 \text{ if } j = 3 \\ 2 \text{ otherwise} \end{cases}$$

Assuming equiprobable inputs derive the expression for the average distortion, $\bar{d}$. State the constrained minimisation problem that you need to solve to derive the rate-distortion function, $R(D)$. Can you solve it?

28. For Qu. 27 what is D_{max} and what are values of p and q that yield $\bar{d} = D_{max}$?

2.12   References

[1] S. Arimoto, An algorithm for calculating the capacity of an arbitrary discrete memoryless channel, *IEEE Trans. Inform. Theory*, IT-18, 14-20, 1972.

[2] R. Blahut, Computation of channel capacity and rate distortion functions, *IEEE Trans. Inform. Theory*, IT-18, 460-473, 1972.

[3] J.-F. Cardoso, Blind signal separation: statistical principles, *Proceedings of the IEEE*, 86(10), 2009-2025, 1998.

[4] T.M. Cover and J.A. Thomas, *Elements of Information Theory*, John Wiley & Sons, New York, 1991.

[5] R.G. Gallager, *Information Theory and Reliable Communication*, John Wiley & Sons, New York, 1968.

[6] T.V.L. Hartley, Transmission of information, *Bell System Tech. J.*, 7, 535-563, 1928.

[7] S. Haykin, *Communication Systems*, John Wiley & Sons, New York, 4th ed., 2001.

[8] J. Jeong, J.C. Gore, and B.S. Peterson, Mutual information analysis of the EEG in patients with Alzheimer's disease, *Clinical Neurophysiology*, 112(5), 827-835, 2001.

[9] Y. Normandin, R. Cardin, and R. De Mori, High-performance connected digit recognition using maximum mutual information, *IEEE Trans. Speech and Audio Processing*, 2(2), 299-311, 1994.

[10] C.E. Shannon, A mathematical theory of communication, *Bell System Tech. J.*, vol. 28, pg 379-423, 623-656, 1948.

[11] C.E. Shannon, Coding theorems for a discrete source with a fidelity criterion, *IRE Nat. Conv. Record*, Part 4, 142-163, 1959.

[12] P. Viola, and W.M. Wells, Alignment by maximization of mutual information, *Int. J. of Comput. Vision*, 24(2), 137-154, 1997.

Chapter 3

Source Coding

3.1 Introduction

An important problem in digital communications and computer storage is the *efficient* transmission and *storage* of information. Furthermore transmission and storage of digital data require the information to be represented in a digital or binary form. Thus there is a need to perform *source coding*, the process of *encoding* the information source *message* to a binary *code word*, and lossless and faithful *decoding* from the binary code word to the original information source *message*. The goal of source coding for digital systems is two-fold:

1. Uniquely map any arbitrary source message to a binary code and back again

2. Efficiently map a source message to the most compact binary code and back again

This is shown in Figure 3.1 for a source transmitting arbitrary text through a binary (digital) communication channel. The source encoder efficiently maps the text to a binary representation and the source decoder performs the reverse operation. It should be noted that the channel is assumed noiseless. In the presence of noise, channel codes are also needed and these are introduced in Chapter 5.

FIGURE 3.1
Noiseless communication system.

We first present the notation and terminology for the general source coding problem and establish the fundamental results and properties of codes for their practical use

105

and implementation. We then show how entropy is the benchmark used to determine the efficiency of a particular coding strategy and then proceed to detail the algorithms of several important binary coding strategies.

Consider a mapping from the source alphabet, S, of size q to the code alphabet, X, of size r and define the source coding process as follows:

DEFINITION 3.1 Source Coding   *Let the information source be described by the source alphabet of size* q, $S = \{s_i : i = 1, 2, \ldots, q\}$, *and define* $X = \{x_j : j = 1, 2, \ldots, r\}$ *as the code alphabet of size* r. *A* source message *of length* n *is an n-length string of symbols from the source alphabet, that is,* $\sigma_i^n = s_{i1} s_{i2} \ldots s_{in}$. *A code word,* $C(.)$, *is a finite-length string of, say,* l *symbols from the code alphabet, that is,* $C(.) = x_{j1} x_{j2} x_{j3} \ldots x_{jl}$. *Encoding is the mapping from a source symbol,* s_i, *to the code word,* $C(s_i)$, *and decoding is the reverse process of mapping a code word,* $C(s_i)$, *to the source symbol,* s_i. *A code table or simply* source code *completely describes the source code by listing the code word encodings of all the source symbols,* $\{s_i \longmapsto C(s_i) : i = 1, 2, \ldots, q\}$.

DEFINITION 3.2 nth Extension of a Code   *The nth extension of a code maps the source messages of length* n, σ_i^n, *which are the symbols from* S^n, *the nth extension of* S, *to the corresponding sequence of code words,* $C(\sigma_i^n) = C(s_{i1}) C(s_{i2}) \ldots C(s_{in})$, *from the code table for source* S.

A source code is identified as:

- a *non-singular code* if all the code words are distinct.

- a *block code of length* $\underline{n}$ if the code words are all of fixed length n.

All practical codes must be *uniquely decodable*.

DEFINITION 3.3 Unique Decodability   *A code is said to be uniquely decodable if, and only if, the nth extension of the code is non-singular for every finite value of n. Informally, a code is said to be uniquely decodable if there are no instances of non-unique (i.e., ambiguous) source decodings for any and all possible code messages.*

NOTE   The following observations can be made:

1. A block code of length n which is non-singular is uniquely decodable

2. It is a non-trivial exercise to prove whether a code is uniquely decodable in general

3. A code is proven to be NOT uniquely decodable if there is at least one instance of a non-unique decoding

The code alphabet of most interest is the *binary code*, $X = \{0, 1\}$, of size $r = 2$ since this represents the basic unit of storage and transmission in computers and digital communication systems. The *ternary code*, $X = \{0, 1, 2\}$, of size $r = 3$ is also important in digital communication systems utilising a tri-level form of line coding. In general we can speak of r-ary codes to refer to code alphabets of arbitrary size r.

EXAMPLE 3.1

Consider the source alphabet, $S = \{s_1, s_2, s_3, s_4\}$, and binary code, $X = \{0, 1\}$. The following are three possible binary source codes.

Source	Code A	Code B	Code C
s_1	0	0	00
s_2	11	11	01
s_3	00	00	10
s_4	11	010	11

We note that:

- Code A is NOT non-singular since $C(s_2) = C(s_4) = 11$.

- Code B is non-singular but it is NOT uniquely decodable since the code sequence 00 can be decoded as either s_3 or $s_1 s_1$, that is $C(s_3) = C(s_1 s_1) = 00$ where $C(s_1 s_1) = C(s_1)C(s_1) = 00$ is from the second extension of the code.

- Code C is a non-singular block code of length 2 and is thus uniquely decodable.

⬚

3.2 Instantaneous Codes

Although usable codes have to be at least uniquely decodable, there is a subclass of uniquely decodable codes, called *instantaneous codes*, that exhibit extremely useful properties for the design and analysis of such codes and the practical implementation of the encoding and decoding processes.

> **DEFINITION 3.4 Instantaneous Codes**   *A uniquely decodable code is said to be instantaneous if it is possible to decode each message in an encoded string without reference to succeeding code symbols.*

A code that is uniquely decodable but not instantaneous requires the past and current code symbols to be buffered at the receiver in order to uniquely decode the source message. An instantaneous code allows the receiver to decode the current code word to the correct source message immediately upon receiving the last code symbol of that code word (e.g., the decoding is performed "on the fly").

EXAMPLE 3.2

Consider the following three binary codes for the source $S = \{s_1, s_2, s_3, s_4\}$:

Source	Code A	Code B	Code C
s_1	0	0	0
s_2	10	01	01
s_3	110	011	011
s_4	1110	0111	111

Codes A, B and C are uniquely decodable (since no instance of a non-unique decoding can be found) but only code A is instantaneous. Why? Consider the following encoded string from code A:

$$\overline{0}\underline{10}\overline{1110}\underline{110}\overline{0}\underline{100}\overline{0}\underline{10} \Rightarrow s_1 s_2 s_4 s_3 s_1 s_2 s_1 s_1 s_2$$

The bars under and over each code symbol highlight each of the encoded source symbols. It is apparent that the "0" code symbol acts as a code word terminator or separator. Hence the moment a "0" is received it represents the last code symbol of the current code word and the receiver can immediately decode the word to the correct source symbol; thus code A is instantaneous.

Now consider the following encoded string from code B:

$$\overline{0110}\underline{1}\overline{0111}\underline{0}\overline{0} \Rightarrow s_3 s_2 s_4 s_1 s_1$$

Here too the "0" acts as a code word separator but, unlike code A, the "0" now represents the first code symbol of the *next* code word. This means the current code word cannot be decoded until the first symbol of the next code word, the symbol "0," is read. For example assume we have received the string 011. We cannot decode this to the symbol s_3 until we see the "0" of the next code word since if the next character we read is in fact a "1" then that means the current code word (0111) is s_4, not s_3. Although we can verify that the code is uniquely decodable (since the "0" acts as a separator) the code is not instantaneous.

Now consider the following encoded string from code C:

$$011111111\ldots$$

Surprisingly we cannot yet decode this sequence! In fact until we receive a "0" or an EOF (end-of-file or code string termination) the sequence cannot be decoded since we do not know if the first code word is 0, 01 or 011. Furthermore once we are in a position to decode the sequence the decoding process is itself not at all as straightforward as was the case for code A and code B. Nevertheless this code is uniquely decodable since the "0" still acts as a separator. ⬜

From Example 3.2 one can see that instantaneous codes are so named because decoding is a very fast and easy process (i.e., instantaneous!). However the practicality of instantaneous codes also lie in the *prefix condition* which allows such codes to be efficiently analysed and designed.

DEFINITION 3.5 Prefix of a Code   *Let $C(s_i) = x_{j1}x_{j2}\ldots x_{jn}$ be a code word of length n. A sub-string of code characters from $C(s_i)$, $x_{j1}x_{j2}\ldots x_{jm}$, where $m < n$, is called a* prefix *of the code word $C(s_i)$.*

DEFINITION 3.6 Prefix Condition   *A necessary and sufficient condition for a code to be instantaneous is that no complete code word of the code be a prefix of some other code word.*

Since the prefix condition is both necessary and sufficient the instantaneous codes are also referred to as *prefix codes* since such codes obey the above prefix condition.

NOTE   Since an instantaneous code is uniquely decodable the prefix condition is a sufficient condition for uniquely decodable codes. However it is not a necessary condition since a code can be uniquely decodable without being instantaneous.

EXAMPLE 3.3

Consider the codes from Example 3.2. We can now immediately state whether the codes are instantaneous or not:

Code A is instantaneous since no code word is a prefix of any other code word (code A obeys the prefix condition). Since it is instantaneous it is also uniquely decodable.

Code B is not instantaneous since $s_1 = 0$ is a prefix of $s_2 = 01$, $s_2 = 01$ is a prefix of $s_3 = 011$, etc. However it is uniquely decodable since the "0" acts as a separator.

Code C is not instantaneous since $s_1 = 0$ is a prefix of $s_2 = 01$, etc. However it is uniquely decodable since the "0" can be used as a separator.

$\Box$

Figure 3.2 graphically depicts the universe of all codes and how the different classes of codes we have described are subsets of one another. That is, the class of all instantaneous codes is a subset of the class of all uniquely decodable codes which in turn is a sub-class of all non-singular codes.

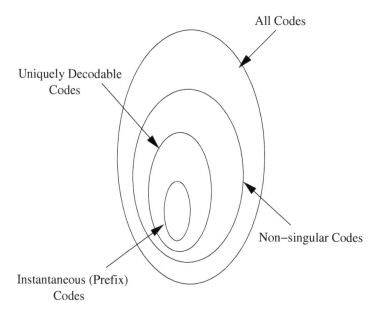

FIGURE 3.2
Classes of codes.

3.2.1 Construction of Instantaneous Codes

The prefix condition not only makes it easy to determine whether a given code is instantaneous or not but it can also be used to systematically design an instantaneous code with specified lengths for the individual code words. The problem can be stated as follows. For a source with q symbols it is required to design the q individual code words with specified code word lengths of $l_1, l_2, \ldots l_q$ such that the code is instantaneous (i.e., it satisfies the prefix condition). To design the code the code word lengths are sorted in order of increasing length. The code words are derived in sequence such that at each step the current code word does not contain any of the other code words as a prefix. A systematic way to do this is to enumerate or count through the code alphabet.

EXAMPLE 3.4

Problem 1: Design an instantaneous binary code with lengths of 3, 2, 3, 2, 2

Solution 1: The lengths are re-ordered in increasing order as 2, 2, 2, 3, 3. For the first three code words of length 2 a count from 00 to 10 is used:

$$00$$
$$01$$
$$10$$

For the next two code words of length 3, we count to 11, form 110, and then start counting from the right most symbol to produce the complete code:

$$00$$
$$01$$
$$10$$
$$110$$
$$111$$

Problem 2: Design an instantaneous ternary code with lengths 2, 3, 1, 1, 2

Solution 2: The code word lengths are re-ordered as 1, 1, 2, 2, 3 and noting that a ternary code has symbols $\{0, 1, 2\}$ we systematically design the code as follows:

$$0$$
$$1$$
$$20$$
$$21$$
$$220$$

Problem 3: Design an instantaneous binary code with lengths 2, 3, 2, 2, 2

Solution 3: The code word lengths are re-ordered as 2, 2, 2, 2, 3 and we immediately see that an instantaneous code cannot be designed with these lengths since:

$$00$$
$$01$$
$$10$$
$$11$$
$$??$$

□

3.2.2   Decoding Instantaneous Codes

Since an instantaneous code has the property that the source symbol or message can
be immediately decoded upon reception of the last code character in the current code
word the decoding process can be fully described by a decoding tree or state machine
which can be easily implemented in logic.

EXAMPLE 3.5

Figure 3.3 is the decoding tree that corresponds to the binary code:

Source	Code
s_1	00
s_2	01
s_3	10
s_4	110
s_5	111

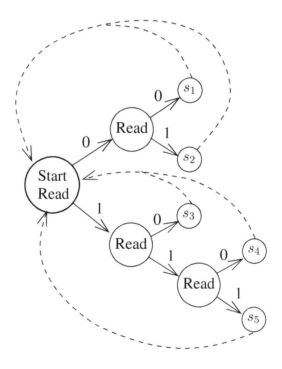

FIGURE 3.3
Decoding tree for an instantaneous code.

The receiver simply jumps to the next node of the tree in response to the current code character and when the leaf node of the tree is reached the receiver produces the corresponding source symbol, s_i, and immediately returns to the root of the tree. Thus there is no need for the receiver to buffer any of the received code characters in order to uniquely decode the sequence. ⬜

3.2.3 Properties of Instantaneous Codes

Property 1 Easy to prove whether a code is instantaneous by inspection of whether the code satisfies the prefix condition.

Property 2 The prefix code permits a systematic design of instantaneous codes based on the specified code word lengths.

Property 3 Decoding based on a decoding tree is fast and requires no memory storage.

Property 4 Instantaneous codes are uniquely decodable codes and where the length of a code word is the main consideration in the design and selection of codes there is no advantage in ever considering the general class of uniquely decodable codes which are not instantaneous.

Property 4 arises because of McMillan's Theorem which will be discussed in the next section.

3.2.4 Sensitivity to Bit Errors

The instantaneous codes we have been discussing are *variable-length codes* since the code words can be of any length. Variable-length codes should be contrasted with *block codes of length n* that restrict code words to be of the same length n. As we will see variable-length codes provide greater flexibility for sources to be encoded more efficiently. However a serious drawback to the use of variable-length codes is their sensitivity to bit or code symbol errors in the code sequence. If a single bit error causes the decoder to interpret a shorter or longer code word than is the case then this will create a synchronisation error between the first bit of the code word generated by the encoder and the root of the decoding tree. Subsequent code words will be incorrectly decoded, including the possibility of *insertion* errors, until the synchronisation is re-established (which may take a very long time). With block codes a single bit error will only affect the current block code and the effect will not propagate to subsequent code words. In fact block codes only suffer this problem initially when the transmission is established and the decoder has to "lock" onto the first bit of the code and where there are unmanaged timing discrepancies or clock skew between the transmitter and receiver.

EXAMPLE 3.6

Consider the following variable-length and block instantaneous binary codes for the source $S = \{s_1, s_2, s_3, s_4\}$.

Source	Variable-length code	Block code
s_1	0	00
s_2	10	01
s_3	110	10
s_4	111	11

Consider the following source sequence:

$$s_1^1\, s_3^2\, s_2^3\, s_4^4\, s_2^5\, s_1^6$$

where s_i^t indicates that symbol s_i is being transmitted at time t. The corresponding variable-length code sequence is:

$$0\overline{11010}\overline{111}\overline{100}$$

Assume the code sequence is now transmitted through an information channel which introduces a bit error in the 2nd bit. The source decoder sees:

$$\overline{00}\overline{1010}\overline{111}\overline{100}$$

and generates the decoded message sequence:

$$s_1^1\, s_1^2\, s_2^3\, s_2^4\, s_4^5\, s_2^6\, s_1^7$$

The single bit error in the encoded sequence produces both a *substitution* error (2nd symbol s_3^2 is substituted by s_1^2) and *insertion* error (s_2^3 is inserted after s_1^2 and subsequent source symbols appear as s_i^{t+1}) in the decoded message sequence and 7 rather than 6 source symbols are produced.

Now assume there is a bit error in the 6th bit. The source decoder sees:

$$0\overline{11011}\overline{111}\overline{100}$$

and generates the decoded message sequence:

$$s_1^1\, s_3^2\, s_4^3\, s_4^4\, s_1^5\, s_1^6$$

The single bit error now causes two isolated substitution errors (3rd symbol s_2^3 substituted by s_4^3 and 5th symbol s_2^5 substituted by s_1^5). Now consider the corresponding block code sequence:

$$s_1^1 \, s_3^2 \, s_2^3 \, s_4^4 \, s_2^5 \, s_1^6 \Rightarrow \underline{00}\overline{\underline{1001}}\,\overline{\underline{110}}\,\underline{100}$$

For errors in both the 2nd and 6th bits, a bit error in the 2nd bit causes a single substitution error (1st symbol s_1^1 substituted by s_2^1) and a bit error in the 6th bit also causes a single substitution error (3rd symbol s_2^3 substituted by s_1^3). In fact any bit errors will only affect the current code word and will not have any effect on subsequent code words, and all errors will be substitution errors.   ☐

From Example 3.6 it is apparent that variable-length codes are very sensitive to bit errors, with errors propagating to subsequent code words and symbols being inserted as well as being substituted. If the main goal of the source coder is to map the source symbols to the code alphabet then a block coding scheme should be used. For example, the ASCII code [1] is a 7-bit (or 8-bit in the case of extended ASCII) block code mapping letters of the English alphabet and keyboard characters to a binary block code of length 7 or 8. Furthermore, a channel coder (see Chapter 5) with an appropriate selection of an error-correcting or error-detecting code (see Chapters 6 to 9) is mandatory to ensure the final code sequence is close to error-free and must always be present when using variable-length codes.

The loss of synchronisation arising from code symbol errors with variable-length codes is a specific example of the more general problem of word synchronisation between the source and receiver which is discussed by Golomb et al. [5].

3.3   The Kraft Inequality and McMillan's Theorem

3.3.1   The Kraft Inequality

If the individual code word lengths are specified there is no guarantee that an instantaneous code can be designed with those lengths (see Example 3.4). The *Kraft Inequality* theorem [8] provides a limitation on the code word lengths for the design of instantaneous codes. Although not of real use in practice (the coding strategies we will later discuss will guarantee codes that are instantaneous) the Kraft Inequality is a precursor to the more important *McMillan's Theorem* [10] which states that where code word lengths are the only consideration, an instantaneous code will always be as good as any uniquely decodable code which is not necessarily instantaneous.

THEOREM 3.1 Kraft Inequality

A necessary and sufficient condition for the existence of an instantaneous code with alphabet size r and q code words with individual code word lengths of $l_1, l_2, \ldots, l_q$ is that the following inequality be satisfied:

$$K = \sum_{i=1}^{q} r^{-l_i} \leq 1 \qquad (3.1)$$

Conversely, given a set of code word lengths that satisfy this inequality, then there exists an instantaneous code with these word lengths.

The proof of the Kraft Inequality is interesting in that it is based on a formal description of how instantaneous codes are constructed (see Section 3.2.1) given the code word lengths, where the Kraft Inequality needs to be satisfied for the code to be successfully constructed.

PROOF   Let s_i be the ith source message or symbol and $C(s_i)$ the corresponding code word of length l_i. The proof requires the code word lengths to be arranged in ascending order, that is, we assume that the code word lengths are arranged such that $l_1 \leq l_2 \leq l_3 \ldots \leq l_q$. The number of possible code words for $C(s_i)$ is r^{l_i}. To ensure the code is instantaneous we need to consider the number of permissible code words for $C(s_i)$ such that the prefix condition is satisfied.

Consider the shortest code word, $C(s_1)$. Then the number of permissible code words for $C(s_1)$ is simply r^{l_1}. Next consider code word $C(s_2)$. The number of possible $C(s_2)$ with $C(s_1)$ as a prefix is given by the expression $r^{l_2-l_1}$ (since with $C(s_1)$ as the prefix the first l_1 symbols are fixed and one can choose the remaining $l_2 - l_1$ symbols arbitrarily). To ensure the prefix condition is satisfied the number of permissible code words for $C(s_2)$ is the number of possible code words, r^{l_2}, less those code words which have $C(s_1)$ as a prefix, $r^{l_2-l_1}$. That is, the number of permissible code words for $C(s_2)$ is $r^{l_2} - r^{l_2-l_1}$. Similarly for $C(s_3)$ the number of permissible code words is the number of possible code words, r^{l_3}, less those code words which have $C(s_2)$ as a prefix, $r^{l_3-l_2}$, and less those code words which have $C(s_1)$ as a prefix, $r^{l_3-l_1}$, that is $r^{l_3} - r^{l_3-l_2} - r^{l_3-l_1}$. For code word, $C(s_i)$, the expression for the number of permissible code words is:

$$r^{l_i} - r^{l_i-l_{i-1}} - r^{l_i-l_{i-2}} - \cdots - r^{l_i-l_1} \qquad (3.2)$$

To be able to construct the code we want to ensure that there is at least one permissible code word for all source messages, $i = 1, 2, \ldots q$. That is we require the following inequalities to be simultaneously satisfied:

$$r^{l_1} \geq 1$$
$$r^{l_2} - r^{l_2-l_1} \geq 1$$

$$\vdots$$

$$r^{l_q} - r^{l_q - l_{q-1}} - \cdots - r^{l_q - l_1} \geq 1 \tag{3.3}$$

By multiplying the ith equation by r^{-l_i} and rearranging we get:

$$r^{-l_1} \leq 1$$
$$r^{-l_2} + r^{-l_1} \leq 1$$
$$\vdots$$
$$r^{-l_q} + r^{-l_{q-1}} + \cdots + r^{-l_1} \leq 1 \tag{3.4}$$

We note that if the last inequality holds then all the preceding inequalities will also hold. Thus the following inequality expression must hold to ensure we can design an instantaneous code:

$$r^{-l_q} + r^{-l_{q-1}} + \cdots + r^{-l_1} \leq 1 \Rightarrow \sum_{i=1}^{q} r^{-l_i} \leq 1 \tag{3.5}$$

which is the Kraft Inequality. $\quad$ ▯

The Kraft Inequality of Equation 3.1 only indicates whether an instantaneous code can be designed from the given code word lengths. It does not provide any indication of what the actual code is, nor whether a code we have designed which satisfies Equation 3.1 is instantaneous, but it does tell us that an instantaneous code with the given code word lengths can be found. Only by checking the prefix condition of the given code can we determine whether the code is instantaneous. However if a code we have designed does *not* satisfy Equation 3.1 then we know that the code is *not* instantaneous and that it will *not* satisfy the prefix condition. Furthermore, we will not be able to find a code with the given code word lengths that is instantaneous. A less apparent property of the Kraft Inequality is that the *minimum* code word lengths for the given alphabet size and number of code words that can be used for designing an instantaneous code is provided by making K as close as possible to 1. Obviously shorter overall code word lengths intuitively yield more efficient codes. This is examined in the next section.

EXAMPLE 3.7

Consider the following binary ($r = 2$) codes

Source	Code A	Code B	Code C
s_1	0	0	0
s_2	100	100	10
s_3	110	110	110
s_4	111	11	11

Code A satisfies the prefix condition and is hence instantaneous. Calculating the Kraft Inequality yields:

$$K = \sum_{i=1}^{4} 2^{-l_i} = 2^{-1} + 2^{-3} + 2^{-3} + 2^{-3} = \frac{7}{8} \le 1$$

which is as expected.

Code B does *not* satisfy the prefix condition since s_4 is a prefix of s_3; hence the code is *not* instantaneous. Calculating the Kraft Inequality yields:

$$K = \sum_{i=1}^{4} 2^{-l_i} = 2^{-1} + 2^{-3} + 2^{-3} + 2^{-2} = 1 \le 1$$

which implies that an instantaneous code is possible with the given code word lengths. Thus a different code can be derived with the same code word lengths as code B which does satisfy the prefix condition. One example of an instantaneous code with the same code word lengths is:

$$0$$
$$110$$
$$111$$
$$10$$

Code C does *not* satisfy the prefix condition since s_4 is a prefix of s_3; hence the code is not instantaneous. Calculating the Kraft Inequality yields:

$$K = \sum_{i=1}^{4} 2^{-l_i} = 2^{-1} + 2^{-2} + 2^{-3} + 2^{-2} = \frac{9}{8} > 1$$

which implies that an instantaneous code cannot be designed with these code word lengths, and hence we shouldn't even try.   ⬚

3.3.2   McMillan's Theorem

As we will discuss in the next section, the use of shorter code word lengths creates more efficient codes. Since the class of uniquely decodable codes is larger than the class of instantaneous codes, one would expect greater efficiencies to be achieved considering the class of all uniquely decodable codes rather than the more restrictive class of instantaneous codes. However, instantaneous codes are preferred over uniquely decodable codes given that instantaneous codes are easier to analyse, systematic to design and can be decoded using a decoding tree (state machine) structure. McMillan's Theorem assures us that we do not lose out if we only consider the class of instantaneous codes.

THEOREM 3.2 McMillan's Theorem

The code word lengths of any uniquely decodable code must satisfy the Kraft Inequality:

$$K = \sum_{i=1}^{q} r^{-l_i} \leq 1 \tag{3.6}$$

Conversely, given a set of code word lengths that satisfy this inequality, then there exists a uniquely decodable code (Definition 3.3) with these code word lengths.

The proof of McMillan's Theorem is presented as it is instructive to see the way it uses the formal definition of unique decodability to prove that the inequality must be satisfied. The proof presented here is based on that from [2].

PROOF   Assume a uniquely decodable code and consider the quantity:

$$\left(\sum_{i=1}^{q} r^{-l_i} \right)^n = \left(r^{-l_1} + r^{-l_2} + \cdots + r^{-l_q} \right)^n \tag{3.7}$$

When written out, the quantity will consist of the q^n terms arising from the nth extension of the code, each of the form:

$$r^{-l_{i1} - l_{i2} - \cdots - l_{in}} = r^{-k} \tag{3.8}$$

where we define $l_{i1} + l_{i2} + \cdots + l_{in} = k$.

Then l_{ij} is the length, l_i, of the jth code word in the nth extension of the code and k is the length of the sequence of code words in the nth extension of the code. Let $l_m = \max_i \{ l_i : i = 1, 2, \ldots, q \}$ be the maximum code word length over the q code words. The minimum code word length is, of course, a length of 1. Then k can assume any value from n to $n l_m$. Let N_k be the number of terms of the form r^{-k}. Then:

$$\left(\sum_{i=1}^{q} r^{-l_i} \right)^n = \sum_{k=n}^{n l_m} N_k r^{-k} \tag{3.9}$$

Thus N_k represents the number of code word sequences in the nth extension of the code with a length of k. If the code is uniquely decodable then the nth extension of the code must be non-singular. That is, N_k must be no greater than r^k, the number of distinct sequences of length k. Thus for any value of n we must have:

$$\left(\sum_{i=1}^{q} r^{-l_i} \right)^n \leq \sum_{k=n}^{n l_m} r^k r^{-k} = \sum_{k=n}^{n l_m} 1$$

$$\leq n l_m - n + 1$$

$$\leq n l_m \tag{3.10}$$

or:

$$\sum_{i=1}^{q} r^{-l_i} \leq (nl_m)^{1/n} \tag{3.11}$$

For $n \geq 1$ and $l_m \geq 1$ we have that $\min_n \left\{ (nl_m)^{1/n} \right\} = \lim_{n \to \infty} \left\{ (nl_m)^{1/n} \right\} = 1$. Since the above inequality has to hold for all values of n, then this will be true if:

$$\sum_{i=1}^{q} r^{-l_i} \leq 1 \tag{3.12}$$

▯

The implication of McMillan's Theorem is that for every non-instantaneous uniquely decodable code that we derive, an instantaneous code with the same code word lengths will always be found since both codes satisfy the same Kraft Inequality. Thus we can restrict ourselves to the class of instantaneous codes since we will not gain any efficiencies based on code word lengths by considering the larger class of uniquely decodable codes.

EXAMPLE 3.8

It is required to design a uniquely decodable ternary ($r = 3$) code with code word lengths 1, 1, 2, 2, 3, 3. Since a uniquely decodable code satisfies the Kraft Inequality by McMillan's Theorem we check whether this is the case. Calculating the Kraft Inequality:

$$K = \sum_{i=1}^{6} 3^{-l_i} = 2(3^{-1}) + 2(3^{-2}) + 2(3^{-3}) = 0.963 \leq 1$$

shows that a uniquely decodable code with these lengths can be found.

BUT the same Kraft Inequality is satisfied for instantaneous codes and since instantaneous codes can be systematically constructed following the procedure described in Section 3.2.1 the following instantaneous code, which is uniquely decodable, is designed:

0
1
20
21
220
221

▯

3.4 Average Length and Compact Codes

3.4.1 Average Length

When considering a collection of possible r-ary codes for the same source a choice needs to be made between the different codes by comparing the performance based on a criteria of interest. For storage and communication purposes the main criterion of interest is the *average length* of a code, where codes with smaller average length are preferred. To calculate the average length the source symbol (or message) probabilities are needed.

DEFINITION 3.7 Average Length *Define $\{P_i : i = 1, 2, \ldots q\}$ as the individual source symbol probabilities for a source with q possible symbols. Define $\{l_i : i = 1, 2, \ldots q\}$ as the length of the corresponding code words for a given source coding. Then the* average length *of the code, L, is given by:*

$$L = \sum_{i=1}^{q} P_i l_i \qquad (3.13)$$

Consider all possible r-ary codes for the same source, S. The number of source symbols, q, and source probabilities, $\{P_i : i = 1, 2, \ldots q\}$, are constant, but the code word lengths $\{l_i : i = 1, 2, \ldots q\}$ vary with each code. The best code or *compact code* will be the one with the smallest average length.

DEFINITION 3.8 Compact Codes *Consider a uniquely decodable code that maps the symbols for a source S to code words from an r-ary code alphabet. The code will be a* compact code *if its average length is less than or equal to the average length of all other uniquely decodable codes for the same source and code alphabet.*

EXAMPLE 3.9

Consider the following two binary codes for the same source. Which code is better?

Source	P_i	Code A	Code B
s_1	0.5	00	1
s_2	0.1	01	000
s_3	0.2	10	001
s_4	0.2	11	01

The average length of code A is obviously $L_A = 2$ bits per symbol. The average length of code B is $L_B = (0.5)1 + (0.1)3 + (0.2)3 + (0.2)2 = 1.8$ bits per symbol.

Code B is better than code A since $L_B < L_A$. But is there another code, call it code C, which has $L_C < L_B$? Or is code B the compact code for this source? And how small can the average length get? ⬚

The fundamental problem when coding information sources and the goal of source coding for data compression is to find compact codes. This obviously requires the individual code word lengths to be made as small as possible, as long as we still end up with a uniquely decodable (i.e., instantaneous) code. Intuitively the average length will be reduced when the shorter length code words are assigned to the most probable symbols and the longer code words to the least probable symbols. This concept is shown by Example 3.9 where code B assigns the shortest code word (of length 1) to the most probable symbol (with probability 0.5). Formally, the problem of searching for compact codes is a problem in constrained optimisation. We state the problem as follows.

REMARK 3.1   Given $\{P_i : i = 1, 2, \ldots q\}$ for a source with q symbols the compact r-ary code is given by the set of integer-valued code word lengths $\{l_i : i = 1, 2, \ldots q\}$ that minimise:

$$L = \sum_{i=1}^{q} P_i l_i \tag{3.14}$$

such that the Kraft Inequality constraint is satisfied:

$$\sum_{i=1}^{q} r^{-l_i} \leq 1 \tag{3.15}$$

⬚

3.4.2   Lower Bound on Average Length

From Chapter 1 the information content of a source is given by its entropy. When using logarithms to base 2 the entropy is measured in units of "(information) bits per source symbol." Similarly when using binary source codes the average length is also measured in units of "(code) bits per source symbol." Since the entropy provides a measure of the intrinsic information content of a source it is perhaps not surprising that, for there to be no losses in the coding process, the average length must be at least the value of the entropy (no loss of information in the coding representation) and usually more to compensate for the inefficiencies arising from the coding process. The following theorem establishes this lower bound on the average length of any possible coding of the source based on the entropy of the source.

THEOREM 3.3

Every instantaneous r-ary code of the source, $S = \{s_1, s_2, \ldots s_q\}$, will have an average length, L, which is at least the entropy, $H_r(S)$, of the source, that is:

$$L \geq H_r(S) \qquad (3.16)$$

with equality when $P_i = r^{-l_i}$ for $i = 1, 2, \ldots q$ where P_i is the probability of source symbol s_i, $H_r(S) = \frac{H(S)}{\log_2 r}$ is the entropy of the source S using logarithms to the base r and $H(S)$ is the entropy of the source S using logarithms to the base 2.

PROOF   Consider the difference between the entropy $H_r(S)$ and the average length L and simplify the expression as follows:

$$H_r(S) - L = \sum_{i=1}^{q} P_i \log_r \frac{1}{P_i} - \sum_{i=1}^{q} P_i l_i$$

$$= \sum_{i=1}^{q} P_i \log_r \frac{1}{P_i} - \sum_{i=1}^{q} P_i \log_r r^{l_i}$$

$$= \frac{1}{\ln r} \sum_{i=1}^{q} P_i \ln \frac{1}{P_i r^{l_i}} \qquad (3.17)$$

Using the inequality $\ln x \leq x - 1$ (with equality when $x = 1$) where $x = \frac{1}{P_i r^{l_i}}$ gives:

$$H_r(S) - L \leq \frac{1}{\ln r} \sum_{i=1}^{q} P_i \left(\frac{1}{P_i r^{l_i}} - 1 \right)$$

$$\leq \frac{1}{\ln r} \left(\sum_{i=1}^{q} \frac{1}{r^{l_i}} - \sum_{i=1}^{q} P_i \right)$$

$$\leq \frac{1}{\ln r} \left(\sum_{i=1}^{q} r^{-l_i} - 1 \right) \qquad (3.18)$$

Since the code is instantaneous the code word lengths will obey the Kraft Inequality so that $\sum_{i=1}^{q} r^{-l_i} \leq 1$ and hence:

$$H_r(S) - L \leq 0 \implies L \geq H_r(S)$$

with equality when $\frac{1}{P_i r^{l_i}} = 1$, or $P_i = r^{-l_i}$ for $\{i = 1, 2, \ldots q\}$.   ▯

The definition of entropy in Chapter 1 now makes practical sense. The entropy is a measure of the intrinsic amount of average information (in r-ary units) and the

average length of the code must be at least equal to the entropy of the code to ensure
no loss in coding (*lossless coding*). Thus the smallest average length, L, for any code
we design will be $L = H_r(S)$. However there is no guarantee that a compact code
for a particular source and code alphabet can be found. Indeed, unless $P_i = r^{-l_i}$ for
$i = 1, 2, \ldots q$ we will always have that $L > H_r(S)$, but the closer L is to $H_r(S)$
then the better or more efficient the code.

DEFINITION 3.9 Code Efficiency   *The efficiency of the code is given by:*

$$\eta = \frac{H_r(S)}{L} \times 100\% \tag{3.19}$$

where if $L = H_r(S)$ the code is 100% efficient.

EXAMPLE 3.10

The entropy for the source in Example 3.9 is:

$$H(S) = 0.5 \log \frac{1}{0.5} + 0.1 \log \frac{1}{0.1} + 2(0.2) \log \frac{1}{0.2} = 1.761 \text{ bits per symbol}$$

Thus for this source we must have $L \geq 1.761$ no matter what instantaneous binary
code we design. From Example 3.9 we have:

- $L_A = 2$ and code A has an efficiency of $\eta = \frac{1.761}{2} = 88\%$.

- $L_B = 1.8$ and code B has an efficiency of $\eta = \frac{1.761}{1.8} = 98\%$.

Since code B is already at 98% efficiency we can probably state with some confidence
that code B is a compact code for source S. And even if it wasn't, at best a compact
code would only provide us with no more than 2% improvement in coding efficiency.

□

Ideally compact codes should exhibit 100% efficiency. This requires that $L = H_r(S)$, and from the condition for equality, $P_i = r^{-l_i}$, it implies $l_i = \log_r \frac{1}{P_i}$
for $i = 1, 2, \ldots q$. The problem lies in the fact that the l_i have to be integer values.
If this is the case then we have a *special source*.

DEFINITION 3.10 Special Source   *A source with symbol probabilities $\{P_i : i = 1, 2, \ldots q\}$ such that $\left\{ \log_r \frac{1}{P_i} : i = 1, 2, \ldots q \right\}$ are integers is a special source
for r-ary codes since an instantaneous code with code word lengths $l_i = \log_r \frac{1}{P_i}$ for
a code alphabet of size r can be designed which is 100% efficient with $L = H_r(S)$.*

EXAMPLE 3.11

Consider the following 4-symbol source:

Source A	P_i
s_1	0.125
s_2	0.25
s_3	0.5
s_4	0.125

We note that the symbol probabilities are of the form $P_i = \left(\frac{1}{2}\right)^{l_i}$ with $l_1 = 3, l_2 = 2$, $l_3 = 1$, and $l_4 = 3$. Thus a 100% efficient compact binary code can be designed with code word lengths of 3, 2, 1, 3 with $L = H(S) = 1.75$. For example:

Source A	Code A
s_1	110
s_2	10
s_3	0
s_4	111

Now consider the following 9-symbol source:

Source B	P_i
s_1	1/9
s_2	1/9
s_3	1/3
s_4	1/27
s_5	1/27
s_6	1/9
s_7	1/9
s_8	1/27
s_9	1/9

We note that the symbol probabilities are of the form $P_i = \left(\frac{1}{3}\right)^{l_i}$ with $l_i = \{2, 2, 1, 3,$ $3, 2, 2, 3, 2\}$ for $i = 1, 2, \ldots, 9$. Thus a 100% efficient ternary code can be designed with code word lengths of 2,2,1,3,3,2,2,3,2 with $L = H_3(S) = 16/9$. $\quad\square$

3.5 Shannon's Noiseless Coding Theorem

3.5.1 Shannon's Theorem for Zero-Memory Sources

Equation 3.16 states that the average length of any instantaneous code is bounded below by the entropy of the source. In this section we show that the average length, L, of a compact code is also bounded above by the entropy plus 1 unit of information (1 bit in the case of a binary code). The lower bound was important for establishing how the efficiency of a source code is measured. Of practical importance is the existence of both a lower and upper bound which, as we will see, leads to the important

Shannon's Noiseless Coding Theorem [11]. The theorem states that the coding efficiency (which will always be less than 100% for compact codes that are not special) can be improved by coding the extensions of the source (message blocks generated by the source) rather than just the source symbols themselves.

THEOREM 3.4

Let L be the average length of a compact r-ary code for the source S. Then:

$$H_r(S) \leq L < H_r(S) + 1 \qquad (3.20)$$

The proof of the theorem as presented makes mention of a possible coding strategy that one may adopt in an attempt to systematically design compact codes. Although such a strategy, in fact, designs codes which are not compact it establishes the theorem which can then be extended to the case of compact codes.

PROOF   Let $S = \{s_1, s_2, \ldots s_q\}$ and consider the sub-optimal coding scheme where the individual code word lengths are chosen according to:

$$\log_r \frac{1}{P_i} \leq l_i < \log_r \frac{1}{P_i} + 1 \qquad (3.21)$$

We can justify that this is a reasonable coding scheme by noting from Theorem 3.3 that a 100% efficient code with $H(S) = L$ is possible if we select $l_i = \log_r \frac{1}{P_i}$ and where $\log_r \frac{1}{P_i}$ does not equal an integer we "round up" to provide integer length assignments, yielding the coding scheme just proposed. Multiplying Equation 3.21 by P_i and summing over i gives:

$$\sum_{i=1}^{q} P_i \log_r \frac{1}{P_i} \leq \sum_{i=1}^{q} P_i l_i < \sum_{i=1}^{q} P_i \log_r \frac{1}{P_i} + \sum_{i=1}^{q} P_i \qquad (3.22)$$

which yields, using the definitions for entropy and average length:

$$H_r(S) \leq L_s < H_r(S) + 1 \qquad (3.23)$$

where L_s is the average length using this sub-optimal coding scheme. Consider a compact code for the same source S with average length L. Then by definition $L \leq L_s$ and from Theorem 3.3 we have $H_r(S) \leq L$. Thus we must also have:

$$H_r(S) \leq L < H_r(S) + 1 \qquad (3.24)$$

$\square$

NOTE One coding scheme resulting in code word lengths given by Equation 3.21 is the *Shannon code* [9]. Due to the sub-optimal nature of this coding scheme it will not be elaborated further upon.

Equation 3.20 indicates the average length of a compact code will be no more than 1 unit away from the average length of a 100% efficient code. However we now show that by taking extensions using Equation 3.20 we can improve the efficiency of the code.

Let $S^n = \{\sigma_1^n, \sigma_2^n, \ldots \sigma_{q^n}^n\}$ be the nth extension of the source $S = \{s_1, s_2, \ldots s_q\}$ where $\sigma_i^n = s_{i1} s_{i2} \ldots s_{in}$. Consider the compact code for S^n with average length L_n. Then from Theorem 3.4 we have that:

$$H_r(S^n) \le L_n < H_r(S^n) + 1 \tag{3.25}$$

The $H_r(S^n)$ can be considered the joint entropy of n independent sources since:

$$P(\sigma_i^n) = P(s_{i1})P(s_{i2}) \ldots P(s_{in}) \tag{3.26}$$

and from Section 1.18 we have the result that $H_r(S^n) = \sum^n H_r(S) = nH_r(S)$ which when substituted into Equation 3.25 and dividing by n yields:

$$H_r(S) \le \frac{L_n}{n} < H_r(S) + \frac{1}{n} \tag{3.27}$$

NOTE The term $\frac{L_n}{n}$ is the average length of the code words per symbol s_i when coding the nth extension S^n which should not be confused with L which is the average length of the code words per symbol s_i when coding the source S. However $\frac{L_n}{n}$ can be used as the average length of a coding scheme for the source S (based on coding the nth extension of the source).

From Equation 3.27 the average length of a compact code for S based on coding the nth extension of S is no more than $\frac{1}{n}$ units away from the average length of a 100% efficient code and this overhead decreases with larger extensions. Intuitively we expect that coding the nth extension of the source will provide more efficient codes, approaching 100% efficiency as $n \to \infty$. We formally state these results in the following theorem.

THEOREM 3.5 Shannon's Noiseless Coding Theorem

Let L be the average length of a compact code for the source S, then:

$$H_r(S) \leq L < H_r(S) + 1$$

Now let L_n be the average length of a compact code for the nth extension of S, then:

$$H_r(S) \leq \frac{L_n}{n} < H_r(S) + \frac{1}{n}$$

and thus:

$$\lim_{n \to \infty} \frac{L_n}{n} = H_r(S)$$

that is, in the limit the code becomes 100% efficient.

NOTE   Since we can't get something for nothing there is a price to be paid in improving coding efficiency by taking extensions. The price is the increased cost of encoding and decoding a code with q^n code words, which represents an exponential increase in complexity.

EXAMPLE 3.12

Taking extensions to improve efficiency is apparent when the source and code alphabets are the same. Consider the case of a binary coding scheme for a binary source. Without extensions we would have the trivial result:

Source	P_i	Compact Code
s_1	0.8	0
s_2	0.2	1

The efficiency of this compact code (which is no code at all!) is $\eta = \frac{H(S)}{L}$ where $L = 1$ bits per symbol and $H(S) = 0.722$ yielding a compact code which is only 72% efficient. Consider taking the second extension of the code where $\sigma_i^2 = s_i s_j$ and $P_i \equiv P(\sigma_i^2) = P(s_i)P(s_j)$ for $i = 1, 2$ and $j = 1, 2$:

Source	P_i	Compact Code
$s_1 s_1$	0.64	0
$s_1 s_2$	0.16	10
$s_2 s_1$	0.16	110
$s_2 s_2$	0.04	111

The compact code for the second extension is as shown in the above table. The average length of this code is $L_2 = (0.64)1 + (0.16)2 + (0.16)3 + (0.04)3 = 1.56$

bits per symbol *pair* and $\frac{L_2}{2} = 0.78$ bits per symbol yielding a compact code for the second extension which is $\eta = \frac{H(S)}{L_2/2} = 92.6\%$ efficient, a significant improvement over 72% efficient. $\square$

3.5.2 Shannon's Theorem for Markov Sources

The statement of Shannon's Noiseless Coding Theorem was proven for zero-memory sources. We now show that a similar statement applies to the general case of Markov sources. Consider an mth order Markov source, S, and the corresponding entropy, $H_r(S)$, based on the mth order Markov model of the source. Define $H_r(\overline{S})$ as the entropy of the *equivalent* zero-memory model of the source (where $\overline{S}$ is the *adjoint source*). Since Equation 3.20 applies to zero-memory sources it also applies to the adjoint source, hence:

$$H_r(\overline{S}) \leq L < H_r(\overline{S}) + 1 \tag{3.28}$$

and also:

$$H_r(\overline{S^n}) \leq L_n < H_r(\overline{S^n}) + 1 \tag{3.29}$$

where $\overline{S^n}$ is the adjoint of the nth extension of the source, S^n (not to be confused with $\overline{S}^n$ which is the nth extension of the adjoint source, $\overline{S}$).

We know from our results from Chapter 1 that $H_r(S^n) \leq H_r(\overline{S^n})$, or equivalently that $H_r(\overline{S^n}) = H_r(S^n) + \epsilon$ (for $\epsilon \geq 0$). We also have that $H_r(S^n) = nH_r(S)$. Substituting these into Equation 3.29 gives:

$$nH_r(S) + \epsilon \leq L_n \leq nH_r(S) + \epsilon + 1 \tag{3.30}$$

and

$$H_r(S) + \frac{\epsilon}{n} \leq \frac{L_n}{n} \leq H_r(S) + \frac{\epsilon}{n} + \frac{1}{n} \tag{3.31}$$

and hence we can see that

$$\lim_{n \to \infty} \frac{L_n}{n} = H_r(S) \tag{3.32}$$

That is, the lower bound on $\frac{L_n}{n}$ is the entropy of the mth order Markov model $H_r(S)$.

Consider a sequence of symbols from an unknown source that we are coding. The entropy of the source depends on the model we use for the source. From Chapter 1 we had that, in general, $H_r(S^m) \leq H_r(S^{m-1})$ where S^m is used to indicate a Markov source of order m. Thus higher order models generally yield lower values for the entropy. And the lower the entropy is, then from Equation 3.32, the smaller the average length we expect as we take extensions to the source. This realisation does not affect how we design the compact codes for extensions to the source, but affects how we calculate the corresponding nth extension symbol probabilities (which are based on the underlying model of the source) and the resulting code efficiency.

RESULT 3.1

When coding the nth extension of a source, the probabilities should be derived based on the "best" or "true" model of the source, and the efficiency of the code should be based on the same model of the source.

EXAMPLE 3.13

Consider the following typical sequence from an unknown binary source:

$$0\ 1\ 0\ 1\ 0\ 0\ 1\ 0\ 1\ 1\ 0\ 1\ 0\ 1\ 0\ 1\ 0\ 1\ 0\ 1$$

If we assume a zero-memory model of the source then we can see that $P(0) = 0.5$, $P(1) = 0.5$ and thus $H(S) = 1$. This implies that the original binary source is 100% efficient.

However the "true" model is the first order Markov model from Figure 3.4.

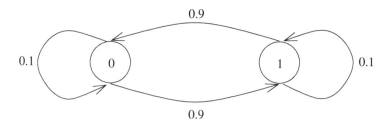

FIGURE 3.4
True model of source.

The "true" entropy is calculated to be $H(S) = 0.469$ and the original binary source is, in fact, only 47% efficient! To improve the efficiency we find a binary code for the second extension. Noting that $P(s_i s_j) = P(s_j | s_i) P(s_i)$ this gives the second extension symbol probabilities:

Source	P_i	Compact Code
$s_1 s_1$	0.05	110
$s_1 s_2$	0.45	0
$s_2 s_1$	0.45	10
$s_2 s_2$	0.05	111

The compact code is shown in the above table. The average length of this code is $L_2 = 1.65$ bits per symbol pair or $\frac{L_2}{2} = 0.825$ bits per symbol which improves the coding efficiency to $\eta = \frac{H(S)}{L_2/2} = 57\%$. ⬜

3.5.3 Code Efficiency and Channel Capacity

Consider the communication system shown in Figure 3.5 for the case of a noiseless channel.

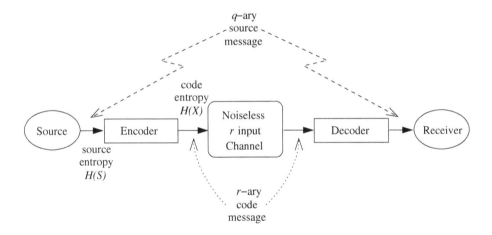

FIGURE 3.5
Noiseless communication system.

The source entropy $H(S)$ is interpreted as the average number of bits transmitted per source symbol. Each source symbol is encoded to a code word of L code symbols on average. Thus the ratio $\frac{H(S)}{L}$ represents the average number of bits transmitted per *code symbol*. By definition this gives the entropy of the encoded sequence or *code entropy*, $H(X)$. That is:

$$H(X) = \frac{H(S)}{L} \tag{3.33}$$

Since $H_r(S) = \frac{H(S)}{\log_2 r}$ and $\eta = \frac{H_r(S)}{L}$ then we have:

$$H(X) = \eta \log_2 r \tag{3.34}$$

Thus if the code is 100% efficient then $H(X) = \log_2 r$ which represents the case of maximum entropy for a source with r symbols. Consider the r-input noiseless channel. From Chapter 2 we established that the mutual information of a noiseless channel is $I(X;Y) = H(X)$, where X and Y are the channel input and output, respectively, and hence for a noiseless channel $C = \max I(X;Y) = \max H(X) = \log_2 r$.

RESULT 3.2

We make the following observations:

(a) A 100% efficient source coding implies maximum code entropy and hence equiprobable code symbols.

(b) A 100% efficient source coding implies maximum or "best" use of the channel.

This result has important implications when we consider the inclusion of a channel code with the source code. As we will see in Chapter 5 one of the key assumptions is that the (binary) inputs to the channel coder are equally likely, and with efficient source coding this will definitely be the case.

EXAMPLE 3.14

Consider the binary source sequence:

$$0\,0\,0\,1\,1\,0\,0\,1\,0\,0$$

Assuming a zero-memory model of the source gives $P(0) = 0.7$, $P(1) = 0.3$ and $H(S) = 0.88$. We find the compact binary code of the second extension of the source is as shown in the following table:

$s_i s_j$	$P(s_i, s_j)$	Compact Code
00	0.49	0
01	0.21	10
10	0.21	110
11	0.09	111

The average length of the code is $\frac{L_2}{2} = 0.905$ and $\eta = \frac{H(S)}{L_2/2} = 0.972$. The code entropy is $H(X) = 0.972$. Consider the corresponding code sequence given the above compact code and initial source sequence:

$$0\ 10\ 110\ 10\,0$$

We assume a zero-memory model and this gives $P(0) = \frac{5}{9}$, $P(1) = \frac{4}{9}$ and $H(X) = 0.99$. Although contrived, this still shows that the code sequence exhibits a higher entropy than the source. ⬚

3.6 Fano Coding

We now describe a coding scheme for the design of instantaneous codes called the *Fano coding* scheme, or *Shannon-Fano code* in reference to the fact that the scheme was independently published by Shannon [11] and Fano [4], that can yield close to optimal codes in most cases. Although mainly of historical interest because of its lack of optimality, Fano coding can be considered the precursor to optimal Huffman coding discussed in the next section as it is based on the same principle but implemented less rigidly.

Fano coding is predicated on the idea that equiprobable symbols should lead to code words of equal length. We describe Fano's method for the design of r-ary instantaneous codes. Consider the source S with q symbols: $\{s_i : i = 1, 2, \ldots q\}$ and associated symbol probabilities $\{P(s_i) : i = 1, 2, \ldots q\}$. Let the symbols be numbered so that $P(s_1) \geq P(s_2) \geq \ldots \geq P(s_q)$. In Fano's method the q symbols are divided or split into r equiprobable or close to equiprobable groups of symbols. Each symbol in the ith group is then assigned the code symbol x_i and this is done for all $i = 1, 2, \ldots r$ groups. Each of the r groups is then further split into r sub-groups (or into less than r sub-groups of one symbol if there are less than r symbols in the group) with each symbol in the ith sub-group having the code symbol x_i appended to it for all $i = 1, 2, \ldots r$ sub-groups. The process is repeated until the groups can be split no further.

For the important case of binary codes, Fano coding splits each group into two equiprobable sub-groups appending a 0 to one group and 1 to the other group. Groups are successively split until no more groups can be split (i.e., each group has only one symbol in it).

EXAMPLE 3.15

Consider the design of a binary code using the Fano coding scheme for a 8-symbol source with the following ordered probability assignments:

$$P(s_1) = 1/2$$
$$P(s_2) = P(s_3) = 1/8$$
$$P(s_4) = P(s_5) = P(s_6) = 1/16$$
$$P(s_7) = P(s_8) = 1/32$$

The regrouping and code symbol assignment in each step of the Fano coding scheme are shown in Figure 3.6. The dashed lines represent the splitting of each set of symbols into two equiprobable groups. For each division a 0 is appended to each symbol in one group and a 1 is appended to each symbol in the other group. The number in front of each dashed line indicates the level of grouping. Thus the level 1 grouping shown in Figure 3.6(a) splits the original source into two groups with one group of probability 1/2 containing $\{s_1\}$ being assigned a 0 and the other group of probability 1/2 containing $\{s_2, s_3, s_4, s_5, s_6, s_7, s_8\}$ being assigned a 1. Each of these groups is split into the level 2 groups as shown in Figure 3.6(b). This continues for each level group until the level 5 grouping of Figure 3.6(e) when there are no more groups left to split is reached. Figure 3.6(e) is the binary Fano code for this source. The average

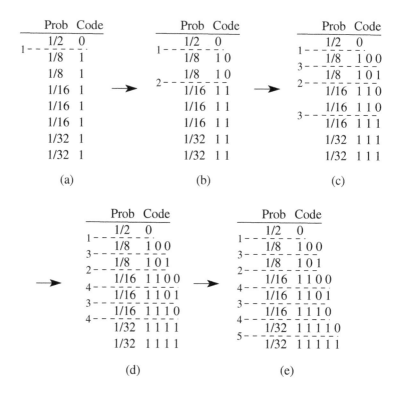

FIGURE 3.6
Binary Fano coding.

length of the Fano code is $L = 2.3125$ bits and since $H(S) = 2.3125$ the Fano code provides the compact code (with efficiency of 100%). This result can also be verified by noting that the source probabilities are such that this is a special source for binary codes.

In Example 3.15 the symbols are successively divided into equiprobable groups of 1/2, 1/4, 1/8, 1/16 and 1/32. In the majority of cases equal groupings are not possible, so closest to equal groupings are used and since there may be many such "quasi" equiprobable groups different Fano codes will result, and not all may yield compact codes.

EXAMPLE 3.16

Consider the design of a binary code using the Fano coding scheme for the 5-symbol source with the following probability assignment:

$$P(s_1) = 4/9$$
$$P(s_2) = P(s_3) = P(s_4) = P(s_5) = P(s_6) = 1/9$$

There are two possible Fano codes depending upon how we perform the level 1 grouping. One grouping creates a group of probability 4/9 containing $\{s_1\}$ and the other group of probability of 5/9 containing $\{s_2, s_3, s_4, s_5, s_6\}$ as shown by Figure 3.7(a). The alternative grouping creates a group of probability 5/9 containing $\{s_1, s_2\}$ and the other group of probability 4/9 containing $\{s_2, s_3, s_4, s_5\}$ as shown by Figure 3.7(b). Both groupings are equally valid. The average lengths of the two Fano codes

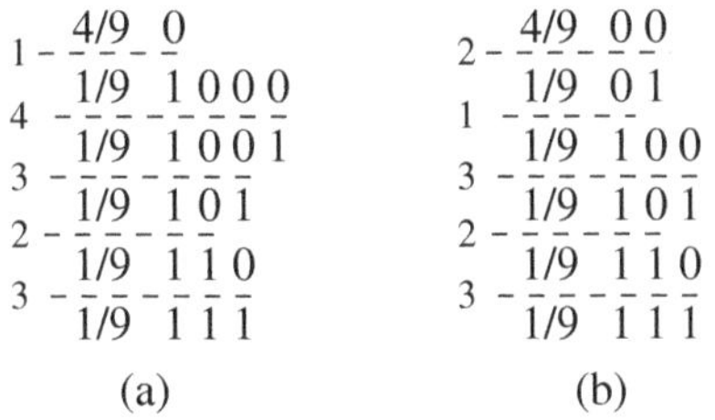

FIGURE 3.7
Different binary Fano codes.

shown in Figure 3.7 are $L_{(a)} = 21/9$ and $L_{(b)} = 22/9$ with code (b) being the compact code. Thus a sub-optimal code may result depending on how the symbols are grouped. ☐

Example 3.16 demonstrates that Fano codes are not necessarily compact and different Fano codes may yield different average lengths. Thus Fano codes are of limited use since such codes cannot be guaranteed to be compact.

3.7 Huffman Coding

3.7.1 Huffman Codes

We now describe an important class of instantaneous codes known as *Huffman codes* which are attributed to the pioneering work of Huffman [7]. Huffman codes arise when using the *Huffman algorithm* for the design of instantaneous codes given the source symbols, s_i, and corresponding source probabilities, $P(s_i)$. The Huffman algorithm attempts to assign each symbol a code word of length proportional to the amount of information conveyed by that symbol. Huffman codes are important because of the following result.

RESULT 3.3
Huffman codes are compact codes. That is, the Huffman algorithm produces a code with an average length, L, which is the smallest possible to achieve for the given number of source symbols, code alphabet and source statistics.

We now proceed to outline the basic principles of the Huffman algorithm and then detail the steps and provide examples for specific cases.

The Huffman algorithm operates by first successively *reducing* a source with q symbols to a source with r symbols, where r is the size of the code alphabet.

DEFINITION 3.11 Reduced Source   *Consider the source S with q symbols: $\{s_i : i = 1, 2, \ldots q\}$ and associated symbol probabilities $\{P(s_i) : i = 1, 2, \ldots q\}$. Let the symbols be renumbered so that $P(s_1) \geq P(s_2) \geq \ldots \geq P(s_q)$. By combining the last r symbols of S, $\{s_{q-r+1}, s_{q-r+2}, \ldots s_q\}$, into one symbol, $\widehat{s}_{q-r+1}$, with probability, $P(\widehat{s}_{q-r+1}) = \sum_{i=1}^{r} P(s_{q-r+i})$, we obtain a new source, termed a reduced source of S, containing only $q - r + 1$ symbols, $\{s_1, s_2, \ldots s_{q-r}, \widehat{s}_{q-r+1}\}$. Call this reduced source S_1. Successive reduced sources S_2 S_3 ... can be formed by a similar process of renumbering and combining until we are left with a source with only r symbols.*

It should be noted that we will only be able to reduce a source to exactly r symbols if the original source has $q = r + \alpha(r - 1)$ symbols where α is a non-negative integer. For a binary $(r = 2)$ code this will hold for any value of $q \geq 2$. For non-binary codes if $\alpha = \frac{q-r}{(r-1)}$ is not an integer value then "dummy" symbols with zero probability are appended to create a source with $q = r + \lceil \alpha \rceil (r - 1)$ symbols, where $\lceil \alpha \rceil$ is the smallest integer greater than or equal to α.

The trivial r-ary compact code for the reduced source with r symbols is then used to design the compact code for the preceding reduced source as described by the following result.

RESULT 3.4

Assume that we have a compact code for the reduced source S_j. Designate the last r symbols from S_{j-1}, $\{s_{q-r+1}, s_{q-r+2}, \ldots s_q\}$, as the symbols which were combined to form the combined symbol, $\widehat{s}_{q-r+1}$ of S_j. We assign to each symbol of S_{j-1}, except the last r symbols, the code word used by the corresponding symbol of S_j. The code words for the last r symbols of S_{j-1} are formed by appending $\{0, 1, \ldots r\}$ to the code word of $\widehat{s}_{q-r+1}$ of S_j to form r new code words.

The Huffman algorithm then operates by back-tracking through the sequence of reduced sources, $S_j, S_{j-1}, \ldots$, designing the compact code for each source, until the compact code for the original source S is designed.

3.7.2 Binary Huffman Coding Algorithm

For the design of binary Huffman codes the Huffman coding algorithm is as follows:

1. Re-order the source symbols in decreasing order of symbol probability.

2. Successively reduce the source S to S_1, then S_2 and so on, by combining the last two symbols of S_j into a combined symbol and re-ordering the new set of symbol probabilities for S_{j+1} in decreasing order. For each source keep track of the position of the combined symbol, $\widehat{s}_{q-1}$. Terminate the source reduction when a two symbol source is produced. For a source with q symbols the reduced source with two symbols will be S_{q-2}.

3. Assign a compact code for the final reduced source. For a two symbol source the trivial code is $\{0, 1\}$.

4. Backtrack to the original source S assigning a compact code for the jth reduced source by the method described in Result 3.4. The compact code assigned to S is the binary Huffman code.

The operation of the binary Huffman coding algorithm is best shown by the following example.

EXAMPLE 3.17

Consider a 5 symbol source with the following probability assignments:

$$P(s_1) = 0.2 \quad P(s_2) = 0.4 \quad P(s_3) = 0.1 \quad P(s_4) = 0.1 \quad P(s_5) = 0.2$$

Re-ordering in decreasing order of symbol probability produces $\{s_2, s_1, s_5, s_3, s_4\}$. The re-ordered source S is then reduced to the source S_3 with only two symbols as shown in Figure 3.8, where the arrow-heads point to the combined symbol created in

S_j by the combination of the last two symbols from S_{j-1}. Starting with the trivial compact code of $\{0, 1\}$ for S_3 and working back to S a compact code is designed for each reduced source S_j following the procedure described by Result 3.4 and shown in Figure 3.8. In each S_j the code word for the last two symbols is produced by taking the code word of the symbol pointed to by the arrow-head and appending a 0 and 1 to form two new code words. The Huffman coding algorithm of Figure 3.8 can be depicted graphically as the binary Huffman coding tree of Figure 3.9. The Huffman coding tree is of similar form to the decoding tree for instantaneous codes shown in Figure 3.3 and can thus be used for decoding Huffman codes. The Huffman code itself is the bit sequence generated by the path from the root to the corresponding leaf node.

	S		S_1	S_2	S_3
s_2	0.4	*1*	0.4 *1*	0.4 *1*	�by0.6 *0*
s_1	0.2	*01*	0.2 *01*	▸0.4 *00*	0.4 *1*
s_5	0.2	*000*	0.2 *000*	0.2 *01*	
s_3	0.1	*0010*	▸0.2 *001*		
s_4	0.1	*0011*			

FIGURE 3.8
Binary Huffman coding table.

The binary Huffman code is:

Symbol	$P(s_i)$	Huffman Code
s_1	0.2	01
s_2	0.4	1
s_3	0.1	0010
s_4	0.1	0011
s_5	0.2	000

The average length of the code is $L = 2.2$ bits/symbol and the efficiency of the Huffman code is $\eta = \frac{H(S)}{L} = \frac{2.122}{2.2} = 96.5\%$. Since the Huffman code is a compact code then any other instantaneous code we may design will have an average length, L_O, such that $L_O \geq L$.   □

From Figure 3.8 in Example 3.17 different reduced sources S_1 and S_2 are possible by inserting the combined symbol $\hat{s}_{q-1}$ in a different position when re-ordering.

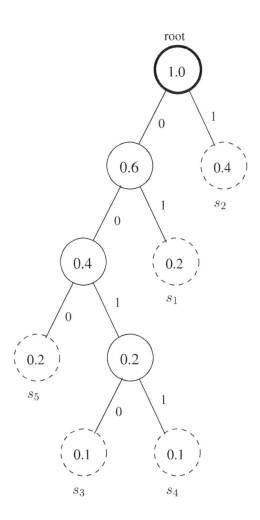

FIGURE 3.9
Binary Huffman coding tree.

This may or may not result in a compact code with different individual code word lengths. Either way the compact code will have the same average length. However the average length *variance*:

$$\sigma^2 = \sum_{i=1}^{q} P_i(l_i - L)^2 \tag{3.35}$$

may be different.

EXAMPLE 3.18

The average length variance for the Huffman code of Example 3.17 is $\sigma^2 = 1.88$. Consider a different Huffman code tree for the same source shown by Figures 3.10 and 3.11.

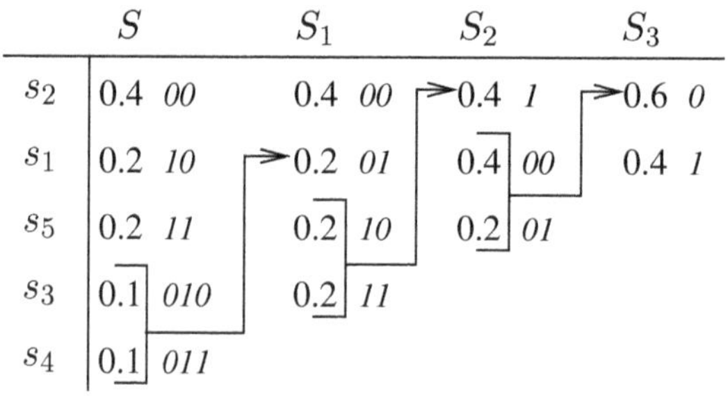

	S	S_1	S_2	S_3
s_2	0.4 *00*	0.4 *00*	0.4 *1*	0.6 *0*
s_1	0.2 *10*	0.2 *01*	0.4 *00*	0.4 *1*
s_5	0.2 *11*	0.2 *10*	0.2 *01*	
s_3	0.1 *010*	0.2 *11*		
s_4	0.1 *011*			

FIGURE 3.10
Different binary Huffman code table.

The binary Huffman code is now:

Symbol	$P(s_i)$	Huffman Code
s_1	0.2	10
s_2	0.4	00
s_3	0.1	010
s_4	0.1	011
s_5	0.2	11

Although this different Huffman code for the same source possesses different individual code word lengths the average length is still $L = 2.2$ bits/symbol; however, the average length variance is now $\sigma^2 = 0.16$.

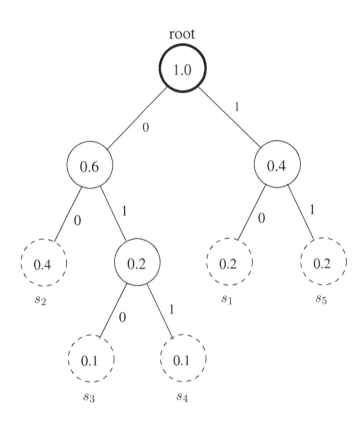

FIGURE 3.11
Different binary Huffman code tree.

The Huffman code from Example 3.18 has a smaller average length variance than the code produced in Example 3.17. Codes with smaller average length variance are preferable since they produce a more constant code bit rate. The following result can be used to ensure the Huffman coding tree produces a compact code with the smallest average length variance.

RESULT 3.5

If the combined symbol, $\widehat{s}_{q-1}$, is placed at the highest available position for that probability assignment when the reduced source is re-ordered then the resulting compact code will possess the smallest average length variance.

3.7.3 Software Implementation of Binary Huffman Coding

The binary Huffman coding algorithm can be implemented in software as a sequence of merge and sort operations on a binary tree. The Huffman coding algorithm is a greedy algorithm that builds the Huffman decoding tree (e.g., Figures 3.9 and 3.11) by initially assigning each symbol as the root node of a single-node tree. The decoding tree is then constructed by successively merging the last two nodes, labeling the edges of the left-right child pairs of the merge operation with a 0 and 1, respectively, and sorting the remaining nodes until only one root node is left. The code word for s_i is then the sequence of labels on the edges connecting the root to the leaf node of s_i. The details of the algorithm including a proof of the optimality of Huffman codes can be found in [3].

3.7.4 r-ary Huffman Codes

For the design of general r-ary Huffman codes the Huffman coding algorithm is as follows:

1. Calculate $\alpha = \frac{q-r}{(r-1)}$. If α is a non-integer value then append "dummy" symbols to the source with zero probability until there are $q = r + \lceil \alpha \rceil (r - 1)$ symbols.

2. Re-order the source symbols in decreasing order of symbol probability.

3. Successively reduce the source S to S_1, then S_2 and so on, by combining the last r symbols of S_j into a combined symbol and re-ordering the new set of symbol probabilities for S_{j+1} in decreasing order. For each source keep track of the position of the combined symbol, $\widehat{s}_{q-r+1}$. Terminate the source reduction when a source with exactly r symbols is produced. For a source with q symbols the reduced source with r symbols will be $S_{\lceil \alpha \rceil}$.

4. Assign a compact r-ary code for the final reduced source. For a source with r symbols the trivial code is $\{0, 1, \ldots, r\}$.

5. Backtrack to the original source S assigning a compact code for the jth reduced source by the method described in Result 3.4. The compact code assigned to S, minus the code words assigned to any "dummy" symbols, is the r-ary Huffman code.

The operation of the r-ary Huffman code is demonstrated in the following example.

EXAMPLE 3.19

We want to design a compact quaternary $(r = 4)$ code for a source with originally 11 $(q = 11)$ symbols and the following probability assignments (re-ordered in decreasing order of probability for convenience):

$$P(s_1) = 0.16 \quad P(s_2) = 0.14 \quad P(s_3) = 0.13 \quad P(s_4) = 0.12 \quad P(s_5) = 0.10$$

$$P(s_6) = 0.10 \quad P(s_7) = P(s_8) = 0.06 \quad P(s_9) = 0.05 \quad P(s_{10}) = P(s_{11}) = 0.04$$

Now $\alpha = 2.33$ and since α is not an integer value we need to append "dummy" symbols so that we have a source with $q = r + \lceil \alpha \rceil (r - 1) = 13$ symbols where $\lceil \alpha \rceil = 3$. Thus we append symbols $\{s_{12}, s_{13}\}$ with $P(s_{12}) = P(s_{13}) = 0.00$. The source S is then reduced to the source S_3 with only r symbols as shown in Figure 3.12, where the arrow-heads point to the combined symbol created in S_j by the combination of the last r symbols from S_{j-1}. Starting with the trivial compact code of $\{0, 1, 2, 3\}$ for S_3 and working back to S a compact code is designed for each reduced source S_j following the procedure described by Result 3.4. The code word for the last r symbols is produced by taking the code word of the symbol pointed to by the arrow-head and appending $\{0, 1, 2, 3\}$ to form r new code words.

The r-ary Huffman code is (ignoring the "dummy" symbols s_{12} and s_{13}):

Symbol	$P(s_i)$	Huffman Code
s_1	0.16	2
s_2	0.14	3
s_3	0.13	00
s_4	0.12	01
s_5	0.10	02
s_6	0.10	03
s_7	0.06	11
s_8	0.06	12
s_9	0.05	13
s_{10}	0.04	100
s_{11}	0.04	101

The average length of the code is $L = 1.78$ quaternary units per symbol and the Huffman code has an efficiency of $\eta = \frac{H_4(S)}{L} = \frac{1.654}{1.78} = 93\%$. ▯

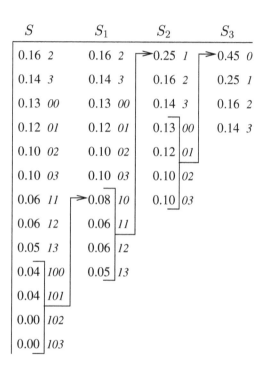

FIGURE 3.12
r-ary Huffman code table.

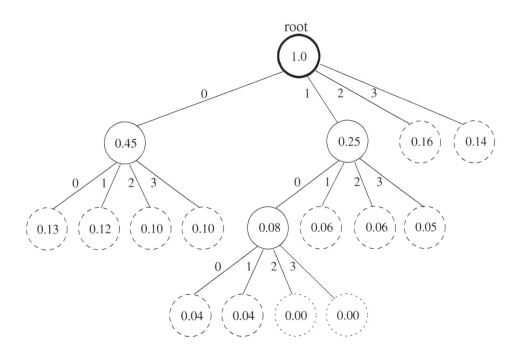

FIGURE 3.13
r-ary Huffman code tree.

3.8 Arithmetic Coding

The Huffman coding method described in Section 3.7 is guaranteed to be optimal in the sense that it will generate a compact code for the given source alphabet and associated probabilities. However a compact code may be anything but optimal if it is not very efficient. Only for special sources, where $l_i = \log_r \frac{1}{P_i} = -\log_r P_i$ is an integer for $i = 1, 2, \ldots q$, will the compact code also be 100% efficient. Inefficiencies are introduced when l_i is a non-integer and it is required to "round-up" to the nearest integer value. The solution to this as a consequence of Shannon's Noiseless Coding Theorem (see Section 3.5) is to consider the nth extension of the source for n large enough and build the Huffman code for all possible blocks of length n. Although this does yield optimal codes, the implementation of this approach can easily become unwieldy or unduly restrictive. Problems include:

- The size of the Huffman code table is q^n, representing an exponential increase in memory and computational requirements.

- The code table needs to be transmitted to the receiver.

- The source statistics are assumed stationary. If there are changes an adaptive scheme is required which re-estimates the probabilities, and recalculates the Huffman code table.

- Encoding and decoding is performed on a per block basis; the code is not produced until a block of n symbols is received. For large n this may require the last segment to be padded with dummy data.

One solution to using Huffman coding on increasingly larger extensions of the source is to directly code the source message to a code sequence using *arithmetic coding* which we will describe here. Rather than deriving and transmitting a code table of size q^n, segmenting the incoming source message into consecutive blocks of length n, and encoding each block by consulting the table, arithmetic coding directly processes the incoming source message and produces the code "on-the-fly." Thus there is no need to keep a code table, making arithmetic coding potentially much more computationally efficient than Huffman coding. The basic concept of arithmetic coding can be traced back to the 1960's; however, it wasn't until the late 1970's and mid 1980's that the method started receiving much more attention, with the paper by Witten et al. [12] often regarded as one of the seminal papers in the area.

Arithmetic coding, like Huffman coding, does require reliable source statistics to be available. Consider the N-length source message $s_{i1}, s_{i2}, \ldots s_{iN}$ where $\{s_i : i = 1, 2, \ldots q\}$ are the source symbols and s_{ij} indicates that the jth character in the message is the source symbol s_i. Arithmetic coding assumes that the probabilities,

$P(s_{ij}|s_{i1}, s_{i2}, \ldots s_{i,j-1})$, for $j = 1, 2, \ldots N$, can be calculated. The origin of these probabilities depends on the underlying source model that is being used (e.g., zero-memory source or mth order Markov model) and how such models are arrived at and probabilities estimated is not discussed here. Rather we assume that the required probabilities are available and concentrate on the encoding and decoding process.

The goal of arithmetic coding is to assign a unique interval along the unit number line or "probability line" $[0, 1)$ of length equal to the probability of the given source message, $P(s_{i1}, s_{i2} \ldots, s_{iN})$, with its position on the number line given by the cumulative probability of the given source message, $Cum(s_{i1}, s_{i2} \ldots, s_{iN})$. However there is no need for direct calculation of either $P(s_{i1}, s_{i2} \ldots, s_{iN})$ or $Cum(s_{i1}, s_{i2} \ldots, s_{iN})$. The basic operation of arithmetic coding is to produce this unique interval by starting with the interval $[0, 1)$ and iteratively subdividing it by $P(s_{ij}|s_{i1}, s_{i2}, \ldots s_{i,j-1})$ for $j = 1, 2, \ldots N$.

The interval subdivision operation of arithmetic coding proceeds as follows. Consider the first letter of the message, s_{i1}. The individual symbols, s_i, are each assigned the interval $[lb_i, hb_i)$ where $hb_i = Cum(s_i) = \sum_{k=1}^{i} P(s_k)$ and $lb_i = Cum(s_i) - P(s_i) = \sum_{k=1}^{i-1} P(s_k)$. That is the length of each interval is $hb_i - lb_i = P(s_i)$ and the end of the interval is given by $Cum(s_i)$. The interval corresponding to the symbol s_{i1}, that is $[lb_{i1}, hb_{i1})$, is then selected. Next we consider the second letter of the message, s_{i2}. The individual symbols, s_i, are now assigned the interval $[lb_i, hb_i)$ where:

$$hb_i = lb_{i1} + Cum(s_i|s_{i1}) * P(s_{i1})$$

$$= lb_{i1} + \left\{ \sum_{k=1}^{i} P(s_k|s_{i1}) \right\} * R \qquad (3.36)$$

$$lb_i = lb_{i1} + \{Cum(s_i|s_{i1}) - P(s_i|s_{i1})\} * P(s_{i1})$$

$$= lb_{i1} + \left\{ \sum_{k=1}^{i-1} P(s_k|s_{i1}) \right\} * R \qquad (3.37)$$

and $R = hb_{i1} - lb_{i1} = P(s_{i1})$. That is the length of each interval is $hb_i - lb_i = P(s_i|s_{i1}) * P(s_{i1})$. The interval corresponding to the symbol s_{i2}, that is $[lb_{i2}, hb_{i2})$, is then selected. The length of the interval corresponding to the message seen so far, s_{i1}, s_{i2}, is $hb_{i2} - lb_{i2} = P(s_{i2}|s_{i1}) * P(s_{i1}) = P(s_{i1}, s_{i2})$. This interval subdivision operation is depicted graphically in Figure 3.14.

We continue in this way until the last letter of the message, s_{iN}, is processed and the final interval of length $hb_{iN} - lb_{iN} = P(s_{i1}, s_{i2} \ldots, s_{iN})$ is produced. Any number that falls within that interval can be chosen and transmitted to identify the interval (and hence the original message) to the receiver. Arithmetic codes are typically binary codes. Thus the binary representation of the number is considered. Optimal coding is assured by selecting a number that requires the minimum number of bits to be transmitted to the receiver (by only transmitting the significant bits and ignoring

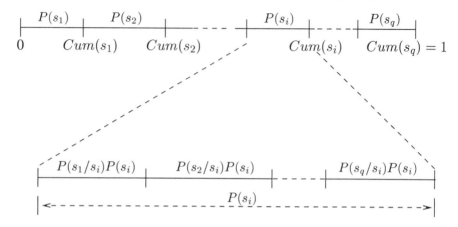

FIGURE 3.14
Interval subdividing operation of arithmetic coding.

the trailing 0's). Since the interval occurs with probability $P(s_{i1}, s_{i2} \ldots, s_{iN})$ then approximately $-\log_2 P(s_{i1}, s_{i2} \ldots, s_{iN})$ bits are needed and where this becomes equality for all message sequences a 100% efficient coding is possible. A simple demonstration of arithmetic coding is given in Example 3.20.

EXAMPLE 3.20

Consider the message s_2, s_2, s_1 originating from the 3-symbol source with the following individual and cumulative probabilities:

Symbol	$P(s_i)$	$Cum(s_i)$
s_1	0.2	0.2
s_2	0.5	0.7
s_3	0.3	1.0

We further assume that the source is zero-memory, thus $P(s_{ij}|s_{i1}, s_{i2}, \ldots s_{i,j-1}) = P(s_{ij})$. As shown by Figure 3.15 initially the probability line $[0, 1)$ is divided into three consecutive, adjoint intervals of $[0, 0.2), [0.2, 0.7)$ and $[0.7, 1.0)$ corresponding to s_1, s_2 and s_3 and the length ratios $0.2 : 0.5 : 0.3$, respectively. The length ratios and interval lengths both correspond to $P(s_i)$. When the first letter of the message, s_2, is received the interval $[0.2, 0.7)$ is selected. The interval $[0.2, 0.7)$ is further divided into three consecutive, adjoint subintervals of length ratios $0.2 : 0.5 : 0.3$, that is $[0.2, 0.3)$, $[0.3, 0.55)$ and $[0.55, 0.7)$. The length ratios correspond to $P(s_i|s_2) = P(s_i)$ and the subinterval lengths are equal to $P(s_i|s_2)P(s_2) = P(s_i)P(s_2) = P(s_i, s_2)$. When the second letter of the message, s_2, is received the subinterval $[0.3, 0.55)$ is selected. The interval $[0.3, 0.55)$ is then further subdivided into the three subintervals $[0.3, 0.35), [0.35, 0.475)$ and $[0.475, 0.55)$ with length ratios $0.2 : 0.5 : 0.3$ and when the third and last letter of the message, s_1, is received the corresponding and final interval $[0.3, 0.35)$ of length $P(s_2)P(s_2)P(s_1) = P(s_2, s_2, s_1) = 0.05$ is selected.

The interval selection process can be summarised by the following table:

Next Letter	Interval
s_2	$[0.2,0.7)$
s_2	$[0.3,0.55)$
s_1	$[0.3,0.35)$

In binary the final interval is $[0.01001100, 0.01011001)$. We need to select a number from within the interval that can be represented with the least number of significant bits. The number 0.0101 falls within the interval and only requires the 4 bits 0101 to be transmitted. We note that we need to transmit $-\log_2 P(s_2, s_2, s_1) = -\log_2(0.05) = 4.32$ bits of information and we have been able to do this with 4 bits. ▯

The above description of arithmetic coding is still incomplete. For example, how does one select the number that falls within the final interval so that it can be transmitted in the least number of bits? Let $[low, high)$ denote the final interval. One scheme described in [6] which works quite well in the majority of cases is to carry out the binary expansion of low and $high$ until they differ. Since $low < high$, at the first place they differ there will be a 0 in the expansion for low and a 1 in the expansion for $high$. That is:

$$low = 0.a_1 a_2 \ldots a_{t-1} 0 \ldots$$
$$high = 0.a_1 a_2 \ldots a_{t-1} 1 \ldots$$

The number $0.a_1 a_2 \ldots a_{t-1} 1$ falls within the interval and requires the least number of bits from any other number within the same interval; so it is selected and transmitted as the t-bit code sequence $a_1 a_2 \ldots a_{t-1} 1$.

3.8.1 Encoding and Decoding Algorithms

We have described how the message is encoded and transmitted but have not yet discussed how the received code is decoded back to the original message. Figure 3.16 lists an algorithm for encoding a message using arithmetic coding to the binary integer code, *value*, and Figure 3.17 lists an algorithm for decoding the received binary integer, *value*, back to the original message.

NOTE   In both the encoding and decoding algorithms mention is made of the *EOF* or end-of-file. Consider Example 3.20. The code sequence 0101 corresponds to the binary fraction 0.0101 and decimal fraction 0.3125. The number 0.3125 not only falls within the final interval for the message s_2, s_2, s_1 but also for the longer

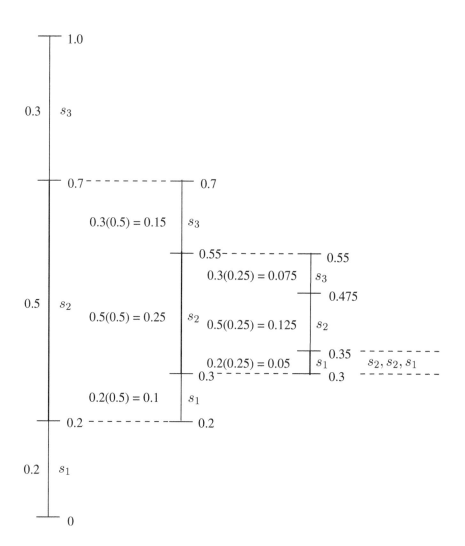

FIGURE 3.15
Arithmetic coding for Example 3.20.

Definitions

1. Let $S = \{s_i : i = 1, 2 \ldots q\}$ denote the q distinct source symbols or letters of the alphabet.

2. Let $s_{i1}, s_{i2} \ldots s_{iN}$ denote the N-length message that is to be encoded where s_{ij} indicates that the jth letter in the message is the symbol s_i.

3. Assume $P(s_{ij}|s_{i1}, s_{i2} \ldots s_{i,j-1})$ for $i = 1, 2 \ldots q$ and $j = 1, 2 \ldots N$ are available or can be estimated from the underlying source model.

Initialisation

- $low = 0.0$
- $high = 1.0$
- $j = 1$

Iteration

While $(input \neq EOF)$ do

1. get next letter of message, s_{ij}

2. $Range = high - low$

3. $lb = \sum_{k=1}^{i-1} P(s_{kj}|s_{i1}, s_{i2} \ldots s_{i,j-1})$

4. $hb = \sum_{k=1}^{i} P(s_{kj}|s_{i1}, s_{i2} \ldots s_{i,j-1}) = lb + P(s_{ij}|s_{i1}, s_{i2} \ldots s_{i,j-1})$

5. $high = low + Range * hb$

6. $low = low + Range * lb$

7. $j = j + 1$

done

Termination

Let $low = 0.a_1 a_2 \ldots a_{t-1} 0 \ldots$ and $high = 0.a_1 a_2 \ldots a_{t-1} 1 \ldots$. Then transmit the code as the binary integer $value = a_1 a_2 \ldots a_{t-1} 1$.

FIGURE 3.16
Encoding algorithm for arithmetic codes.

Definitions

1. Let $S = \{s_i : i = 1, 2 \ldots q\}$ denote the q distinct source symbols or letters of the alphabet.

2. Let $value$ denote the binary integer that is to be decoded to the original message $s_{i1}, s_{i2} \ldots s_{iN}$, where s_{ij} indicates that the jth letter in the message is the symbol s_i.

3. Assume $P(s_{ij}|s_{i1}, s_{i2} \ldots s_{i,j-1})$ for $i = 1, 2 \ldots q$ and $j = 1, 2 \ldots N$ are available or can be estimated from the underlying source model.

Initialisation

- $low = 0.0$

- $high = 1.0$

- $j = 1$

- get $value$ and store $value$ as the binary fraction $.value0\ldots.$

Iteration

Repeat

1. Start at $i = 0$ and repeat:

    (a) $i = i + 1$

    (b) $lb = \sum_{k=1}^{i-1} P(s_{kj}|s_{i1}, s_{i2} \ldots s_{i,j-1})$

    (c) $hb = \sum_{k=1}^{i} P(s_{kj}|s_{i1}, s_{i2} \ldots s_{i,j-1})$

    until $lb \leq \frac{value - low}{high - low} \leq hb$

2. output s_{ij} as the symbol s_i

3. $Range = high - low$

4. $high = low + Range * hb$

5. $low = low + Range * lb$

6. $j = j + 1$

until EOF is reached

FIGURE 3.17
Decoding algorithm for arithmetic codes.

length messages s_2, s_2, s_1, s_2 and s_2, s_2, s_1, s_2, s_2 and so on. Thus there is a need for the decoder to know when to stop decoding. Possible solutions to this problem include:

- Defining a special or extra symbol called EOF which is placed at the end of each message.

- Transmitting the length of the message, N, along with the code.

- Encoding and decoding messages in fixed-sized blocks and only transmitting the size of the block to the decoder at the beginning of the code sequence.

A demonstration of the encoding and decoding algorithms depicted in Figures 3.16 and 3.17 is provided by Examples 3.21 and 3.22.

EXAMPLE 3.21

Consider a zero-memory source with the following source alphabet and probabilities:

Symbol	$P(s_i)$	$Cum(s_i)$	$[lb, hb)$
a	6/15	6/15	$[0.0, 6/15)$
b	2/15	8/15	$[6/15, 8/15)$
d	2/15	10/15	$[8/15, 10/15)$
$\_$	4/15	14/15	$[10/15, 14/15)$
$.$	1/15	15/15	$[14/15, 1.0)$

Suppose the message $bad\_dab.$ is generated by the source where . is the EOF. Applying the encoding algorithm of Figure 3.16, where $P(s_{ij}|s_{i1}, s_{i2}, \ldots s_{i,j-1}) = P(s_{ij})$, yields the following:

Next letter	$Range$	$[lb, hb)$	$[low, high)$
Init	-	-	$[0.0, 1.0)$
b	1.0	$[6/15, 8/15)$	$[0.4, 0.533333333)$
a	0.133333333	$[0.0, 6/15)$	$[0.4, 0.453333333)$
d	0.053333333	$[8/15, 10/15)$	$[0.428444444, 0.435555556)$
$\_$	0.007111111	$[10/15, 14/15)$	$[0.433185185, 0.435081481)$
d	0.001896296	$[8/15, 10/15)$	$[0.434196543, 0.434449383)$
a	0.000252840	$[0.0, 6/15)$	$[0.434196543, 0.434297679)$
b	0.000101136	$[6/15, 8/15)$	$[0.434236998, 0.434250482)$
$.$	0.000013485	$[14/15, 1.0)$	$[0.434249583, 0.434250482)$

The binary representation of the $low = 0.434249583$ and $high = 0.434250482$ is:

$$low = 0.01101111001010101111\ldots$$
$$high = 0.01101111001010110000\ldots$$

and hence the integer code that uniquely identifies the interval that is transmitted is the 16-bit $value = 0110111100101011$.

☐

EXAMPLE 3.22

Consider decoding the received code $value = 0110111100101011$ produced by the encoding process in Example 3.21. Applying the decoding algorithm of Figure 3.17 given the source alphabet and probabilities from Example 3.21 to the stored value of $value = 0.01101111001010110 = 0.434249878$ yields the following:

$[low, high)$	$Range$	$\frac{value-low}{high-low}$	$s_i : [lb, hb)$
$[0.0, 1.0)$	1.0	0.434249878	$b : [6/15, 8/15)$
$[0.4, 0.533333333)$	0.133333333	0.256874149	$a : [0, 6/15)$
$[0.4, 0.453333333)$	0.053333333	0.642185373	$d : [8/15, 10/15)$
$[0.428444444, 0.435555556)$	0.007111111	0.816389094	$\_ : [10/15, 14/15)$
$[0.433185185, 0.435081481)$	0.001896296	0.561459102	$d : [8/15, 10/15)$
$[0.434196543, 0.434449383)$	0.000252840	0.210943262	$a : [0, 6/15)$
$[0.434196543, 0.434297679)$	0.000101136	0.527358154	$b : [6/15, 8/15)$
$[0.434236998, 0.434250482)$	0.000013485	0.955186157	$. : [14/15, 1)$

and the decoded message is $bad\_dab.$ where . is the EOF used to terminate the decoding process.

☐

3.8.2 Encoding and Decoding with Scaling

The encoding and decoding algorithms depicted in Figures 3.16 and 3.17 iteratively reduce the length of the interval with each successive letter from the message. Thus with longer messages, smaller intervals are produced. If the interval range becomes too small, then underflow is a very real problem. In Example 3.21 the interval range is already down to 1.011e-04 after only the 8th letter in the message and in the binary representation of the final $[low, high)$ interval the first (most significant) 16 bits are identical. With modern 64-bit CPUs the low and $high$ will become indistinguishable when the first 64 bits are identical. The practical implementation of the encoding and decoding algorithms requires an added scaling operation. The scaling operation rescales the current $[low, high)$ based on the pseudo-algorithm depicted in Figure 3.18.

The scaling operation not only solves the underflow problem but permits transmission of the coded $value$ before the encoding operation is complete. Thus a long message block can be encoded without having to wait until the end of the message before transmitting the coded $value$. Similarly, the decoder can begin decoding the

---

Rescale

If $low = 0.a_1 a_2 \ldots a_{t-1} 0 l_1 l_2 \ldots$ and $high = 0.a_1 a_2 \ldots a_{t-1} 1 h_1 h_2 \ldots$ then rescale to:

$$low = 0.0 l_1 l_2 \ldots$$
$$high = 0.1 h_1 h_2 \ldots$$

If <u>encoding</u> then $value = value + a_1 a_2 \ldots a_{t-1}$ (initially $value$ is the NULL string), where $+$ is the string concatenation operator.
If <u>decoding</u> then $value = (value*2^{t-1}) \bmod 1$, where $\bmod 1$ returns the fractional component of the product.

FIGURE 3.18
Scaling algorithm to avoid underflow.

most significant bits of the coded *value* (which are transmitted first by the encoder when rescaling the intervals) before all the bits have been received. Examples 3.23 and 3.24 illustrate the use of the scaling algorithm of Figure 3.18 for the encoding and decoding of arithmetic codes.

EXAMPLE 3.23

The encoding from Example 3.21 is now repeated with the additional rescaling operation of Figure 3.18 to prevent underflow from occurring. The encoding with scaling is shown in the following table where $[low, high)_r$ indicates that the low and $high$ are expressed in base r:

Next letter	*Range*	$[lb, hb)$	$[low, high)_{10}$	$[low, high)_2$	*output*
Init	-	-	[0.0, 1.0)	[0.0, 1.0)	
b	1.0	[6/15, 8/15)	[.4, .533333)	[.01100110, .10001000)	
a	.133333	[0, 6/15)	[.4, .453333)	[**.01100110**, **.01110100**)	
(rescale)			[.2, .626667)	[.00110..., .10100...)	**011**
d	.426667	[8/15, 10/15)	[.427556, .484444)	[**.01101101**, **.01111100**)	
(rescale)			[.420444, .875556)	[.01101..., .11100...)	**011**
$_$	.455111	[10/15, 14/15)	[.723852, .845215)	[**.10111001**, **.11011000**)	
(rescale)			[.447704, .690430)	[.0111001..., .1011000...)	**1**
d	.242726	[8/15, 10/15)	[.577158, .609521)	[**.10010011**, **.10011100**)	
(rescale)			[.234520, .752335)	[.0011..., .1100...)	**1001**
a	.517815	[0, 6/15)	[.234520, .441646)	[**.00111100**, **.01110001**)	
(rescale)			[.469040, .883292)	[.0111100..., .1110001...)	**0**
b	.414252	[6/15, 8/15)	[.634741, .689975)	[**.10100010**, **.10110000**)	
(rescale)			[.077928, .519797)	[.00010..., .10000...)	**101**
$.$	.441869	[14/15, 1)	[.490339, .519797)	[**.01111101**, **.10000101**)	
(terminate)					**1**

When the last or EOF letter of the message is read a final 1 is transmitted. Concatenating the intermediate bits that were output yields the final $value = 0110111100101011$

which is the same *value* as in Example 3.21.

☐

EXAMPLE 3.24

The decoding from Example 3.22 is repeated with the additional rescaling operation of Figure 3.18. The decoding processing with scaling is shown in the following table:

$\frac{value-low}{high-low}$	s_i	$[low, high)_{10}$	$[low, high)_2$	*Range*	*value*
-	-	[0.0, 1.0)	[0.0, 1.0)	1.0	.434250
.434250	b	[.400000,.533333)	[.01100110, .10001000)	.133333	.434250
.256874	a	[.4, .453333)	[.**01**100110, .**01**110100)		
(rescale)		[.2, .626667)	[.001100..., .10100...)	.426667	.473999
.642185	d	[.427556, .484444)	[.**01**101101, .**01**111100)		
(rescale)		[.420444, .875556)	[.01101..., .11100...)	.455111	.791992
.816389	-	[.723852, .845215)	[.**10**111001, .**11**011000)		
(rescale)		[.447704, .690430)	[.0111001..., .1011000...)	.242726	.583984
.561459	d	[.577158, .609521)	[.**10010**011, .**10011**100)		
(rescale)		[.234520, .752335)	[.0011..., .1100...)	.517815	.343750
.210944	a	[.234520, .441646)	[.**00**111100, .**01**110001)		
(rescale)		[.469040, .883292)	[.0111100..., .1110001...)	.414252	.687500
.527360	b	[.634741, .689975)	[.**10**100010, .**10**110000)		
(rescale)		[.077928, .519797)	[.00010..., .10000...)	.441870	.500000
.955197	.				

☐

NOTE The scaling operation not only expands the interval but repositions it around 0.5. That is, the rescaled interval has *low* < 0.5 and *high* > 0.5. Thus, rescaling is needed whenever *low* and *high* are both less than or greater than 0.5. One pathological condition that may arise and defeat the scaling operation is when *low* is just below 0.5 and *high* is just above 0.5. Consider:

$$low = 0.0111111111111$$
$$high = 0.1000000000001$$

The interval cannot be rescaled and the interval $Range = high - low = 0$ and $high = 0.5$ when the CPU bit-size is exceeded and underflow occurs.

3.8.3 Is Arithmetic Coding Better Than Huffman Coding?

Shannon's noiseless coding theorem states that if one designs a compact binary code, using Huffman coding, for the nth extension of a source then the average length of the code, $\frac{L_n}{n}$, is no greater than $H(S) + \frac{1}{n}$. A similar theorem outlined in [6] states that for the arithmetic coding scheme just described the average length is also no greater than $H(S) + \frac{1}{n}$ when encoding a message of length n. Thus Huffman coding and arithmetic coding exhibit similar performance in theory. However the arithmetic code encoding and decoding algorithms with scaling can be implemented for messages of any length n. Huffman coding for the nth extension of a source, however, becomes computationally prohibitive with increasing n since the computational complexity is of the order q^n.

3.9 Higher-order Modelling

Shannon's Noiseless Coding Theorem discussed in Section 3.5 states that code efficiency can be improved by coding the nth extension of the source. Huffman coding is then used to derive the compact code for S^n, the nth extension of the source. The arithmetic codes from Section 3.8 directly code the n-length message sequence to the code word. In both cases source statistics are needed to provide the required conditional and joint probabilities which we obtain with the appropriate model. But what model do we use? Result 3.1 alluded to the requirement that we use the "true" model for the source, except that in practice the order of the "true" model is not known, it may be too high to be computationally feasible or, indeed, the "true" model is not a Markov model of any order! We now formally define what we mean by the different orders of modelling power.

> **DEFINITION 3.12 kth-order Modelling**
> *In kth-order modelling the joint probabilities $P(s_{i1}, s_{i2}, \ldots s_{i,k+1})$ of all possible $(k+1)$-length message sequences are given. Furthermore:*
>
> - *Since*
>
> $$P(s_{i1}, s_{i2}, \ldots s_{im}) = \sum_{m+1} \sum_{m+2} \cdots \sum_{k+1} P(s_{i1}, \ldots s_{im}, s_{i,m+1}, \ldots s_{i,k+1})$$
>
> $$(3.38)$$
>
> *then the joint probabilities for the mth-order model, where $m \leq k + 1$, are also given.*
>
> - *Since*
>
> $$P(s_{i,m+1} | s_{i1}, s_{i2}, \ldots s_{im}) = \frac{P(s_{i1}, s_{i2}, \ldots, s_{im}, s_{i,m+1})}{P(s_{i1}, s_{i2}, \ldots, s_{im})} \qquad (3.39)$$
>
> *then the stationary state and conditional probabilities for the mth order Markov model, where $m < k + 1$, are also provided.*

We need to consider what effect the model order has on the Huffman and arithmetic coding schemes we have discussed. In Chapter 1 it was shown that higher-order Markov models yield lower values of entropy, and since the lower bound on the average length of a code is the entropy, higher-order modelling would be expected to produce more efficient codes (see Result 3.1). Thus it is not only important to consider the implementation of the encoder and decoder for both the Huffman and arithmetic algorithms but also how the source statistics are calculated.

Ideally for encoding a n-length message or nth extension of a source the highest-order model of order $k = (n - 1)$ should be used. Thus coding of longer messages and source extensions will indirectly benefit from higher orders of modelling power which, from Shannon's Noiseless Coding Theorem and Result 3.1, will produce more efficient codes as the source entropy decreases. Intuitively, longer messages and extensions are able to better exploit any redundant or repetitive patterns in the data.

3.9.1   Higher-order Huffman Coding

With higher-order Huffman coding we consider the case of deriving the Huffman code for the nth extension of the source, S^n, using a kth order model of the source, S.

- If $n \leq k + 1$ then the joint probabilities $P(s_{i1}, s_{i2}, \ldots s_{in})$ of the $(n - 1)$th order (Markov) model can be used.

- If $n > k + 1$ then the product of the joint kth order probabilities can be used.

In practice the $(k + 1)$-gram model probabilities $P(s_{i1}, s_{i2}, \ldots s_{i,k+1})$ are directly estimated from the frequency counts:

$$P(s_{i1}, s_{i2}, \ldots s_{i,k+1}) = \frac{C(s_{i1}, s_{i2}, \ldots s_{i,k+1})}{\sum C(s_{i1}, s_{i2}, \ldots s_{i,k+1})} \qquad (3.40)$$

where $C(s_{i1}, s_{i2}, \ldots s_{i,k+1})$ is the number of times the $(k + 1)$-length message, $s_{i1}, s_{i2}, \ldots s_{i,k+1}$, is seen and $\sum C(s_{i1}, s_{i2}, \ldots s_{i,k+1})$ is the sum of *all* $(k + 1)$-length messages that have been seen so far.

When building the Huffman coding tree only the relative order of the nth extension probabilities is important. Thus a fixed scaling of the probabilities can be applied without affecting the Huffman coding algorithm. Thus practical implementations use the model counts, $C(s_{i1}, s_{i2}, \ldots s_{i,k+1})$, directly instead of the model probabilities, $P(s_{i1}, s_{i2}, \ldots s_{i,k+1})$.

EXAMPLE 3.25

After observing a typical 100 length binary message the following 2nd order model counts are produced:

$$C(000) = 47$$
$$C(001) = 12$$
$$C(010) = 7$$
$$C(011) = 8$$
$$C(100) = 12$$
$$C(101) = 3$$
$$C(110) = 7$$
$$C(111) = 4$$

The binary Huffman code for the 3rd extension of the source is to be designed based on the above 2nd order model. The binary Huffman code tree is shown in Figure 3.19 where the counts are directly used for construction of the tree.

The Huffman code is given in the following table:

3rd extension	Huffman code
000	1
001	010
010	0010
011	0000
100	011
101	00011
110	0011
111	00010

□

3.9.2 Higher-order Arithmetic Coding

When performing arithmetic coding on a message of length n we require knowledge of the conditional probabilities $P(s_{ij}|s_{i1}, s_{i2}, \ldots s_{i,j-1})$ for $j = 1, 2, \ldots n$ as shown by Figure 3.16. Assuming the kth order model is given then:

- If $n \leq k + 1$ then $P(s_{ij}|s_{i1}, s_{i2}, \ldots s_{i,j-1}) = \frac{P(s_{i1}, s_{i2}, \ldots s_{i,j-1}, s_{ij})}{P(s_{i1}, s_{i2}, \ldots s_{i,j-1})}$ for $j = 1, 2, \ldots n$ where the joint probabilities of the $(n-1)$th order model can be used.

- If $n > k+1$ then for $j = k+2, k+3, \ldots n$ we have $P(s_{ij}|s_{i1}, s_{i2}, \ldots s_{i,j-1}) = P(s_{ij}|s_{i,j-k}, s_{i,j-k+1}, \ldots s_{i,j-1}) = \frac{P(s_{i,j-k}, s_{i,j-k+1}, \ldots s_{i,j-1}, s_{ij})}{P(s_{i,j-k}, s_{i,j-k+1}, \ldots s_{i,j-1})}$.

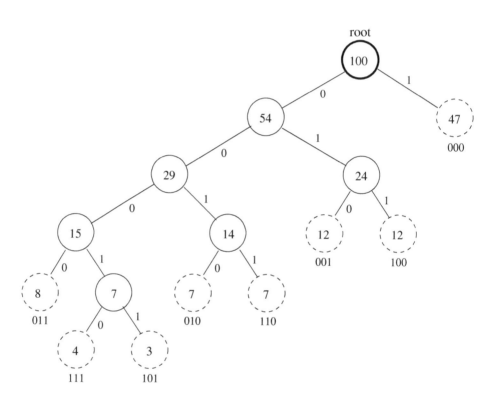

FIGURE 3.19
3rd order Huffman coding tree.

The kth order probabilities are derived from the respective frequency counts of the data seen thus far using Equation 3.40. Unlike Huffman coding, arithmetic coding requires the absolute probabilities so a count of both $C(s_{i1}, s_{i2}, \ldots s_{i,k+1})$ and $\sum C(s_{i1}, s_{i2}, \ldots s_{i,k+1})$ are needed.

EXAMPLE 3.26

Consider performing arithmetic coding on the message 01011 based on the 2nd order model counts from Example 3.25. From the counts the 3rd extension, 2nd extension and single binary probabilities can be calculated as shown in Figure 3.20.

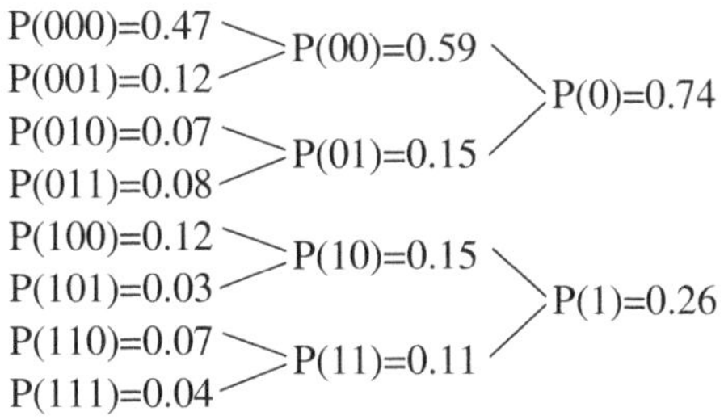

$$
\begin{array}{l}
P(000){=}0.47 \\
P(001){=}0.12
\end{array} {\Large\rangle} P(00){=}0.59
$$
$$
\begin{array}{l}
P(010){=}0.07 \\
P(011){=}0.08
\end{array} {\Large\rangle} P(01){=}0.15
$$
$$
P(0){=}0.74
$$
$$
\begin{array}{l}
P(100){=}0.12 \\
P(101){=}0.03
\end{array} {\Large\rangle} P(10){=}0.15
$$
$$
\begin{array}{l}
P(110){=}0.07 \\
P(111){=}0.04
\end{array} {\Large\rangle} P(11){=}0.11
$$
$$
P(1){=}0.26
$$

FIGURE 3.20
Joint probabilities up to 3rd order.

Applying the encoding algorithm of Figure 3.16 where $P(s_{i2}|s_{i1}) = \frac{P(s_{i1},s_{i2})}{P(s_{i1})}$ and $P(s_{ij}|s_{i,j-2}, s_{i,j-1}) = \frac{P(s_{i,j-2},s_{i,j-1},s_{ij})}{P(s_{i,j-2},s_{i,j-1})}$ yields:

$s_{ij}|s_{i,j-1}, s_{i,j-2}$	Range	$P(s_{ij}|s_{i1}, s_{i2}, \ldots s_{i,j-1})$	$[lb, hb)$	$[low, high)$
-	-	-	-	$[0.0, 1.0)$
0/-	1.0	$P(0) = 0.74$	$[0.0, 0.74)$	$[0.0, 0.74)$
1/0	0.74	$P(1|0) = \frac{P(01)}{P(0)} = 0.203$	$[0.797, 1.0)$	$[0.590, 0.740)$
0/01	0.15	$P(0|01) = \frac{P(010)}{P(01)} = 0.467$	$[0.0, 0.467)$	$[0.590, 0.660)$
1/10	0.070	$P(1|10) = \frac{P(101)}{P(10)} = 0.200$	$[0.800, 1.0)$	$[0.646, 0.660)$
1/01	0.014	$P(1|01) = \frac{P(011)}{P(01)} = 0.533$	$[0.467, 1.0)$	$[0.6525, 0.6600)$

The corresponding interval sub-division of the probability line $[0.0, 1.0)$ is shown in Figure 3.21.

The final interval for the message 01011 is $[0.6525, 0.6600)$ which in binary is $[0.101001110000, 0.101010001111)$ and hence the code word that is transmitted is 10101. It should be noted that for this example the same number of bits is used for both the code word and the message. $\quad\square$

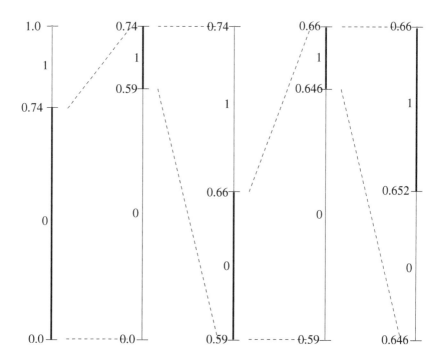

FIGURE 3.21
Subdivision of probability line.

3.10 Exercises

1. For the following binary codes, determine:

   (a) Whether the code is *uniquely decodable*. If not, exhibit two source messages with the same code.

   (b) Whether the code is *instantaneous*. If not, can you design an instantaneous code with the same lengths of code words?

	code A	code B	code C	code D	code E	code F	code G	code H
s_1	000	0	0	0	0	0	01	1010
s_2	001	01	10	10	10	100	011	001
s_3	010	011	110	110	1100	101	10	101
s_4	011	0111	1110	1110	1101	110	1000	0001
s_5	100	01111	11110	1011	1110	111	1100	1101
s_6	101	011111	111110	1101	1111	001	0111	1011

2. Consider a block code based on an r-symbol code alphabet and designed for a source with q possible messages. Derive the expression for the lower bound on the block code word length.

3. We are going to devise a code for the decimal digits $\{0\text{-}9\}$ using a binary code. Analysis shows that we should use the shortest codes for 0 and 1. If we code:

   $$\text{digit } 0 \text{ to code } 00$$

   $$\text{digit } 1 \text{ to code } 11$$

   find the minimum code word length for the remaining digits $\{2\text{-}9\}$ assuming we want them to be the same code word length.

4. For the following code word lengths:

Code	Lengths
A	2 2 2 4 4 4
B	1 1 2 3 3
C	1 1 2 2 2 2

   (a) Can an instantaneous binary code be formed? If so, give an example of such a code.

   (b) Can an instantaneous ternary code be formed? If so, give an example of such a code.

(c) If neither a binary nor ternary code can be formed find the smallest number of code symbols that will allow a code to be formed. Give an example
of such a code.

*5. Find all possible combinations of the individual code word lengths l_1, l_2, l_3
when coding the messages s_1, s_2, s_3 which result in uniquely decodable binary
codes with words not more than 3 bits long (i.e., $l_i \leq 3$).

6. A 6-symbol source has the following statistics and suggested *binary* and *ternary*
codes:

s_i	P_i	Code A	Code B
s_1	0.3	0	00
s_2	0.2	10	01
s_3	0.1	1110	02
s_4	0.1	1111	10
s_5	0.2	1100	11
s_6	0.1	1101	12

(a) What is the *efficiency* of binary code A?

(b) What is the *efficiency* of ternary code B?

(c) Can you design a more efficient binary code and, if so, what is the efficiency of your code?

(d) Can you design a more efficient ternary code and, if so, what is the efficiency of your code?

(e) Which is the most efficient code: binary or ternary?

7. Consider the following information source:

Symbol	s_1	s_2	s_3	s_4	s_5	s_6
P_i	0.1	0.1	0.45	0.05	0.2	0.1

(a) What is $H(S)$?

(b) Derive a *compact binary code* using *Huffman coding*. What is the *average length* of your code?

(c) Can you find another compact code with different individual code word
lengths? If so, what is the average length of this code?

(d) What is the efficiency of the code?

(e) What coding strategy can be used to improve the efficiency?

8. Consider the following source:

Symbol	s_1	s_2	s_3	s_4	s_5
P_i	0.1	0.2	0.2	0.4	0.1

Design a compact binary code for the source with minimum variance between code word lengths. What is the efficiency of your code? Now design the compact binary code with maximum variance and compare.

9. Derive the compact ternary code using Huffman coding for the following source and compute the efficiency:

Symbol	s_1	s_2	s_3	s_4	s_5	s_6	s_7	s_8
P_i	0.07	0.4	0.05	0.2	0.08	0.05	0.12	0.03

10. Consider the following binary source:

Symbol	s_1	s_2
P_i	0.1	0.9

(a) Find $H(S)$.

(b) What is the compact (indeed trivial!) binary code for this source? Compare the average length L_1 with $H(S)$. What is the efficiency?

(c) Repeat (b) for S^n (the nth extension of S) for $n = 2, 3$. Find L_n/n and compare this with $H(S)$. What is happening?

11. A binary source has the following 2nd extension probabilities:

Symbol	P_i
00	0.75
01	0.1
10	0.1
11	0.05

(a) Design the source encoder by devising a compact code for the source.

(b) What is the efficiency of the code you designed in (a)?

(c) What is the entropy of the source encoder *output* (the code entropy), $H(X)$? Compare your answer with the source entropy, $H(S)$. Will this always be the case?

*12. Suppose a long binary message contains half as many 1's as 0's. Find a binary Huffman coding strategy which uses:

(a) at most 0.942 bits per symbol.

(b) at most 0.913 bits per symbol.

13. What is the efficiency of a binary source S in which 1 has a probability of 0.85? Find an extension of S with efficiency at least 95%.

*14. A binary source emits 0 with probability 3/8. Find a ternary coding scheme that is at least 90% efficient.

15. A source emits symbol x a third of the time and symbol y the rest of the time. Devise a ternary coding scheme which is:

(a) at most 0.65 ternary units per symbol

(b) at most 0.60 ternary units per symbol

(c) at most 0.55 ternary units per symbol

What is the efficiency of the code you devised for the above cases?

16. You are required to design a ternary source encoder that is at least 97% efficient. Observations of the ternary source over a certain time interval reveal that symbol s_1 occurred 15 times, s_2 occurred 6 times and s_3 occurred 9 times. Design the coding scheme. What is the efficiency of your code?

17. A binary message has been modeled as a 1st order Markov source:

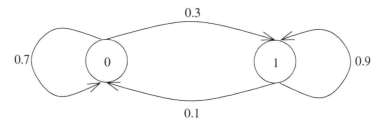

(a) Derive a binary compact code for the 2nd extension of the source. What is the efficiency of your code?

(b) Repeat (a) for the 3rd extension of the source.

18. A binary source emits an equal number of 0's and 1's but emits the same symbol as the previous symbol twice as often as emitting a different symbol than the previous symbol. Derive a binary encoding which is at least 95% efficient.

19. The output of an unknown binary information source transmitting at 100 bps (bits per second) produces a different output (from the previous output) three times as often as producing the same output. Devise a binary source coding strategy for each of the following cases, so that the channel bit rate is:

(a) at most 95 bps.

(b) at most 80 bps.

*20. Analysis of a binary information source reveals that the source is three times more likely to emit the same bit if the last two bits were the same; otherwise it is equally likely to produce a 0 or a 1.

(a) Design a coding scheme which is at least 85% efficient.

(b) Design a coding scheme which is at least 90% efficient.

*21. Consider the source of Qu. 10:

Symbol	s_1	s_2
P_i	0.1	0.9

(a) Perform arithmetic coding on the sequence $s_2 s_1 s_2$ to produce the coded output. Compare your answer with the Huffman code of the 3rd extension of the source from 10(c).

(b) Repeat (a) for the sequence $s_1 s_1 s_1$

(c) Repeat (a) for the sequence $s_1 s_2 s_1$

(d) From the above cases, is arithmetic coding of a 3 letter message comparable to, superior to or inferior to Huffman coding of the corresponding 3rd extension of the source?

NOTE Do not use any scaling. To convert from base 10 to base 2 (and vice versa) you will need a number base converter that can deal with non-integer values. One such tool available from the Internet is http://www.math.com/students/converters/source/base.htm.

22. Perform arithmetic decoding of your coded output values from Qu. 21(a),(b),(c) and confirm that the original 3 letter messages are produced.

23. Repeat Qu. 21 but this time use scaling.

24. Repeat Qu. 22 but this time use scaling.

*25. Huffman codes are guaranteed to be instantaneous codes. Can the same statement be made about arithmetic codes?

26. Consider Example 3.26:

(a) Perform arithmetic coding of the sequence 011 and compare with the corresponding Huffman code from Example 3.25.

(b) Repeat (a) for the sequence 000.

(c) Repeat (a) for the sequence 001.

(d) From the above cases, is arithmetic coding comparable to, superior to or inferior to Huffman coding?

27. Consider Example 3.26:

(a) Perform arithmetic decoding of the coded binary integer value 10101 assuming the original message was 3 bits long.

(b) Repeat (a) but for the coded binary integer value of 1.

*28. After observing a typical sample of a binary source the following 2nd order model counts are produced:

$$C(000) = 20$$
$$C(001) = 10$$
$$C(010) = 5$$
$$C(011) = 15$$
$$C(100) = 15$$
$$C(101) = 10$$
$$C(110) = 20$$
$$C(111) = 5$$

(a) Derive the binary Huffman code table for the 3rd extension of the source such that there is minimum variance in the individual code word lengths. What is the average length of your code?

(b) Perform arithmetic coding on the following 3 letter messages and compare with the corresponding Huffman code:

  i. 110
  ii. 111
  iii. 100

3.11   References

[1] ASCII, retrieved January 25, 2002 from
     http://www.webopedia.com/TERM/A/ASCII.html

[2] N. Abramson, *Information Theory and Coding*, McGraw-Hill, New York, 1953.

[3] T.H. Cormen, C.E. Leiserson, and R.L. Rivest, *Introduction to Algorithms*, MIT Press, Cambridge, MA, 1990.

[4] R.M. Fano, *Transmission of Information*, MIT Press, Cambridge, MA, 1961.

[5] S.W. Golomb, R.E. Peile, and R.A. Scholtz, *Basic Concepts in Information Theory and Coding*, Plenum Press, New York, 1994.

[6] D. Hankerson, G.A. Harris, and P.D. Johnson, Jr., *Introduction to Information Theory and Data Compression*, CRC Press, Boca Raton, FL, 1998.

[7] D.A. Huffman, A method for the construction of minimum redundancy codes, *Proc. IRE*, 40, 1098-1101, 1952.

[8] L.G. Kraft, A device for quantizing, grouping and coding amplitude modulated pulses, *M.S. thesis*, Dept. of E.E., MIT, Cambridge, MA, 1949.

[9] J. C. A. van der Lubbe, *Information Theory*, Cambridge University Press, London, 1997.

[10] B. McMillan, Two unequalities implied by unique decipherability", *IRE Trans. Inform. Theory*, IT-2, 115-116, 1956.

[11] C.E. Shannon, A mathematical theory of communication, *Bell System Tech. J.*, 28, 379-423, 623-656, 1948.

[12] I.H. Witten, R.M. Neal and J.G. Cleary, Arithmetic coding for compression, *Communications of the ACM*, 30(6), 520-540, 1987.

[13] I.H. Witten, A. Moffat, and T.C. Bell, *Managing Gigabytes*, Morgan Kaufmann Publishers, San Francisco, 2nd ed., 1999.

Chapter 4

Data Compression

4.1 Introduction

We have seen that entropy or information content is a measure of predictability or redundancy. In situations where there is redundancy in a body of information, it should be possible to adopt some form of coding which exploits the redundancy in order to reduce the space which the information occupies. This is the idea underlying approaches to data compression.

In the previous chapter, we have looked at encoding methods that assume that successive characters in a sequence are independent or that the sequence was generated by a Markov source. In this chapter, we extend these ideas to develop data compression techniques that use the additional redundancy that arises when there are relationships or correlations between neighbouring characters in the sequence. This additional redundancy makes it possible to achieve greater compression, though at the cost of extra computational effort.

EXAMPLE 4.1

Suppose we have a message that consists of only four letters, say A, B, C, D. To measure the information content of such a message, it is convenient to code the letters as binary digits, and count the total number of digits to give an estimate of the information content in bits.

If there is no redundancy in the sequence, so that the letters occur at random, then we must use two binary digits to code each letter, say 00 for A, 01 for B, 10 for C and 11 for D. A sequence such as

$$BADDACDCADCCABBC$$

will be coded as

$$01001111001011100011101000010110.$$

The information content will be two bits per letter.

Suppose instead that the occurrence of letters is not completely random, but is constrained in some way, for example, by the rule that A is followed by B or C with equal probability (but never by D) , B is followed by C or D with equal probability, C is followed by D or A with equal probability and D is followed by A or B with equal probability. We can encode a sequence that satisfies these rules by using two binary digits to encode the first letter as above, and then using one binary digit to encode each successive letter. If the preceding letter is A, we can use 0 to encode the following letter if it is B and 1 to encode it if it is C, and so on. In this case, a sequence such as

$$BDBCACDBCDBDBDACDA$$

will be coded as

$$0111011010011110100$$

and the information content will be $(n + 1)$ bits if there are n letters in the sequence. If we had used the coding that we used for the completely random sequence, it would be coded as

$$011101100010110110110111011100101100$$

which is approximately twice as long.

The redundancy in the second sequence has enabled us to reduce the number of binary digits used to represent it by half.

⬚

The example above illustrates how structure can be used to remove redundancy and code sequences efficiently. In the following sections we will look at various kinds of structure in sequence and image data and the ways in which they can be used for data compression.

4.2 Basic Concepts of Data Compression

We have seen in the previous chapter that simple coding schemes can be devised based on the frequency distribution of characters in an alphabet. Frequently occurring characters are assigned short codes and infrequently occurring characters are assigned long codes so that the average number of code symbols per character is minimised.

Repetition is a major source of redundancy. In cases where single characters occur many times in succession, some form of *run-length coding* can be an effective data compression technique. If there are strings of characters or some other patterns that occur frequently, some form of *dictionary coding* may be effective.

In some cases, it may be possible to create a statistical model for the data and use the predictions of the model for compression purposes. Techniques that use a single model to compress all kinds of data are known as *static compression techniques*. Such techniques are likely to be inefficient, because of the mismatch between the model and the actual statistics of the data. As an alternative, *semi-adaptive compression techniques* construct a model for each set of data that is encoded and store or transmit the model along with the compressed data. The overhead involved in storing or transmitting the model can be significant. For this reason, semi-adaptive techniques are rarely used. The best compression algorithms use *adaptive compression techniques*, in which the model is built up by both the compressor and decompressor in the course of the encoding or decoding process, using that part of the data that has already been processed.

When sequences of characters representing text or similar material have to be compressed, it is important that the original material is reproduced exactly after decompression. This is an example of *lossless compression*, where no information is lost in the coding and decoding. When images are compressed, it may be permissible for the decompressed image not to have exactly the same pixel values as the original image, provided the difference is not perceptible to the eye. In this case, some form of *lossy compression* may be acceptable. This involves a loss of information between the coding and decoding processes.

4.3   Run-length Coding

Run-length coding is a simple and effective means of compressing data in which it is frequently the case that the same character occurs many times in succession. This may be true of some types of image data, but it is not generally true for text, where it is rare for a letter of the alphabet to occur more than twice in succession.

To compress a sequence, one simply replaces a repeated character with one instance of the character followed by a count of the number of times it occurs. For example, the sequence

$$AABBBBCCCDCCDDDBBBBBAAAA$$

could be replaced by

$$A2B4C3D1C2D3B5A4$$

reducing the number of characters from 24 to 16. To decompress the sequence, each combination of a character and a count is replaced by the appropriate number of characters.

Protocols need to be established to distinguish between the characters and the counts in the compressed data. While the basic idea of run-length coding is very simple, complex protocols can be developed for particular purposes. The standard for facsimile transmission developed by the International Telephone and Telegraph Consultative Committee (CCITT) (now the International Telecommunications Union) [4] involves such protocols.

4.4   The CCITT Standard for Facsimile Transmission

Facsimile machines have revolutionised the way in which people do business. Sending faxes now accounts for a major part of the traffic on telephone lines. Part of the reason why facsimile machines gained wide acceptance quickly may be due to the fact that they integrated into the existing telecommunications infrastructure very easily. However, a major factor in their gaining acceptance was the adoption of the world-wide standard proposed by the CCITT, which made it possible for every facsimile machine to communicate with every other facsimile machine, as they all complied with the standard.

The CCITT Group 3 standard is used to send faxes over analogue telephone lines. It specifies two run-length coding operations that are used to compress the data that represents the image that is being transmitted. As the standard was intended primarily for the transmission of pages of text, it only deals with binary images, where a single bit is used to specify whether a pixel is black or white.

The Group 3 standard is designed to transmit A4-sized pages (210 mm by 297 mm). The image is broken into horizontal scan lines which are 215 mm long and contain 1,728 pixels. The scan lines are either 0.26 mm apart for images transmitted at the standard resolution or 0.13 mm apart for images transmitted at high resolution. This means that the information content of a page is just over 2 million bits at standard resolution and over 4 million bits at high resolution.

The images are coded one line of pixels at a time, with a special end-of-line character being used to ensure that the lines are not confused. The standard specifies two run-length coding algorithms, a one-dimensional algorithm that codes a single line of pixels in isolation, and a two-dimensional one that codes a line of pixels in terms of the differences between it and the preceding line of pixels.

The one-dimensional algorithm is used to code the first line of each page. It treats the line of pixels as a succession of runs of white and black pixels in alternation,

Run length	Run of white pixels	Run of black pixels
0	00110101	0000110111
1	000111	010
2	0111	11
3	1000	10
4	1011	011
5	1100	0011
6	1110	0010
7	1111	00011
8	10011	000101

Table 4.1   Huffman codes of the pixel run lengths.

and codes the number of pixels in each run into a number of bits using a modified Huffman coding technique. It assumes that the first run of pixels is white; if this is not the case, it transmits the code for zero. Separate Huffman codes are used for the numbers of black and white pixels, as these have different statistics.

The Huffman codes were developed from a set of test documents. They provide representations for run lengths from 0 to 63, and for multiples of 64 up to 1728. Run lengths up to 63 are represented by a single code. Run lengths of 64 or more are represented by concatenating the code for the greatest multiple of 64 that is less than the run length with the code for the remaining number of pixels in the run. The following example illustrates the coding process.

EXAMPLE 4.2

Table 4.1 shows the Huffman codes for run lengths from 0 to 8. Using them, a line of pixels that starts

$$WWWWBBBBBBBWWWBBBBBB...$$

where W denotes a white pixel and B denotes a black pixel would be encoded as

$$1011\ 00011\ 1000\ 0010\ ...$$

where the spaces have been introduced to show where each code word ends. The number of bits has been reduced from 21 to 17; this is not a great reduction, but substantial compression can be achieved when the run lengths are over 64.   ☐

The two-dimensional code describes the differences between the runs of pixels in the line being encoded with the runs in the preceding line. Where a run of pixels in the current line ends within three pixels of the end of a run of pixels of the same colour in the preceding line, this is encoded. Otherwise, the number of pixels in the current run is coded using the one-dimensional code. There is also provision for ignoring runs in the previous line. Details of the codes used in the two-dimensional algorithm can be found in the standard [4].

If there is a transmission error, continued use of the two-dimensional coding algorithm can cause it to propagate down the page. For this reason, the one-dimensional algorithm is used to code every second line in standard resolution transmissions and every fourth line in high resolution transmissions, with the intervening lines being encoded using the two-dimensional code. This limits the extent to which errors can propagate.

The decoding process simply generates runs of pixels with the appropriate run lengths to re-create the lines of pixels.

There is also a Group 4 standard that is intended for sending faxes over digital telephone lines. It makes provision for the compression of greyscale and colour images as well as binary images.

4.5   Block-sorting Compression

The *Block-sorting Compression* technique is designed for compressing texts where the letters that precede each letter form a *context* for that letter and letters often appear in the same context. It was introduced by Burrows and Wheeler [1] and has been implemented in the *bzip2* compression program [2].

The technique works by breaking the text up into blocks and rearranging the letters in each block so that the resulting sequence can be compressed efficiently using runlength coding or some other simple technique. For decompression, the rearranged sequence is recovered and the original sequence is restored. What makes the technique work is the way in which the original sequence can be restored with a minimum of effort.

The technique is best discussed by way of an example. Suppose we have a block of fifteen letters, say, *the fat cat sat*. We list the fifteen cyclic permutations of the block, separating the last letter of the sequence from the rest, to get:

```
the_fat_cat_sa   t
he_fat_cat_sat   t
e_fat_cat_satt   h
_fat_cat_satth   e
fat_cat_satthe   _
at_cat_satthe_   f
t_cat_satthe_f   a
_cat_satthe_fa   t
cat_satthe_fat   _
at_satthe_fat_   c
t_satthe_fat_c   a
_satthe_fat_ca   t
satthe_fat_cat   _
atthe_fat_cat_   s
tthe_fat_cat_s   a
```

where the underscore character has been used to represent the spaces between the words. The strings in the left hand column are the contexts of the letters in the right hand column. We now sort these permutations in alphabetical order of the contexts:

```
at_cat_satthe_   f
atthe_fat_cat_   s
at_satthe_fat_   c
_satthe_fat_ca   t
_cat_satthe_fa   t
the_fat_cat_sa   t
t_satthe_fat_c   a
fat_cat_satthe   _
t_cat_satthe_f   a
_fat_cat_satth   e
tthe_fat_cat_s   a
satthe_fat_cat   _
cat_satthe_fat   _
he_fat_cat_sat   t
e_fat_cat_satt   h
```

where the ordering is from right to left and the underscore character precedes the alphabetic characters in the ordering.

The sequence of letters in the right hand column is the required rearrangement. The sequence the_fat_cat_sat has become fscttta_aea__th. In this case, three of the t's and two of the underscore characters appear together in the rearranged sequence. In larger blocks, there will be more long runs of the same character. The rearranged sequence can be compressed with a run-length coder or a *move-to-front* coder.

(A move-to-front coder is a simple way of generating a code based on a probability distribution. A frequency distribution of the letters is constructed by passing through the sequence and counting the number of times each letter occurs. After each letter is counted, the order of letters in the frequency distribution is changed by moving the current letter to the top of the list. In the end, the list is in approximate order of decreasing frequency, with the more frequently occurring letters appearing near the top of the list. The letters are then encoded using short codes for letters at the top of the list and longer codes for the less frequently occurring letters lower down the list.)

To restore the sequence to the original order, we need to know where the sequence starts and which letter follows which. In the example above, the first letter is the second last letter of the rearranged sequence. The order in which letters follow each other can be found by aligning the last letter in the left hand column of the table above with the letter in the right hand column. This matching can be accomplished very simply by sorting the rearranged sequence into alphabetical order. If we do this, we get:

```
_ f
_ s
_ c
a t
a t
a t
c a
e _
f a
h e
s a
t _
t _
t t
t h
```

To reconstruct the original sequence, we start with the letter in the right hand column of the second-last row of the table, knowing that this is the first letter of the original sequence. This is the last t in the right hand column. To find the letter that follows it, we look for the last t in the left hand column. The letter in the right hand column of that row is the following letter, h. There is only one h in the first column, and it is followed by e. We can trace our way through the whole sequence in this way and construct the original text.

In this coding technique, we use the information about structure that is contained in contexts to eliminate redundancy and achieve compression. This is a very powerful approach to compression, and has been used in a number of other ways, as we shall see in the next sections.

4.6 Dictionary Coding

Texts such as books, newspaper articles and computer programs usually consist of words and numbers separated by punctuation and spaces. An obvious way of compressing these is to make a list of all the words and numbers appearing in the text, and then converting the text to a sequence of indices which point to the entries in the list. (The list of words is referred to as the *dictionary*. This is perhaps a different usage of the word from the common one, where a dictionary lists both words and their meanings. *Lexicon* might be a better name, but the methods that are based on this idea are generally referred to as *dictionary coding methods*.) This can be a simple and powerful means of data compression, but it has some drawbacks.

First, it is suitable primarily for text and not for images or other data. It would probably not work well on purely numerical data. Second, both the compressor and the decompressor need to have access to the dictionary. This can be achieved if the dictionary is stored or transmitted along with the compressed text, but this could require large amounts of overhead, especially for small texts. Alternatively, a large dictionary could be constructed and held by both the compressor and decompressor and used for all texts. This is likely to be inefficient, as any given text will use only a small proportion of the words in the dictionary.

It is therefore better to construct a dictionary for each text. The techniques that we describe in this section enable both the compressor and decompressor to do this in a way that does not require any overhead for the transmission or storage of the dictionary. The trick is to use the part of the text that has already been compressed or decompressed to construct the dictionary.

This trick was proposed by Ziv and Lempel ([8], [9]). It has been used in a number of algorithms which have been implemented in the compression software that is commonly available today.

The first algorithm was proposed by Ziv and Lempel in 1977 [8], and will be referred to as *LZ77*. In this algorithm, the compressed text consists of a sequence of triples, with each triple consisting of two numbers, m and n, and a character, c. The triple, (m, n, c), is interpreted as an instruction indicating that the next $m + 1$ characters match the m characters located n characters back plus the next character, c. For example, the triple $(237, 12, q)$ is interpreted to mean: *the next 12 characters match the 12 characters found 237 characters back in the text, copy them from there and then append a "q" to the text.*

The compressor scans the text, trying to match the next characters to a sequence of characters earlier in the text. If it succeeds, it creates the appropriate triple. If it fails, it creates a triple in which the numbers are zero, indicating this failure. The decompressor reverses this process, appending sequences of characters from the section of the text that has already been decompressed and following them with the characters in the triples.

EXAMPLE 4.3

The text *the fat cat sat on the mat.* can used to illustrate the LZ77 algorithm. It is too short for the algorithm to compress it effectively, but the repetitions in the text make it good for the purposes of illustration.

We will use the underscore character to represent the space character, so the text is

the_fat_cat_sat_on_the_mat.

When we start the coding, there is no earlier text to refer to, so we have to code the first nine characters by themselves. The compressed text begins:

$$(0,0,t)\ (0,0,h)\ (0,0,e)\ (0,0,\_)\ (0,0,f)\ (0,0,a)\ (0,0,t)\ (0,0,\_)\ (0,0,c)$$

At this point we can now refer back to the sequence at_, so we can code the next eight characters in two triples:

$$(4,3,s)\ (4,3,o)$$

The next two characters have to be coded individually:

$$(0,0,n)\ (0,0,\_)$$

For the last seven characters, we can make two references back into the text:

$$(19,4,m)\ (11,2,.)$$

The references in this example are shown diagrammatically Figure 4.1.   ▯

Implementations of the LZ77 algorithm usually set limits on the amount of text that is searched for matching sequences and on the number of matching characters copied at each step. There are also minor variations that have been used to improve its efficiency. For example, the *gzip* [3] program uses a hash table to locate possible starting points of matching sequences.

The algorithm proposed by Ziv and Lempel in 1978 [9], which will be referred to as *LZ78*, builds a set of sequences that is used as a dictionary as the compression and decompression progress. The compressed text consists of references to this dictionary instead of references to positions in the text. The sequences can be stored in a tree structure, which makes it easy to carry out the matching. In this algorithm, the compressed text consists of a sequence of pairs, with each pair (i, c) interpreted as

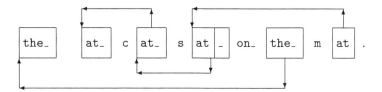

FIGURE 4.1
Diagrammatic representation of references in the LZ77 algorithm.

the ith item in the dictionary followed by the character c. The compression process adds an entry to the dictionary at each stage. The decompression process builds the same dictionary in a similar way to the compression algorithm using the original text that is being generated from the coded sequence. The success of this operation relies on the fact that references to the dictionary are made after that part of the dictionary has been built.

EXAMPLE 4.4

To compress the text

the_fat_cat_sat_on_the_mat.

using LZ78, both the compressor and decompressor begin with a dictionary consisting of the empty sequence. This is item number zero in the dictionary.

The compressed text consists of pairs consisting of a dictionary reference and a character. If there is no matching item in the dictionary, item 0 is referenced and the next character given. Thus the first five characters will be coded as

$$(0, t)\ (0, h)\ (0, e)\ (0, \_)\ (0, f)\ (0, a)$$

and the following items will be added to the dictionary:

Item 1: t

Item 2: h

Item 3: e

Item 4: _

Item 5: f

Item 6: a

The next character is already in the dictionary, so it is coded as

$$(1, \_)$$

and

Item 7: t_

is added to the dictionary.

The next character has not occurred before, so it is coded as

$$(0, c)$$

and

Item 8: c

is added to the dictionary.

The next two characters are coded as

$$(6, t)$$

and

Item 9: at

is added to the dictionary.

The steps above and the rest of the coding are summarised in Table 4.2.

☐

EXAMPLE 4.5

The compression capabilities of LZ78 can be better appreciated by the following sequence of text

ababababababababababab.

where the coding is summarised in Table 4.3. It should be noted that at each stage the longest sequence in the dictionary that has been so far is selected. Thus it is evident from the table that the LZ78 algorithm builds successively longer string patterns as more of the sequence is processed achieving a corresponding improvement in compression.

Item sequence	Code	Item number	Current sequence
t	(0,t)	1	t
h	(0,h)	2	th
e	(0,e)	3	the
_	(0,_)	4	the_
f	(0,f)	5	the_f
a	(0,a)	6	the_fa
t_	(1,_)	7	the_fat_
c	(0,c)	8	the_fat_c
at	(6,t)	9	the_fat_cat
_s	(4,s)	10	the_fat_cat_s
at_	(9,_)	11	the_fat_cat_sat_
o	(0,o)	12	the_fat_cat_sat_o
n	(0,n)	13	the_fat_cat_sat_on
_t	(4,t)	14	the_fat_cat_sat_on_t
he	(2,e)	15	the_fat_cat_sat_on_the
_m	(4,m)	16	the_fat_cat_sat_on_the_m
at.	(9,.)	17	the_fat_cat_sat_on_the_mat.

Table 4.2   Example of coding using LZ78.

Item sequence	Code	Item number	Current sequence
a	(0,a)	1	a
b	(0,b)	2	ab
ab	(1,b)	3	abab
aba	(3,a)	4	abababa
ba	(2,a)	5	abababab a
bab	(5,b)	6	ababababab
abab	(4,b)	7	ababababababab
abab.	(7,.)	8	abababababababababab.

Table 4.3   Another example of coding using LZ78.

Characters	Code	Item number	Item sequence	Current sequence
t	116	128	th	t
h	104	129	he	th
e	101	130	e_	the
_	32	131	_f	the_
f	102	132	fa	the_f
a	97	133	at	the_fa
t	116	134	t_	the_fat
_	32	135	_c	the_fat_
c	99	136	ca	the_fat_c
at	133	137	at_	the_fat_cat
_	32	138	_s	the_fat_cat_
s	115	139	sa	the_fat_cat_s
at_	137	140	at_o	the_fat_cat_sat_
o	111	141	on	the_fat_cat_sat_o
n	110	142	n_	the_fat_cat_sat_on
_	32	143	_t	the_fat_cat_sat_on_
th	128	144	the	the_fat_cat_sat_on_th
e_	130	145	e_m	the_fat_cat_sat_on_the_
m	109	146	ma	the_fat_cat_sat_on_the_m
at	133	147	at.	the_fat_cat_sat_on_the_mat
.	46			the_fat_cat_sat_on_the_mat.

Table 4.4   Example of coding using LZW.

⬜

Welch [7] proposed a variation of LZ78 in which the compressor and decompressor begin with the set of characters in the ASCII alphabet and extend this to sequences of two or more characters. This means that no characters are transmitted, only their codes. The algorithm codes the longest sequence of characters from the current position in the text that already exists in the dictionary and forms a new dictionary entry comprising the just coded sequence plus the next character. This algorithm is known as *LZW*. It is the algorithm implemented in the UNIX utility *compress*.

EXAMPLE 4.6

To compress the text

```
the_fat_cat_sat_on_the_mat.
```

the compressor and decompressor begin with a dictionary consisting of 127 items. The characters t, h, e and space have ASCII values of 116, 104, 101 and 32, respectively; so these are the first four codes. Table 4.4 summarises the coding process, showing the items that are added to the dictionary at each stage.

Characters	Code	Item number	Item sequence	Current sequence
a	97	128	ab	a
b	98	129	ba	ab
ab	128	130	aba	abab
aba	130	131	abab	abababa
ba	129	132	bab	ababababa
bab	132	133	baba	ababababab
abab	131	134	ababa	abababababababab
abab	131	135	abab.	ababababababababab
.	46			abababababababababab.

Table 4.5 Another example of coding using LZW.

In this case, the LZW algorithm outputs more code words than the corresponding LZ78. However it should be noted that for LZW the code words are transmitted as 8 bit numbers whereas LZ78 will require more than 8 bits to transmit both the character and corresponding index into the dictionary.

⬚

EXAMPLE 4.7

Compression of the text:

$$\texttt{abababababababababab.}$$

from Example 4.5 is summarised in Table 4.5. The coded sequence 97, 98, 128, 130, 129, 132, 131, 131, 46 is transmitted in $9 \times 8 = 72$ bits whereas the original sequence takes up $21 \times 7 = 147$ bits. This represents a compression ratio of at least 50 %.

⬚

4.7 Statistical Compression

Statistical Compression or *Model-based Coding* techniques use probabilistic models which predict the next character in the data on the basis of probability distributions that are constructed adaptively from the data. The compressor builds up its models as it compresses the data, while the decompressor builds up identical models as it decompresses the data. Figure 4.2 is a block diagram of a statistical compression system. The system consists of a compressor and a decompressor, each with its own predictor.

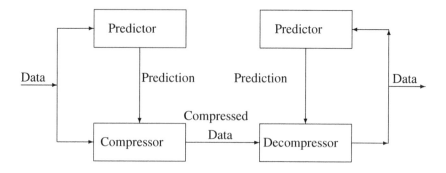

FIGURE 4.2
Block diagram of a statistical compression system.

As each character enters the compressor, its predictor generates a probability distribution which is used by the compressor to encode the character, using Huffman coding or arithmetic coding. The predictor then updates its models.

When the encoded character enters the decompressor, its predictor generates the same probability distribution as was used to encode the character, which the decompressor uses to decode it. The decoded character is then used by the decompressor's predictor to update its models.

As long as the predictors start off with the same models and update them as each character is processed, the models used by the compressor and decompressor will remain synchronised.

Any predictor that makes predictions in the form of a probability distribution over the set of possible characters can be used in a statistical compressor. For example, n-gram models can be used.

4.8   Prediction by Partial Matching

Prediction by Partial Matching (PPM) is the most commonly used statistical compression technique. It uses a collection of n-gram models to provide the required probability distributions.

A PPM compressor or decompressor builds up n-gram models for values of n from 1 to some maximum, usually 3 or 4. These n-gram models consist of frequency counts for each of the n-grams that have appeared in the text so far. The n-gram models are used to estimate the probability of the next character given the previous $n - 1$ characters.

The reason why it is necessary to have a sequence of n-gram models instead of just one is that the sequence of characters comprising the previous $n - 1$ characters followed by the next character may not have appeared before in the text. When this occurs, the n-gram model cannot be used to estimate the probability of the next character. This problem is usually dealt with by implementing an *escape mechanism*, where an attempt is made to estimate the probability using the n-gram models with successively smaller contexts until one is found for which the probability is non-zero.

In the encoding process, the contexts in the n-gram model with the largest contexts that match the current context are found. If the frequency of the n-gram that is made up of the current context followed by the next character in the text has a frequency of zero, a special character, called the *escape character*, is encoded and the contexts of the next largest n-gram model are examined. This process continues until a non-zero frequency is found. When this happens, the corresponding probability distribution is used to encode the next character, using arithmetic coding. The frequency counts of all the n-gram models maintained by the compressor are then updated.

In the decoding process, the decoder selects the appropriate n-gram model on the basis of the number of escape characters that precede the coded character. The probability distribution derived from that n-gram model is used to decode the character, and the frequency counts of all the n-gram models maintained by the decompressor are updated.

There are a number of variations on the basic PPM scheme. Most of them differ in the formulae that they use to convert the frequency counts in the n-gram models to probabilities. In particular, the probability that should be assigned to the escape character may be computed in a number of different ways.

PPM compression has been shown to achieve compression ratios as good as or better than the dictionary-based methods and block-sorting compression. It has also been shown that given a dictionary encoder, it is possible to construct a statistical encoder that achieves the same compression.

4.9 Image Coding

The compression techniques that have been discussed above were designed for the compression of text and similar data, where there is a sequence of characters which makes the data essentially one-dimensional. Images, however, are inherently two-dimensional and there are two-dimensional redundancies that can be used as the basis of compression algorithms.

Nevertheless, the one-dimensional techniques have been shown to work quite effectively on image data. A two-dimensional array of pixel values can be converted to a one-dimensional array by concatenating successive rows of pixels. Text compression methods can be applied to this one-dimensional array, even though the statistics of the pixel values will be very different from the statistics of characters in texts. Compression programs such as *compress* and *zip* have been found to reduce image files to about half their uncompressed size.

The *Graphics Interchange Format* (GIF) has been in use since 1987 as a standard format for image files. It incorporates a compression scheme to reduce the size of the image file. The GIF uses a palette of 256 colours to describe the pixel values of each image. The palette represents a selection of colours from a large colour space and is tailored to each image. Each pixel in the image is given an 8-bit value that specifies one of the colours in the palette. The image is converted to a one-dimensional array of pixel values and the LZW algorithm is applied to this array.

The GIF is being supplanted by the *Portable Network Graphics* (PNG) format. This uses the *gzip* compression scheme instead of LZW. It also has a number of other improvements, including preprocessing filters that are applied to each row of pixels before compression and extended colour palettes.

Images also come in various colour formats. The simplest are the *binary images*, where each pixel is black or white, and each pixel value can be specified by a single bit. *Greyscale images* use integer values to specify a range of shades of grey from black to white. The most common practice is to use eight-bit integers to specify greyscale values, except in medical imaging where twelve bits or sixteen bits may be used to achieve the desired resolution in grey levels. *Colour images* usually require three numbers to specify pixel values. A common format is to use eight-bit integers to specify red, green and blue values, but other representations are also used. *Multispectral images*, generated by systems that collect visible, infrared or ultraviolet light, may generate several integer values for each pixel.

The different colour formats have their own statistical characteristics. Binary images usually consist of runs of white pixels alternating with runs of black pixels. Run-length encoding is an effective means of compressing such images, and the CCITT Group standard for facsimile transmissions described above is an example of run-length coding applied to binary images.

Greyscale images are less likely to have runs of pixels with the same greyscale values. However, there may be similarities between successive rows of the image which will make dictionary-based methods of statistical compression effective. Colour images and multispectral images can be compressed as three or more separate greyscale images. There may be correlations between the component images in these cases, but it is difficult to use the resulting redundancy to improve compression performance.

One-dimensional compression techniques cannot make use of the redundancies that arise from the two-dimensional structure of images if they are applied to single rows or columns of pixels. It is possible, however, to implement dictionary-based compression or PPM compression using two-dimensional contexts. Figure 4.3 shows a

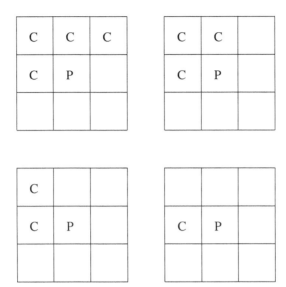

FIGURE 4.3
Two-dimensional contexts of a pixel for PPM compression.

set of contexts consisting of four pixels, three pixels, two pixels or one pixel that could be used in a PPM compressor. The pixels marked "C" represent the context pixels used to predict the centre pixel (marked "P"). The positions of the pixels that are used to predict the centre pixel are such that these pixels would be encoded and decoded before the pixel in the centre. The full contexts would not be available for pixels at the edge of the image; so the escape mechanism would have to be invoked.

Another way in which to take advantage of the two-dimensional structure of images is to use *pyramid coding*. This generates a set of approximations to the image by aggregating adjacent pixel values. A common practice is to divide the image into 2×2 subimages so that each image is half as wide and half as high as the one from which it is derived. The approximation process is repeated to form a sequence of images which may end with an image consisting of a single pixel.

Reversing the process creates a sequence of images that show increasing detail. This is a useful feature for applications where images are downloaded from a remote source. The sequence of images can be displayed to give an indication of the progress of the download operation.

Generating the sequence of images actually increases the number of pixel values required to specify the image. Compression techniques can be used to reduce the total size of the images to less than the size of the original image.

EXAMPLE 4.8

Figure 4.4 illustrates the construction of a 4-level image pyramid, starting with an 8-pixel × 8-pixel array. Each image is divided into an array of 2-pixel × 2-pixel subimages. The pixel values in each of these subimages are used to compute the pixel value of the corresponding image at the next level down. There are a number of ways of doing this; in Figure 4.4 the median of the four pixel values (rounded to the nearest integer) is taken.

The image pyramid contains a total of 85 pixel values, in comparison to the 64 pixel values in the original image. To make it possible to achieve some compression, we can replace the pixel values at each level with the differences between the pixel values and the value of the corresponding pixel in the level below.

Figure 4.5 shows the differences. It is possible to recreate the original image by starting with the single-pixel image and adding and subtracting the differences at each level. Taking differences can reduce the range of pixel values in the images. In the original image, the maximum pixel value is 189 and the minimum pixel value is 55, a range of 134. In the corresponding difference image, the maximum difference is +40 and the minimum difference is −37, a range of 77. The range of pixel values in the 4 × 4 image is 99, while the range of differences is 47. In both these cases, the range has been reduced by about half; so it should be possible to code the differences using one less bit per difference than it would take to code the pixel values.

$\Box$

Full use is made of the two-dimensional structure of images in *transform coding methods*. A two-dimensional transform is used to concentrate the information content of an image into particular parts of the transformed image, which can then be coded efficiently. The most common transform used for this purpose is the *Discrete Cosine Transform* (DCT).

To use the DCT for transform coding, the image is broken up into square subimages, usually 8 pixels by 8 pixels or 16 pixels by 16 pixels. The pixel values in these arrays are then subjected to the two-dimensional DCT, which produces a square array of transform coefficients. Even though the pixel values are integers, the transform coefficients are real numbers.

The DCT acts to concentrate the information contained in the array of pixels into a corner of the transform array, usually at the top left corner. The largest transform coefficients appear in this corner, and the magnitude of the coefficients diminishes diagonally across the array so that the transform coefficients near the bottom right corner are close to zero.

If the transform coefficients are scanned into a one-dimensional array using a zig-zag pattern like that shown in Figure 4.6, the zero coefficients come at the end of the array. They can be thresholded and discarded. The remaining coefficients can then be coded using Huffman coding or arithmetic coding.

185	124	124	124	124	74	74	74
189	127	127	127	127	74	74	74
187	121	121	121	121	71	71	71
173	110	110	110	110	66	66	66
155	97	97	97	97	64	64	64
137	88	88	88	88	66	66	66
119	75	75	75	75	59	59	59
104	66	66	66	66	55	55	55

156	126	99	74
147	116	90	68
117	92	77	65
90	70	62	57

| 136 | 82 |
| 91 | 64 |

FIGURE 4.4
Construction of a 4-level image pyramid.

+29	-32	-2	-2	+25	-25	0	0
+33	-29	+1	+1	+28	-25	0	0
+40	-26	+5	+5	+31	-19	+3	+3
+26	-37	-6	-6	+20	-24	-2	-2
+38	-20	+5	+5	+20	-13	-1	-1
+20	-29	-4	-4	+11	-11	+1	+1
+29	-15	+5	+5	+13	-3	+2	+2
+14	-24	-4	-4	+4	-7	-2	-2

+20	-10	+17	-8
+11	-20	+8	-14
+26	+1	+13	+1
-1	-21	-2	-7

+50	-4
+5	-22

86

FIGURE 4.5
Differences in the 4-level image pyramid.

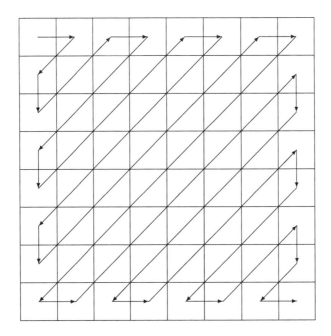

FIGURE 4.6
Zig-zag scanning pattern for encoding DCT coefficients.

To reverse the compression process, the coefficients are restored to an array, the remaining coefficients are set to zero and the inverse DCT is applied. This results in a lossy reconstruction, the amount of information lost depending on the number of DCT coefficients that were set to zero. Differences in the way neighbouring subimages are treated can also result in the reconstructed image exhibiting discontinuities across the boundaries of the subimages.

The use of the DCT in transform coding can result in very high compression ratios, at the cost of some loss of fidelity. The JPEG standard [5] for still image compression (named after its creators, the Joint Photographic Experts Group) uses DCT-based transform coding. The original JPEG standard is being superceded by the JPEG 2000 standard, which adds many features.

For compressing sequences of images, such as those generated by video cameras, a further level of redundancy can be exploited because each image is very similar to the one the precedes it, except when there is an abrupt change of scene. This is the two-dimensional analogue of the process in the CCITT Group 3 standard where lines are coded in terms of the differences between them and the preceding line. The MPEG standard [6] for video compression (devised by the Motion Picture Experts Group) uses this to compress video sequences. The MPEG standard also includes algorithms for coding of digitized audio data.

4.10   Exercises

1. The sequence

$$BADCABAEBABCABAD$$

   is generated by a process that follows the following rules: (1) A may be followed by B, C, D or E; (2) B may be followed by A or C; (3) C is always followed by A; (4) D may be followed by A, B, C or E; (5) E may be followed by B or C; and (6) the sequence must begin with B or D. Devise an encoding scheme that uses these rules to encode the sequence in as few bits as possible. How many bits are required to encode the sequence? How many bits would be required if the occurrence of the letters in the sequence was random?

2. The following table describes the behaviour of a Markov source with four states $\sigma_1, \sigma_2, \sigma_3, \sigma_4$ and alphabet $\{A, B, C, D\}$.

Initial state	Final state	Probability of transition	Output symbol
σ_1	σ_1	0.25	A
σ_1	σ_2	0.25	B
σ_1	σ_3	0.25	C
σ_1	σ_4	0.25	D
σ_2	σ_1	0.50	A
σ_2	σ_4	0.50	B
σ_3	σ_2	0.50	C
σ_3	σ_4	0.50	D
σ_4	σ_3	0.50	A
σ_4	σ_4	0.50	B

Given that the source always begins in state σ_1, devise an efficient way of coding the output of the source using the alphabet $\{0, 1\}$.

3. Use the one-dimensional run-length coding algorithm of the CCITT Group 3 standard and the Huffman codes in Table 4.1 to code the following sequences of pixels

$$WWWWWBBBBBBWWWWWWWBBBBBWWWW$$

$$BBBBWWWWWWWBBBWWWWWBBBBBBB$$

where W denotes a white pixel and B denotes a black pixel. How much compression is achieved by the coding?

4. Apply the block-sorting process to the string `hickory_dickory_dock`. (The underscore characters _ denote spaces between the words.)

5. The string `bbs__aaclaakpee_abh` is the result of block-sorting a string of sixteen letters and three spaces (denoted by the underscore character, _). Given that the first letter is the last b in the sequence and that the underscore character precedes the letters in the sorting order, restore the sequence to the original order.

6. Use the LZ77 algorithm to compress and decompress the string

   `she_sells_seashells_on_the_seashore.`

7. Use the LZ78 algorithm to compress and decompress the string

   `she_sells_seashells_on_the_seashore.`

8. Use the LZW algorithm to compress and decompress the string

   `she_sells_seashells_on_the_seashore.`

9. Figure 4.7 shows pixel values in an 8×8 image. Construct a four-level image pyramid from these values and compute the differences in the pyramid.

142	142	142	142	0	0	0	0
142	142	142	142	0	0	0	0
142	142	142	142	0	0	0	0
142	142	142	142	0	0	0	0
0	0	0	0	142	142	142	142
0	0	0	0	142	142	142	142
0	0	0	0	142	142	142	142
0	0	0	0	142	142	142	142

FIGURE 4.7
Pixel values in an 8×8 image.

10. Apply the two-dimensional Discrete Cosine Transform to the pixel values
    shown in Figure 4.7. Rearrange the output of the transform in a one-dimensional
    array using the zig-zag scanning pattern shown in Figure 4.6. How many of
    the transform coefficients are small enough to be ignored? Set the values of the
    last fifteen transform coefficients to zero and apply the inverse Discrete Cosine
    Transform to recover the image. How much have the pixel values changed?

4.11   References

[1] M. Burrows and D. J. Wheeler, A Block-sorting Lossless Data Compression
Algorithm, Technical Report 124, Digital Equipment Corporation, Palo Alto, CA,
1994.

[2] The bzip2 and libbzip2 official home page,
http://sources.redhat.com/bzip2/index.html

[3] The *gzip* home page, `http://www.gzip.org`

[4] The International Telecommuications Union,
`http://www.itu.int/home/index.html`

[5] The JPEG home page, `http://www.jpeg.org`

[6] The MPEG home page, `http://mpeg.telecomitalialab.com`

[7] T. A. Welch, A technique for high performance data compression, *IEEE Comput.,* 17(6), 8–20, 1984.

[8] J. Ziv and A. Lempel, A universal algorithm for sequential data compression, *IEEE Trans. Inform. Theory,* IT-23(3), 337–343, 1977.

[9] J. Ziv and A. Lempel, Compression of individual sequences via variable rate coding, *IEEE Trans. Inform. Theory,* IT-24(5), 530–536, 1978.

Chapter 5

Fundamentals of Channel Coding

5.1 Introduction

In Chapter 2 we discussed the discrete memoryless channel as a model for the transmission of information from a transmitting source to a receiving destination. The mutual information, $I(A; B)$, of a channel with input A and output B measures the amount of information a channel is able to convey about the source. If the channel is noiseless then this is equal to the information content of the source A, that is $I(A; B) = H(A)$. However in the presence of noise there is an uncertainty or equivocation, $H(A|B)$, that reduces the mutual information to $I(A; B) = H(A) - H(A|B)$. In practice the presence of noise manifests itself as random errors in the transmission of the symbols. Depending on the information being transmitted and its use, random errors may be tolerable (e.g., occasional bit errors in the transmission of an image may not yield any noticeable degradation in picture quality) or fatal (e.g., with Huffman codes discussed in Chapter 3, any bit errors in the encoded variable-length output will create a cascade of errors in the decoded sequence due to the loss of coding synchronisation). Furthermore the errors tolerable at low error rates of 1 in a million (or a probability of error of 10^{-6}) may become intolerable at higher error rates of 1 in 10 or 1 in a 100. When the errors introduced by the information channel are unacceptable then channel coding is needed to reduce the error rate and improve the reliability of the transmission.

The use of channel coders with source coders in a modern digital communication system to provide efficient and reliable transmission of information in the presence of noise is shown in Figure 5.1. Our discussion of channel codes will be principally concerned with *binary block codes*, that is, both the channel coder inputs and outputs will be in binary and fixed-length block codes will be used. Since a digital communication system uses a binary channel (most typically a BSC) and the source coder will encode the source to a binary code, then the intervening channel coder will code binary messages to binary codes. The operation of a channel coder is provided by the following definition.

DEFINITION 5.1 Channel Coding   *The* channel encoder *separates or segments the incoming bit stream (the output of the source encoder) into equal length blocks of L binary digits and maps each L-bit message block into an N-bit* code word *where N > L and the extra N − L* check bits *provide the required error protection. There are M = 2^L messages and thus 2^L code words of length N bits. The* channel decoder *maps the received N-bit word to the most likely code word and inversely maps the N-bit code word to the corresponding L-bit message.*

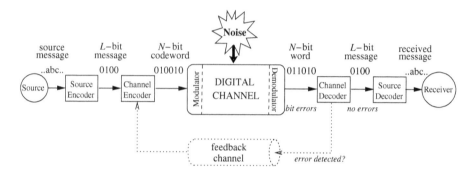

FIGURE 5.1
Noisy communication system.

The channel coder performs a simple mapping operation from the input L-bit message to the corresponding N-bit code word where $N > L$. The rationale behind this mapping operation is that the redundancy introduced by the extra $N - L$ bits can be used to detect the presence of bit errors or indeed identify which bits are in error and correct them (by inverting the bits). How does this work? Since $N > L$ then that means there are $2^N - 2^L$ received words of length N bits that are not code words (where 2^N is the space of all words of length N-bits and 2^L is the subset corresponding to the code words). The key idea is that a bit error will change the code word to a non-code word which can then be detected. It should be obvious that this is only possible if $N > L$.

The channel decoder has the task of detecting that there has been a bit error and, if possible, correcting the bit error. The channel decoder can resolve bit errors by two different systems for error-control.

DEFINITION 5.2 Automatic-Repeat-Request (ARQ)   *If the channel decoder performs* error detection *then errors can be detected and a feedback channel from the channel decoder to the channel encoder can be used to control the retransmission of the code word until the code word is received without detectable errors.*

> **DEFINITION 5.3 Forward Error Correction (FEC)** *If the channel decoder performs* error correction *then errors are not only detected but the bits in error can be identified and corrected (by bit inversion).*

In this chapter we present the theoretical framework for the analysis and design of channel codes, define metrics that can be used to evaluate the performance of channel codes and present Shannon's Fundamental Coding Theorem, one of the main motivators for the continuing research in the design of ever more powerful channel codes.

5.2 Code Rate

A price has to be paid to enable a channel coder to perform error detection and error correction. The extra $N - L$ bits require more bits to be pushed into the channel than were generated from the source coder, thus requiring a channel with a higher bit-rate or bandwidth.

Assume that the source coder generates messages at an average bit rate of n_s bits per second, that is, 1 bit transmitted every $T_s = \frac{1}{n_s}$ seconds. If the channel encoder maps each L-bit message (of information) into an N-bit code word then the channel bit rate will be $n_c = \frac{N}{L} n_s$ and since $N > L$ there will be more bits transmitted through the channel than bits entering the channel encoder. Thus the channel must transmit bits faster than the source encoder can produce.

> **DEFINITION 5.4 Code Rate (Channel Codes)** *In Chapter 2 the general expression for the code rate was given as:*
>
> $$R = \frac{H(M)}{n} \tag{5.1}$$
>
> *For the case of channel coding we assume M equally likely messages where $M = 2^L$ and each message is transmitted as the code word of length N. Thus $H(M) = \log M = L$ and $n = N$ yielding the code rate for channel codes:*
>
> $$R = \frac{L}{N} = \frac{n_s}{n_c} = \frac{T_c}{T_s} \tag{5.2}$$

The code rate, R, measures the relative amount of information conveyed in each code word and is one of the key measures of channel coding performance. A higher value for R (up to its maximum value of 1) implies that there are fewer redundant $N - L$ check bits relative to the code word length N. The upside in a higher value for the code rate is that more message information is transmitted with each code word since for fixed N this implies larger values of L. The downside is that with fewer check bits and redundancy a higher code rate will make it more difficult to cope with transmission errors in the system. In Section 5.7 we introduce Shannon's Fundamental Theorem which states that R must not exceed the channel capacity C for error-free transmission.

In some systems the message rate, n_s, and channel rates, n_c, are fixed design parameters and for a given message length L the code word length is selected based on $N = \lfloor Ln_c/n_s \rfloor$, that is, the largest integer less than Ln_c/n_s. If Ln_c/n_s is not an integer then "dummy" bits have to be transmitted occasionally to keep the message and code streams synchronised with each other.

EXAMPLE 5.1

Consider a system where the message generation rate is $n_s = 3$ bps and the channel bit rate is fixed at $n_c = 4$ bps. This requires a channel coder with an overall $R = \frac{3}{4}$ code rate.

If we choose to design a channel coder with $L = 3$ then obviously we must choose $N = 4$ without any loss of synchronisation since Ln_c/n_s is an integer value. The code rate of this channel coder is $R = \frac{3}{4}$.

Let us now attempt to design a channel coder with $L = 5$. Then $Ln_c/n_s = 6\frac{2}{3}$ and $N = 6$ and the channel coder has an apparent $R = \frac{5}{6}$ code rate. However this system will experience a cumulative loss of synchronisation due to the unaccounted for gap of $\frac{2}{3}$ bit per code word transmitted. To compensate for this, 2 "dummy" bits need to be transmitted for every 3 code word transmissions. That is, for every 3 message blocks of $L \times 3 = 15$ bits duration the channel encoder will transmit 3 code words of $N \times 3 = 18$ bits duration plus 2 "dummy" bits, that is, 20 bits in total for an overall $R = \frac{15}{20} = \frac{3}{4}$ code rate. This is depicted in Figure 5.2.

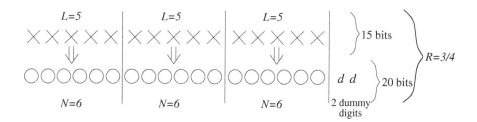

FIGURE 5.2

Channel encoder mapping from message bits (represented by the X's) to the code bits (represented by the circles), including dummy digits for Example 5.1.

5.3   Decoding Rules

When a code word is transmitted through a channel it may be subject to bit errors due to the presence of noise. Thus the received word may be different from the transmitted code word and the receiver will need to make a decision on which code word was transmitted based on some form of decoding rule. The form of decoding rule we adopt will govern how channel codes are used and the level of error protection they provide. We begin the discussion by providing the framework for discussing decoding rules in general and examine two specific decoding rules of interest.

Let $\mathbf{a}_i = a_{i1} a_{i2} \dots a_{iN}$ be the ith N-bit code word (for $1 \le i \le M$) that is transmitted through the channel and let $\mathbf{b} = b_1 b_2 \dots b_N$ be the corresponding N-bit word produced at the output of the channel. Let $\mathbf{B}_M$ represent the set of M valid N-bit code words and let the complement of this set be $\mathbf{B}_M^c$, the set of remaining N-bit binary sequences which are not code words. Then $\mathbf{B}_N = \mathbf{B}_M \cup \mathbf{B}_M^c$ is the set of all possible N-bit binary sequences and $\mathbf{b} \in \mathbf{B}_N$ whereas $\mathbf{a}_i \in \mathbf{B}_M$. The channel decoder has to apply a *decoding rule* on the received word $\mathbf{b}$ to decide which code word $\mathbf{a}_i$ was transmitted, that is, if the decision rule is $D(.)$ then $\mathbf{a}_i = D(\mathbf{b})$. Let $P_N(\mathbf{b}|\mathbf{a}_i)$ be the probability of receiving $\mathbf{b}$ given $\mathbf{a}_i$ was transmitted. For a discrete memoryless channel this probability can be expressed in terms of the channel probabilities as follows:

$$P_N(\mathbf{b}|\mathbf{a}_i) = \prod_{t=1}^{N} P(b_t|a_{it}) \tag{5.3}$$

Let $P_N(\mathbf{a}_i)$ be the a priori probability for the message corresponding to the code word $\mathbf{a}_i$. Then by Bayes' Theorem the probability that message $\mathbf{a}_i$ was transmitted given $\mathbf{b}$ was received is given by:

$$P_N(\mathbf{a}_i|\mathbf{b}) = \frac{P_N(\mathbf{b}|\mathbf{a}_i) P_N(\mathbf{a}_i)}{P_N(\mathbf{b})} \tag{5.4}$$

If the decoder decodes $\mathbf{b}$ into the code word $\mathbf{a}_i$ then the probability that this is correct is $P_N(\mathbf{a}_i|\mathbf{b})$ and the probability that it is wrong is $1 - P_N(\mathbf{a}_i|\mathbf{b})$. Thus to minimise the error the code word $\mathbf{a}_i$ should be chosen so as to maximise $P_N(\mathbf{a}_i|\mathbf{b})$. This leads to the *minimum-error decoding rule*.

DEFINITION 5.5 Minimum-Error Decoding Rule   *We choose:*

$$D_{ME}(\mathbf{b}) = \mathbf{a}^* \qquad (5.5)$$

where $\mathbf{a}^* \in \mathbf{B}_M$ *is such that:*

$$P_N(\mathbf{a}^*|\mathbf{b}) \geq P_N(\mathbf{a}_i|\mathbf{b}) \ \forall i \qquad (5.6)$$

Using Equation 5.4 and noting that $P_N(\mathbf{b})$ *is independent of the code word* $\mathbf{a}_i$ *the condition simplifies to:*

$$P_N(\mathbf{b}|\mathbf{a}^*)P_N(\mathbf{a}^*) \geq P_N(\mathbf{b}|\mathbf{a}_i)P_N(\mathbf{a}_i) \ \forall i \qquad (5.7)$$

requiring both knowledge of the channel probabilities and channel input probabilities. This decoding rule guarantees minimum error in decoding.

An alternative decoding rule, the *maximum-likelihood decoding rule*, is based on maximising $P_N(\mathbf{b}|\mathbf{a}_i)$, the likelihood that $\mathbf{b}$ is received given $\mathbf{a}_i$ was transmitted. This decoding rule is not necessarily minimum error but it is simpler to implement since the channel input probabilities are not required.

DEFINITION 5.6 Maximum-Likelihood Decoding Rule   *We choose:*

$$D_{ML}(\mathbf{b}) = \mathbf{a}^* \qquad (5.8)$$

where $\mathbf{a}^* \in \mathbf{B}_M$ *is such that:*

$$P_N(\mathbf{b}|\mathbf{a}^*) \geq P_N(\mathbf{b}|\mathbf{a}_i) \ \forall i \qquad (5.9)$$

requiring only knowledge of the channel probabilities. This decoding rule does not guarantee minimum error in decoding.

NOTE   The maximum-likelihood decoding rule is the same as the minimum-error decoding rule when the channel input probabilities are equal.

Assume we have a decoding rule $D(.)$. Denote $\mathbf{B}_i$ as the set of all possible received words $\mathbf{b}$ such that $D(\mathbf{b}) = \mathbf{a}_i$, that is the set of all N-bit words that are decoded to the code word $\mathbf{a}_i$, and define its complement as $\mathbf{B}_i^c$. Then the probability of a decoding error given code word $\mathbf{a}_i$ was transmitted is given by:

$$P(E|\mathbf{a}_i) = \sum_{\mathbf{b} \in \mathbf{B}_i^c} P_N(\mathbf{b}|\mathbf{a}_i) \qquad (5.10)$$

and the overall *probability of decoding error* is:

$$P(E) = \sum_{i=1}^{M} P(E|\mathbf{a}_i)P(\mathbf{a}_i) \qquad (5.11)$$

EXAMPLE 5.2

Consider a BSC with the following channel probabilities:

$$\mathbf{P} = \begin{bmatrix} 0.6 \ 0.4 \\ 0.4 \ 0.6 \end{bmatrix}$$

and channel encoder with $(L, N) = (2, 3)$, that is using a channel code with $M = 2^2 = 4$ code words of $N = 3$ bits in length. Assume the code words transmitted into the channel, and their corresponding probability of occurrence, are:

Code word	$P_N(\mathbf{a}_i)$
$\mathbf{a}_1 = (000)$	0.4
$\mathbf{a}_2 = (011)$	0.2
$\mathbf{a}_3 = (101)$	0.1
$\mathbf{a}_4 = (110)$	0.3

Say the received word at the output of the channel is $\mathbf{b} = 111$. If we apply the maximum likelihood decoding rule of Equation 5.9 then we choose the $\mathbf{a}_i$ that maximises $P_N(\mathbf{b}|\mathbf{a}_i)$. Calculating these probabilities for all of the possible $M = 4$ code words gives:

$$P_N(\mathbf{b}|\mathbf{a}_1) = P_N(111|000) = P(1|0) \times P(1|0) \times P(1|0) = 0.064$$
$$P_N(\mathbf{b}|\mathbf{a}_2) = P_N(111|011) = P(1|0) \times P(1|1) \times P(1|1) = 0.144$$
$$P_N(\mathbf{b}|\mathbf{a}_3) = P_N(111|101) = P(1|1) \times P(1|0) \times P(1|1) = 0.144$$
$$P_N(\mathbf{b}|\mathbf{a}_4) = P_N(111|110) = P(1|1) \times P(1|1) \times P(1|0) = 0.144$$

from which we have that code words $\mathbf{a}_2, \mathbf{a}_3$ and $\mathbf{a}_4$ are equally likely to have been transmitted in the maximum likelihood sense if $\mathbf{b} = 111$ was received. For the purposes of decoding we need to make a decision and choose one of the $\mathbf{a}_2, \mathbf{a}_3$ and $\mathbf{a}_4$, and we choose $D_{ML}(\mathbf{b} = 111) = \mathbf{a}_2$. If we apply the minimum error decoding rule of Equation 5.7 we choose the $\mathbf{a}_i$ that maximises $P_N(\mathbf{b}|\mathbf{a}_i)P_N(\mathbf{a}_i)$. Calculating these probabilities for all of the $M = 4$ code words and using the provided a priori $P(\mathbf{a}_i)$ gives:

$$P_N(\mathbf{b}|\mathbf{a}_1)P_N(\mathbf{a}_1) = 0.064 \times 0.4 = 0.0256$$
$$P_N(\mathbf{b}|\mathbf{a}_2)P_N(\mathbf{a}_2) = 0.144 \times 0.2 = 0.0288$$
$$P_N(\mathbf{b}|\mathbf{a}_3)P_N(\mathbf{a}_3) = 0.144 \times 0.1 = 0.0144$$
$$P_N(\mathbf{b}|\mathbf{a}_4)P_N(\mathbf{a}_4) = 0.144 \times 0.3 = 0.0432$$

from which we have that code word $\mathbf{a}_4$ minimises the error in decoding the received word $\mathbf{b} = 111$, that is $D_{ME}(\mathbf{b} = 111) = \mathbf{a}_4$.   ⬜

Although the minimum error decoding rule guarantees minimum error it is rarely used in practice, in favour of maximum likelihood decoding, due to the unavailability of the a priori probabilities, $P_N(\mathbf{a}_i)$. Since with arbitrary sources and/or efficient source coding (see Section 3.5.3) one can assume, without serious side effects, that messages are equiprobable then the use of maximum likelihood decoding will usually lead to minimum error.

5.4   Hamming Distance

An important parameter for analysing and designing codes for robustness in the presence of errors is the number of bit errors between the transmitted code word, $\mathbf{a}_i$, and the received word, $\mathbf{b}$, and how this relates to the number of bit "errors" or differences between two different code words, $\mathbf{a}_i$ and $\mathbf{a}_j$, This measure is provided by the Hamming distance on the space of binary words of length N. The properties of the Hamming distance, attributed to R.W. Hamming [3] who established the fundamentals of error detecting and error correcting codes, are instrumental in establishing the operation and performance of channel codes for both error detection and error correction.

DEFINITION 5.7 Hamming Distance   *Consider the two N-length binary words* $\mathbf{a} = a_1 a_2 \ldots a_N$ *and* $\mathbf{b} = b_1 b_2 \ldots b_N$. *The* Hamming *distance between* $\mathbf{a}$ *and* $\mathbf{b}$, $d(\mathbf{a}, \mathbf{b})$, *is defined as the number of bit positions in which* $\mathbf{a}$ *and* $\mathbf{b}$ *differ. The Hamming distance is a metric on the space of all binary words of length N since for arbitrary words,* $\mathbf{a}$, $\mathbf{b}$, $\mathbf{c}$, *the Hamming distance obeys the following conditions:*

   1. $d(\mathbf{a}, \mathbf{b}) \geq 0$ *with equality when* $\mathbf{a} = \mathbf{b}$

   2. $d(\mathbf{a}, \mathbf{b}) = d(\mathbf{b}, \mathbf{a})$

   3. $d(\mathbf{a}, \mathbf{b}) + d(\mathbf{b}, \mathbf{c}) \geq d(\mathbf{a}, \mathbf{c})$ *(triangle inequality)*

EXAMPLE 5.3

Let $N = 8$ and let:

$$\mathbf{a} = 11010001$$
$$\mathbf{b} = 00010010$$
$$\mathbf{c} = 01010011$$

Then have that $d(\mathbf{a}, \mathbf{b}) = 4$ since $\mathbf{a}$ differs from $\mathbf{b}$ in the following 4 bit locations: $a_1 \neq b_1, a_2 \neq b_2, a_7 \neq b_7, a_8 \neq b_8$. Similarly, $d(\mathbf{b}, \mathbf{c}) = 2$ since $\mathbf{b}$ and $\mathbf{c}$ differ by 2 bits, bits 2 and 8, and $d(\mathbf{a}, \mathbf{c}) = 2$ since $\mathbf{a}$ and $\mathbf{c}$ also differ by 2 bits, bits 1 and 7. We can also verify the triangle inequality since:

$$d(\mathbf{a}, \mathbf{b}) + d(\mathbf{b}, \mathbf{c}) \geq d(\mathbf{a}, \mathbf{c}) \;\Rightarrow\; 4 + 2 \geq 2$$

□

5.4.1 Hamming Distance Decoding Rule for BSCs

Consider the maximum likelihood decoding rule where the code word, $\mathbf{a}_i$, which maximises the conditional probability $P_N(\mathbf{b}|\mathbf{a}_i)$ (or likelihood), has to be found. Consider a BSC with bit error probability q. Let the Hamming distance $d(\mathbf{b}, \mathbf{a}_i) = D$ represent the number of bit errors between the transmitted code word $\mathbf{a}_i$ and received word $\mathbf{b}$. Then this gives the following expression for the conditional probability:

$$P(\mathbf{b}|\mathbf{a}_i) = (q)^D (1 - q)^{N-D} \tag{5.12}$$

If $q < 0.5$ then Equation 5.12 is maximised when $\mathbf{a}_i$ is chosen such that $d(\mathbf{b}, \mathbf{a}_i)$ is minimised.

RESULT 5.1 Hamming Distance Decoding Rule
The binary word, $\mathbf{b}$, of length N is received upon transmission of one of the M possible N-bit binary code words, $\mathbf{a}_i \in \mathbf{B}_M$, through a BSC. Assuming the maximum likelihood decoding rule we choose the most likely code word as follows:

- **if $\mathbf{b} = \mathbf{a}_i$ for a particular i then** *the code word $\mathbf{a}_i$ was sent*

- **if $\mathbf{b} \neq \mathbf{a}_i$ for any i, we find the code word $\mathbf{a}^* \in \mathbf{B}_M$ which is closest to $\mathbf{b}$ in** *the Hamming sense:*
$$d(\mathbf{a}^*, \mathbf{b}) \leq d(\mathbf{a}_i, \mathbf{b}) \;\forall i \tag{5.13}$$

  – **if** *there is only one candidate $\mathbf{a}^*$ **then** the t-bit error, where $t = d(\mathbf{a}^*, \mathbf{b})$, is corrected and $\mathbf{a}^*$ was sent*

  – **if** *there is more than one candidate $\mathbf{a}^*$**then** the t-bit error, where $t = d(\mathbf{a}^*, \mathbf{b})$, can only be detected*

EXAMPLE 5.4

Consider the following channel code:

Message $(L = 2)$	Code word $(N = 3)$
00	000
01	001
10	011
11	111

There are $M = 2^L = 4$ messages and 4 corresponding code words. With $N = 3$ length code words there are $2^N = 8$ possible received words, 4 of these will be the correct code words and 4 of these will be non-code words. If the received word, $\mathbf{b}$, belongs to the set of code words $\{000, 001, 011, 111\}$ then the Hamming distance decoding rule would imply that we decode the received word as the code word (i.e., $\mathbf{a}^*$ is the same as $\mathbf{b}$). That is if $\mathbf{b} = 000$ then $\mathbf{a}^* = 000$, and so on. If the received word, $\mathbf{b}$, belongs to the set of non-code words $\{010, 100, 101, 110\}$ then the Hamming distance decoding rule would operate as follows:

$\mathbf{b} = b_1 b_2 b_3$	Closest code word	Action
010	000 (b_2 in error), 011 (b_3 in error)	1-bit error detected
100	000 (b_1 in error)	1-bit error corrected
101	001 (b_1 in error), 111 (b_2 in error)	1-bit error detected
110	111 (b_3 in error)	1-bit error corrected

$\square$

5.4.2   Error Detection/Correction Using the Hamming Distance

An important indication of the error robustness of a code is the Hamming distance between two different code words, $\mathbf{a}_i$ and $\mathbf{a}_j$. From Example 5.4 it is apparent that errors are detected when the received word, $\mathbf{b}$, is equi-distant from more than one code word and errors are corrected when the received word is closest to only one code word. Both forms of error robustness rely on there being sufficient distance between code words so that non-code words can be detected and even corrected. The analysis for specific behaviour of a code to errors is tedious and unproductive. The most useful result is when we consider the general error robustness of a code. That is, whether a code can detect or correct *all* errors up to t-bits no matter where the errors occur and in which code words. To this end we need to define the following important measure of a code's error performance.

DEFINITION 5.8 Minimum Distance of a Code   *The minimum distance of a block code $\mathcal{K}_n$, where $\mathcal{K}_n$ identifies the set of length n code words, is given by:*

$$d(\mathcal{K}_n) = \min \{d(\mathbf{a}, \mathbf{b}) \,|\, \mathbf{a}, \mathbf{b} \in \mathcal{K}_n \text{ and } \mathbf{a} \neq \mathbf{b}\} \qquad (5.14)$$

that is, the smallest Hamming distance over all pairs of distinct code words.

---

t-bit error detection

A block code $\mathcal{K}_n$ is said to detect *all* combinations of up to t errors provided that for each code word $\mathbf{a}_i$ and each received word $\mathbf{b}$ obtained by corrupting up to t bits in $\mathbf{a}_i$, the resulting $\mathbf{b}$ is not a code word (and hence can be detected via the Hamming distance decoding rule). This is an important property for ARQ error control schemes where codes for error detection are required.

RESULT 5.2 Error Detection Property

A block code, $\mathcal{K}_n$, detects up to t errors if and only if its minimum distance is greater than t:

$$d(\mathcal{K}_n) > t \qquad (5.15)$$

PROOF Let $\mathbf{a}^*$ be the code word transmitted from the block code $\mathcal{K}_n$ and $\mathbf{b}$ the received word. Assume there are t bit errors in the transmission. Then $d(\mathbf{a}^*, \mathbf{b}) = t$. To detect that $\mathbf{b}$ is in error it is sufficient to ensure that $\mathbf{b}$ does not correspond to any of the $\mathbf{a}_i$ code words, that is, $d(\mathbf{b}, \mathbf{a}_i) > 0 \; \forall i$. Using the triangle inequality we have that for any code word $\mathbf{a}_i$:

$$d(\mathbf{a}^*, \mathbf{b}) + d(\mathbf{b}, \mathbf{a}_i) \geq d(\mathbf{a}^*, \mathbf{a}_i) \text{ or } d(\mathbf{b}, \mathbf{a}_i) \geq d(\mathbf{a}^*, \mathbf{a}_i) - d(\mathbf{a}^*, \mathbf{b})$$

To ensure that $d(\mathbf{b}, \mathbf{a}_i) > 0$ we must have that $d(\mathbf{a}^*, \mathbf{a}_i) - d(\mathbf{a}^*, \mathbf{b}) > 0$ or $d(\mathbf{a}^*, \mathbf{a}_i) > d(\mathbf{a}^*, \mathbf{b})$. Since $d(\mathbf{a}^*, \mathbf{b}) = t$ we get the final result that:

$$d(\mathbf{a}^*, \mathbf{a}_i) > t \; \forall i$$

which is guaranteed to be true if and only if $d(\mathcal{K}_n) > t$. $\square$

Figure 5.3 depicts the two closest code words, $\mathbf{a}_i$ and $\mathbf{a}_j$, at a distance of $t + 1$ (i.e., $d(\mathcal{K}_n) = t+1$). The hypersphere of distance $t+1$ drawn around each code word may touch another code word but no code word falls within the hypersphere of another code word since this would violate $d(\mathcal{K}_n) = t + 1$. Clearly any received word $\mathbf{b}$ of distance $< t$ from any code word will fall within the hypersphere of radius $t + 1$ from one of more code words and hence be detectable since it can never be mistaken for a code word.

t-bit error correction

A block code $\mathcal{K}_n$ is said to correct *all* combinations of up to t errors provided that for each code word $\mathbf{a}_i$ and each received word $\mathbf{b}$ obtained by corrupting up to t bits in $\mathbf{a}_i$, the Hamming distance decoding rule leads uniquely to $\mathbf{a}_i$. This is an important property for FEC error control schemes where codes for error correction are required.

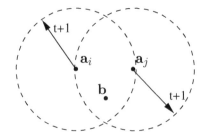

FIGURE 5.3
Diagram of t-bit error detection for $d(\mathcal{K}_n) = t + 1$.

RESULT 5.3 Error Correction Property
A block code, $\mathcal{K}_n$, corrects up to t errors if and only if its minimum distance is greater than 2t:

$$d(\mathcal{K}_n) > 2t \qquad\qquad (5.16)$$

PROOF   Let $\mathbf{a}^*$ be the code word transmitted from the block code $\mathcal{K}_n$ and $\mathbf{b}$ the received word. Assume there are t bit errors in the transmission. Then $d(\mathbf{a}^*, \mathbf{b}) = t$. To detect that $\mathbf{b}$ is in error and ensure that the Hamming distance decoding rule uniquely yields $\mathbf{a}^*$ (the error can be corrected) then it is sufficient that $d(\mathbf{a}^*, \mathbf{b}) < d(\mathbf{a}_i, \mathbf{b})\ \forall i$. Using the triangle inequality we have that for any code word $\mathbf{a}_i$:

$$d(\mathbf{a}^*, \mathbf{b}) + d(\mathbf{b}, \mathbf{a}_i) \geq d(\mathbf{a}^*, \mathbf{a}_i) \text{ or } d(\mathbf{b}, \mathbf{a}_i) \geq d(\mathbf{a}^*, \mathbf{a}_i) - d(\mathbf{a}^*, \mathbf{b})$$

To ensure that $d(\mathbf{a}^*, \mathbf{b}) < d(\mathbf{a}_i, \mathbf{b})$, or $d(\mathbf{b}, \mathbf{a}_i) > d(\mathbf{a}^*, \mathbf{b})$, we must have that $d(\mathbf{a}^*, \mathbf{a}_i) - d(\mathbf{a}^*, \mathbf{b}) > d(\mathbf{a}^*, \mathbf{b})$ or $d(\mathbf{a}^*, \mathbf{a}_i) > 2d(\mathbf{a}^*, \mathbf{b})$. Since $d(\mathbf{a}^*, \mathbf{b}) = t$ we get the final result that:

$$d(\mathbf{a}^*, \mathbf{a}_i) > 2t\ \forall i$$

which is guaranteed to be true if and only if $d(\mathcal{K}_n) > 2t$.   ☐

Figure 5.4 depicts the two closest code words, $\mathbf{a}_i$ and $\mathbf{a}_j$, at a distance of $2t + 1$ (i.e., $d(\mathcal{K}_n) = 2t + 1$). The hyperspheres of distance t drawn around each code word do not touch each other. Clearly any received word $\mathbf{b}$ of distance $\leq t$ from any code word will only fall within the hypersphere of radius t from that code word and hence can be corrected.

EXAMPLE 5.5

The even-parity check code has $N = L + 1$ and is created by adding a parity-check bit to the original L-bit message such that the resultant code word has an even number of 1's. We consider the specific case for $L = 2$ where the code table is:

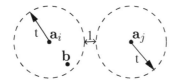

FIGURE 5.4
Diagram of t-bit error correction for $d(\mathcal{K}_n) = 2t + 1$.

Message	Code word
00	000
01	011
10	101
11	110

By computing the Hamming distance between all pairs of distinct code words we find that $d(\mathcal{K}_N) = 2$, that is, no two code words are less than a distance of 2 from each other. By Result 5.2 since $d(\mathcal{K}_N) = 2 > t = 1$ the even-parity check code is able to detect all single-bit errors. It can be shown that $d(\mathcal{K}_N) = 2$ for any length even-parity check code. ▯

EXAMPLE 5.6

The repetition code is defined for single-bit messages ($L = 1$) where the message bit is repeated ($N - 1$) times (for N odd) and then the errors are corrected by using a majority vote decision rule (which is functionally equivalent to the Hamming distance decoding rule but more efficiently implemented). Consider the $N = 3$ repetition code:

Message	Code word
0	000
1	111

where the message is repeated twice to form a code word of length $N = 3$. The minimum distance of the code is, by inspection, $d(\mathcal{K}_3) = 3$, and from Result 5.2, $d(\mathcal{K}_3) > t$ with $t = 2$, and this code can detect all double-bit errors. Furthermore by Result 5.3, $d(\mathcal{K}_3) > 2t$ with $t = 1$, and this code can also correct all single-bit errors. The decoder can be designed for either 2-bit error detection or 1-bit error correction. The Hamming decoding rule as described will perform 1-bit error correction. To see how the Hamming and majority vote decoding rules work to provide 1-bit error correction we consider what happens when any of the 8 possible words are received:

Received word	Closest code word	Majority bit	Message?
000	000	0	0
001	000	0	0
010	000	0	0
011	111	1	1
100	000	0	0
101	111	1	1
110	111	1	1
111	111	1	1

The majority vote decoding rule simply selects the bit that occurs most often in the received word (the majority bit). With N odd then this will be guaranteed to always produce a clear majority (decision). From the above table if, for example, the received word is $\mathbf{b} = 011$ then we have two 1's, one 0 and the majority bit is 1, and hence the message is 1. It can be easily seen that the majority vote selects the code word that is closest to the received word and hence is equivalent to the Hamming distance decoding rule.

For an N-bit repetition code it can be shown that $d(\mathcal{K}_N) = N$ and hence will be able to perform $\lfloor \frac{N}{2} \rfloor$-bit error correction, where the operator $\lfloor x \rfloor$ is the largest integer less than or equal to x, or detect up to $N - 1$ errors if only error detection is considered.

⬚

EXAMPLE 5.7

Consider the code from Example 5.4:

Message ($L = 2$)	Code word ($N = 3$)
00	000
01	001
10	011
11	111

We see that $d(\mathcal{K}_3) = 1$ since the first code word 000 is of distance 1 from the second code word 001. This implies that from Results 5.2 and 5.3 this code cannot detect all single-bit errors, let alone correct any errors. In Example 5.4 the Hamming distance decoding rule was either providing 1-bit detection or 1-bit correction. The discrepancy is explained by noting that Results 5.2 and 5.3 are restricted to codes which are able to detect or correct *all* t-bit errors. With this in mind it is obvious this code cannot detect a single-bit error in bit 3 of the code word 000 since this would produce the code word 001 and not be detected.

⬚

EXAMPLE 5.8

Consider the following code $\mathcal{K}_6$ for $L = 3$-bit messages:

Message ($L = 3$)	Code word ($N = 6$)
000	000000
001	001110
010	010101
011	011011
100	100011
101	101101
110	110110
111	111000

By computing the distance between all pairs of distinct code words, requiring $\binom{8}{2} =$ 28 computations of the Hamming distance, we find that $d(\mathcal{K}_6) = 3$ and this code can correct all 1-bit errors. For example, if $\mathbf{a} = 010101$ is sent and $\mathbf{b} = 010111$ is received then the closest code word is $\mathbf{a} = 010101$ and the most likely single-bit error in b_5 is corrected. This will always be the case no matter which code word is sent and which bit is in error; up to all single-bit errors will be corrected. However the converse statement is not true. All possible received words of length $N = 6$ do not simply imply a code word (no errors) or a code word with a single-bit error. Consider $\mathbf{b} = 111111$. The closest code words are $\mathbf{a} = 110110$, $\mathbf{a} = 101101$ and $\mathbf{a} = 011011$, implying that a 2-bit error has been detected. Thus a code for t-bit error correction may sometimes provide greater than t-bit error detection. In FEC error control systems where there is no feedback channel this is undesirable as there is no mechanism for dealing with error detection.

⬚

5.5 Bounds on M, Maximal Codes and Perfect Codes

5.5.1 Upper Bounds on M and the Hamming Bound

The $d(\mathcal{K}_N)$ for a particular block code $\mathcal{K}_N$ specifies the code's error correction and error detection capabilities as given by Results 5.2 and 5.3. Say we want to design a code, $\mathcal{K}_N$, of length N with minimum distance $d(\mathcal{K}_N)$. Is there a limit (i.e., upper bound) on the number of code words (i.e., messages), M, we can have? Or say we want to design a code of length N for messages of length L. What is the maximum error protection (i.e., maximum $d(\mathcal{K}_N)$) that we can achieve for a code with these parameters? Finally, say we want to design a code with messages of length L that is capable of t-bit error correction. What is the smallest code word length, N, that we can use? The answer to these important design questions is given by the parameter $B(N, d(\mathcal{K}_N))$ defined below.

DEFINITION 5.9 Upper Bound on M

For a block code $\mathcal{K}_N$ of length N and minimum distance $d(\mathcal{K}_N)$ the maximum number of code words, and hence messages, M, to guarantee that such a code can exist is given by:

$$M = B(N, d(\mathcal{K}_N)) \tag{5.17}$$

RESULT 5.4

The following are elementary results for $B(N, d(\mathcal{K}_N))$:

1. $B(N, 1) = 2^N$

2. $B(N, 2) = 2^{N-1}$

3. $B(N, 2t + 1) = B(N + 1, 2t + 2)$

4. $B(N, N) = 2$

where $B(N, 2t+1) = B(N+1, 2t+2)$ indicates that if we know the bound for <u>odd</u> $d(\mathcal{K}_N) \equiv 2t+1$ then we can obtain the bound for the next <u>even</u> $d(\mathcal{K}_N)+1 \equiv 2t+2$, and vice versa.

EXAMPLE 5.9

Using the relation $B(N, 2t + 1) = B(N + 1, 2t + 2)$ with $t = 0$ gives $B(N, 1) = B(N+1, 2)$. If we are given that $B(N, 1) = 2^N$ then this also implies $B(N+1, 2) = 2^N$, and making the substitution $N' = N + 1$ and then replacing N' by N yields $B(N, 2) = 2^{N-1}$. $\Box$

The proof of Result 5.4 can be found in [6]. Additional results for $B(N, d(\mathcal{K}_N))$ can be found by considering the important *Hamming* or *sphere-packing bound* stated in the following theorem:

THEOREM 5.1 Hamming Bound

If the block code of length N is a t-bit error correcting code, then the number of code words, M, must the satisfy the following inequality:

$$M \leq \frac{2^N}{\sum_{i=0}^{t} \binom{N}{i}} \tag{5.18}$$

which is an upper bound (the Hamming *bound) on the number of code words.*

PROOF Let $V(N, t)$ be defined as the number of words of length N that are within a Hamming distance of t from a code word $\mathbf{a}_j$. There will be $\binom{N}{i}$ such words at a Hamming distance of i, and by adding up the number of words of Hamming distance i for $i = 0, 1, 2, \ldots t$ we obtain:

$$V(N, t) = \sum_{i=0}^{t} \binom{N}{i} \qquad (5.19)$$

An alternative interpretation which is sometimes useful is to consider $V(N, t)$ as the volume of the hypersphere of radius t centred at $\mathbf{a}_j$ in the space of all N-length words.

Since the code is a t-bit error correcting code then for any pair of code words, $\mathbf{a}_j$ and $\mathbf{a}_k$, this implies $d(\mathbf{a}_j, \mathbf{a}_k) > 2t$ and thus no word will be within a Hamming distance of t from more than one code word. Consider all M code words and the set of words that are within a Hamming distance of t, $V(N, t)$, from each code word, that is $N_t = M.V(N, t)$. To guarantee that no word is within a Hamming distance of t from more than one code word we must ensure that the possible number of distinct sequences of length N, 2^N, be at least N_t, that is:

$$M \sum_{i=0}^{t} \binom{N}{i} \leq 2^N \qquad (5.20)$$

which proves the Hamming bound. This is shown in Figure 5.5 where it is evident that, since for t-bit error correction the hyperspheres of length t around each code word must not touch, this can only happen if the space of all words of length N is at least the sum of all the hyperspheres.

It should be noted that the condition for equality with the Hamming bound occurs when all words of length N reside within one of the hyperspheres of radius t, that is, when each word of length N is a code word or of distance t or less from a code word. $\square$

We know from Result 5.3 that for a t-bit error correcting code $d(\mathcal{K}_N) > 2t$. Thus a t-bit error correcting code will have a minimum distance of $d(\mathcal{K}_N) = 2t+1$ or more. Consider the problem of finding the upper bound for a code with $d(\mathcal{K}_N) = 2t + 1$ (i.e., $B(N, 2t + 1)$). Since $d(\mathcal{K}_N) > 2t$ this is a t-bit error correcting code and hence subject to the Hamming bound from Theorem 5.1 we expect that:

$$M = B(N, 2t + 1) \leq \frac{2^N}{\sum_{i=0}^{t} \binom{N}{i}} \qquad (5.21)$$

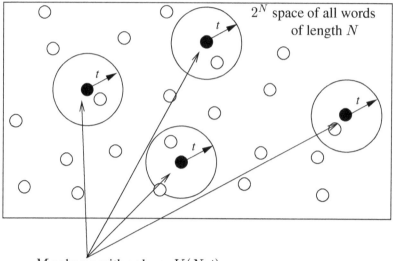

M spheres with volume $V(N,t)$

FIGURE 5.5
Proof of Hamming bound for t-bit error correcting codes. Solid dots represent code words; open dots represent all other words of length N.

EXAMPLE 5.10

Consider the important case of 1-bit error correcting codes where $d(\mathcal{K}_N) = 3$. The Hamming bound for 1-bit error correction is:

$$M \leq \frac{2^N}{\binom{N}{0} + \binom{N}{1}} = \frac{2^N}{N+1}$$

and hence the upper bound on the number of messages with $d(\mathcal{K}_N) = 3$ is:

$$B(N,3) \leq \frac{2^N}{N+1}$$

Using $B(N, 2t+1) = B(N+1, 2t+2)$ for $t = 1$ then gives:

$$B(N,4) \leq \frac{2^{N-1}}{N}$$

which is the upper bound on the number of messages for a code with $d(\mathcal{K}_N) = 4$. ☐

It should be noted that the Hamming bound of Theorem 5.1 is a necessary but not sufficient condition. That is, if we find values of N, M and t that satisfy Equation 5.18 this is not sufficient to guarantee that such a code actually exists.

EXAMPLE 5.11

Consider designing a 1-bit error correcting code ($t = 1$) using code words of length $N = 4$. We can satisfy Equation 5.18 with $M = 3$. However no code with $M = 3$ code words of length $N = 4$ and a minimum distance of at least $d(\mathcal{K}_4) = 2t + 1 = 3$ can actually be designed. In fact the maximum possible value of M is 2, that is, $B(4, 3) = 2$. ⬚

5.5.2 Maximal Codes and the Gilbert Bound

The code $\mathcal{K}_3 = \{000, 011\}$ has $d(\mathcal{K}_3) = 2$ and contains $M = 2$ code words of length $N = 3$ implying $L = 1$ and a code rate $R = 1/3$. The code can be made more efficient (i.e., higher code rate) by augmenting it to the code $\mathcal{K}_3 = \{000, 011, 101, 110\}$ which is also $d(\mathcal{K}_3) = 2$ but contains $M = 4$ code words of length $N = 3$ implying $L = 2$ and a code rate $R = 2/3$. The second code is more efficient than the first code without sacrificing the minimum distance. We can define such codes as follows:

DEFINITION 5.10 Maximal Codes
A code $\mathcal{K}_N$ of length N and minimum distance $d(\mathcal{K}_N)$ with M code words is said to be maximal *if it is not part of, or cannot be augmented to, another code $\mathcal{K}_N$ of length N, minimum distance $d(\mathcal{K}_N)$ but with $M + 1$ code words. It can be shown that a code is maximal if and only if for all words $\mathbf{b}$ of length N there is a code word $\mathbf{a}_i$ such that $d(\mathbf{b}, \mathbf{a}_i) < d(\mathcal{K}_N)$.*

Thus maximal codes are more efficient in that for the same $d(\mathcal{K}_N)$ they provide the maximum code rate possible.

EXAMPLE 5.12

The code $\mathcal{K}_3 = \{000, 011\}$ with $d(\mathcal{K}_3) = 2$ mentioned above is not maximal since for word $\mathbf{b} = 101$, $d(\mathbf{a}_1, \mathbf{b}) = d(000, 101) = 2$ and $d(\mathbf{a}_1, \mathbf{b}) = d(011, 101) = 2$ and thus there is no code word such that $d(\mathbf{b}, \mathbf{a}_i) < 2$. ⬚

If code $\mathcal{K}_N$ is a maximal code then it satisfies the *Gilbert bound* first proposed by Gilbert [2].

THEOREM 5.2 Gilbert Bound

If a block code $\mathcal{K}_N$ of length N is a maximal code with minimum distance $d(\mathcal{K}_N)$ then the number of code words, M, must satisfy the following inequality:

$$M \geq \frac{2^N}{\sum_{i=0}^{d(\mathcal{K}_N)-1} \binom{N}{i}} \tag{5.22}$$

which provides a lower bound (the Gilbert bound) on the number of code words.

PROOF   If code $\mathcal{K}_N$ is a maximal code then each word of length N must be of distance $d(\mathcal{K}_N) - 1$ or less from at least one code word. The number of words within a Hamming distance of $d(\mathcal{K}_N) - 1$ from a code word $\mathbf{a}_j$ is given by Equation 5.19:

$$V(N, d(\mathcal{K}_N) - 1) = \sum_{i=0}^{d(\mathcal{K}_N)-1} \binom{N}{i} \tag{5.23}$$

Consider all M code words and the set of words that are within a Hamming distance of $d(\mathcal{K}_N) - 1$ from each code word. If a code is maximal then to ensure that all possible distinct sequences of length N, 2^N, are guaranteed to reside within a distance $d(\mathcal{K}_N)$ from at least one code word we must have that 2^N be no greater than the total number of words, $M.V(N, d(\mathcal{K}_N) - 1)$, that are within a distance $d(\mathcal{K}_N)$ of at least one code word, that is:

$$M \sum_{i=0}^{d(\mathcal{K}_N)-1} \binom{N}{i} \geq 2^N \tag{5.24}$$

which proves the Gilbert bound. This is shown in Figure 5.6 where it is evident that to ensure that each word of length N falls within one of the hyperspheres of radius $d(\mathcal{K}_N) - 1$ surrounding a code word the space of all words of length N must be no greater than the sum of the volumes of all the hyperspheres.

The condition for equality with the Gilbert bound occurs when the hyperspheres of radius $d(\mathcal{K}_N) - 1$ are filled uniquely with words of length N. That is, each word of length N will be of distance less than $d(\mathcal{K}_N)$ from precisely one code word.   ▯

EXAMPLE 5.13

Consider the code $\mathcal{K}_3 = \{000, 011\}$ with $d(\mathcal{K}_3) = 2$ just discussed. The Gilbert

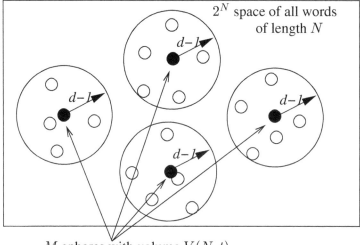

FIGURE 5.6
Proof of Gilbert bound for a code with minimum distance d. Solid dots represent code words; open dots represent all other words of length N.

bound is satisfied since:

$$M = 2 < \frac{2^N}{\sum_{i=0}^{d(\mathcal{K}_N)-1} \binom{N}{i}} = \frac{8}{4} = 2$$

However we know that this code is not maximal. Hence the Gilbert bound is a necessary but not sufficient condition for maximal codes. □

5.5.3 Redundancy Requirements for t-bit Error Correction

Instead of asking how many errors a code can correct, we can ask how much redundancy needs to be present in a code in order for it to be able to correct a given number of errors. Let $r = N - L$ be the number of check bits (the redundancy) that are added by the channel coder. The Hamming bound from Equation 5.18 provides the following lower bound on the redundancy needed for a t-bit error correcting code:

$$r \geq \log_2(V(N, t)) \text{ or } 2^r \geq V(N, t) \tag{5.25}$$

where $V(N, t)$ is given by Equation 5.19. The Gilbert bound from Equation 5.22 provides an upper bound on the redundancy if the code is to be maximal. For a t-bit error correcting code the minimum distance must be at least $d(\mathcal{K}_N) = 2t + 1$, and thus:

$$r \leq \log_2(V(N, 2t)) \text{ or } 2^r \leq V(N, 2t) \tag{5.26}$$

EXAMPLE 5.14

Suppose we want to construct a code with $L = 12$ which is capable of 3-bit error correction. How many check bits are needed for a maximal code? Define $r = N - 12$ and $t = 3$. A computation of 2^r together with $V(N, 3)$ and $V(N, 6)$ for a range of values of N is shown by the table below:

N	$r = N - 12$	$V(N, 3)$	2^r	$V(N, 6)$
22	10	1794	1024	100302
23	11	2048	2048	133102
24	12	2325	4096	174465
25	13	2626	8192	226109
26	14	2952	16384	289992
27	15	3304	32768	368344
28	16	3683	65536	463688
29	17	4090	131072	578863
30	18	4526	262144	717056
31	19	4992	524288	881816
32	20	5489	1048576	1077097
33	21	6018	2097172	1307274

We see that we need between 11 and 20 check bits in order to design a maximal code to correct three errors. Since both the Hamming and Gilbert bounds are necessary but not sufficient then all we can say is that a code may exist and if it does it will have between 11 and 20 check bits. □

5.5.4   Perfect Codes for t-bit Error Correction

A block code with L-bit length messages will allocate $M = 2^L$ words of length N-bits (where $N > L$) as the code words. Assume the code is designed for t-bit error correction. The space of all words of length N, 2^N, will fall into one of the following two mutually exclusive sets:

Set C The code words themselves and all the words formed by corrupting each code word by all combinations of up to t-bit errors. That is, all the words falling uniquely within one of the hyperspheres of radius t centred at each code word. Since the code can correct t-bit errors these words are guaranteed to be of distance t or less from precisely one of the code words and the errors can be corrected.

Set $\overline{C}$ Non-code words of distance greater than t from one or more code words. That is, all the words which fall outside the hyperspheres of radius t surrounding each code word. Since such words may be equidistant to more than one code word errors may only be detected.

If all the words of length N belong to **Set C** then the code is a *perfect code*.

DEFINITION 5.11 Perfect Codes   *Perfect codes for t-bit error correction possess the following properties:*

1. *All received words are either a code word or a code word with up to t-bit errors which, by definition, can be corrected. Thus there is no need to handle detectable errors.*

2. *The code is maximal.*

3. *The code provides maximum code rate (and minimum redundancy) for t-bit error correction.*

4. *The minimum distance is $2t + 1$.*

A code is defined as a perfect code for t-bit errors provided that for every word $\mathbf{b}$ of length N, there exists precisely one code word of distance t or less from $\mathbf{b}$. This implies equality with the Hamming bound. That is, a perfect code satisfies:

$$M = \frac{2^N}{\sum_{i=0}^{t} \binom{N}{i}} \Rightarrow 2^{N-L} = \sum_{i=0}^{t} \binom{N}{i} \tag{5.27}$$

EXAMPLE 5.15

Consider the $(N, L) = (6, 3)$ block code from Example 5.8:

Message ($L = 3$)	Code word ($N = 6$)
000	000000
001	001110
010	010101
011	011011
100	100011
101	101101
110	110110
111	111000

This is a code for 1-bit error correction since $d(\mathcal{K}_6) = 3$. However this is not a perfect code for 1-bit error correction since:

- $2^{N-L} = 2^3 = 8 \neq \sum_{i=0}^{1} \binom{6}{i} = \binom{6}{0} + \binom{6}{1} = 7$

- the received word 111111 is not at a distance of 1 from any of the code words and using the Hamming decoding rule it is at a distance of 2 from three candidate code words: 011011, 101101 and 110110. Thus there is a 2-bit error detected.

▯

EXAMPLE 5.16

Most codes will not be perfect. Perfect codes for 1-bit error correction form a family of codes called the Hamming codes, which are discussed in Section 7.7. An example of the $(N, L) = (7, 4)$ Hamming code is given by:

Message	Code word	Message	Code word
0000	0000000	1000	1000011
0001	0001111	1001	1001100
0010	0010110	1010	1010101
0011	0011001	1011	1011010
0100	0100101	1100	1100110
0101	0101010	1101	1101001
0110	0110011	1110	1110000
0111	0111100	1111	1111111

This is a code for 1-bit error correction since $d(\mathcal{K}_7) = 3$. This is also a perfect code for 1-bit error correction since:

- $2^{N-L} = 2^3 = 8 = \sum_{i=0}^{1} \binom{7}{i} = \binom{7}{0} + \binom{7}{1} = 8$

- any 7-bit received word will be of distance 1 from precisely one code word and hence 1-bit error correction will always be performed, even if there has been, say, a 2-bit error. Consider received word 0000011 as the code word 0000000 with a 2-bit error. It will be decoded incorrectly as the code word 1000011 with a 1-bit error. It should be noted that a 1-bit error is more likely than a 2-bit error and the Hamming decoding rule chooses the most likely code word.

▯

5.6   Error Probabilities

There are two important measures for a channel code:

1. The code rate. Codes with higher code rate are more desirable.

2. The probability of error. Codes with lower probability of error are more desirable. For error correcting codes the *block error probability* is important. For error detecting codes the *probability of undetected error* is important.

There is usually a trade-off between the code rate and probability of error. To achieve a lower probability of error it may be necessary to sacrifice code rate. In the next section Shannon's Fundamental Coding Theorem states what limits are imposed when designing channel codes. In this section we examine the different error probabilities and the tradeoffs with code rate.

5.6.1   Bit and Block Error Probabilities and Code Rate

The *bit error probability*, $P_{eb}(\mathcal{K}_n)$, is the probability of bit error between the message and the decoded message. For a BSC system without a channel coder we have $P_{eb}(\mathcal{K}_n) = q$. Otherwise for a channel coder with $L > 1$ the calculation of the bit error probability is tedious since all combinations of message blocks, and the corresponding bit errors, have to be considered.

The *block error probability*, $P_e(\mathcal{K}_n)$, is the probability of a decoding error by the channel decoder. That is, it is the probability that the decoder picks the wrong code word when applying the Hamming distance decoding rule. For the case of $L = 1$ the bit error and block error probabilities are the same; otherwise they are different. For $L > 1$ the calculation of the block error probability is straightforward if we know the minimum distance of the code.

EXAMPLE 5.17

Consider Code $\mathcal{K}_6$ from Example 5.8. Say message 000 is transmitted as code word 000000. Two bit errors occur in the last two bits and the received word is 000011. The channel decoder will then select the nearest code word which is 100011. Thus a block error has occurred. If this always occurs then $P_e(\mathcal{K}_n) = 1$. However the code word 100011 is decoded as message 100 which has only the first bit wrong and the remaining two bits correct when compared with the original message. If this always occurs then $P_b(\mathcal{K}_n) = 0.333$.   ⬚

The block error probability is used to assess and compare the performance of codes since it is easier to calculate and provides the worst-case performance.

RESULT 5.5 Block Error Probability for a t-bit Error Correcting Code

$$P_e(\mathcal{K}_n) = 1 - \sum_{i=0}^{t} \binom{n}{i} q^i (1-q)^{n-i} \qquad (5.28)$$

PROOF The probability of exactly t-bit errors in a word of length n is $\binom{n}{t} q^t (1-q)^{n-t}$. A t-bit error correcting code is able to handle up to t-bit errors. Thus the probability of no decoding error is $P_c(\mathcal{K}_n) = \sum_{i=0}^{t} \binom{n}{i} q^i (1-q)^{n-i}$ and the probability of error is $P_e(\mathcal{K}_n) = 1 - P_c(\mathcal{K}_n)$. ☐

For some codes there is a direct trade-off between the block error probability and the code rate as shown in the following example.

EXAMPLE 5.18

Assume a BSC with $q = 0.001$. Then for the no channel code case $P_b(\mathcal{K}_1) = 0.001 = 1 \times 10^{-3}$ but at least we have $R = 1$.

Consider using a $N = 3$ repetition code:

Message	Code word
0	000
1	111

Since $d(\mathcal{K}_3) = 3$ this is a 1-bit error correcting code and

$$P_e(\mathcal{K}_3) = 1 - \sum_{i=0}^{t} \binom{n}{i} q^i (1-q)^{n-i}$$

$$= 1 - \binom{3}{0}(0.999)^3 - \binom{3}{1}(0.999)^2(0.001)^1$$

$$= 3 \times 10^{-6}$$

Since $L = 1$ then $P_b(\mathcal{K}_3) = P_e(\mathcal{K}_3) = 3 \times 10^{-6}$. We see that the block error (or bit error) probability has decreased from 1×10^{-3} to 3×10^{-6}, but the code rate has also decreased from 1 to $\frac{1}{3}$.

Now consider using a $N = 5$ repetition code:

Message	Code word
0	00000
1	11111

Since $d(\mathcal{K}_5) = 5$ this is a 2-bit error correcting code and

$$P_e(\mathcal{K}_5) = 1 - \sum_{i=0}^{t} \binom{n}{i} q^i (1-q)^{n-i}$$

$$= 1 - \binom{5}{0} (0.999)^5 - \binom{5}{1} (0.999)^4 (0.001)^1 - \binom{5}{2} (0.999)^3 (0.001)^2$$

$$= 1 \times 10^{-8}$$

We now see that the block error (or bit error) probability has decreased from 1×10^{-3} to 3×10^{-6} to 1×10^{-8}, but the code rate has also decreased from 1 to $\frac{1}{3}$ to $\frac{1}{5}$, respectively. Thus there is a clear trade-off between error probability and code rate with repetition codes. $\Box$

5.6.2 Probability of Undetected Block Error

An undetected block error occurs if the received word, $\mathbf{b}$, is a different code word than was transmitted. Let $\{\mathbf{a}_i \in \mathbf{B}_M : i = 1, 2, \ldots M\}$ and assume that code word $\mathbf{a}_i$ is transmitted and $\mathbf{b} = \mathbf{a}_j$ is received where $j \neq i$, that is, the received word is also a code word, but different to the transmitted code word. The Hamming decoding rule will, of course, assume there has been no error in transmission and select $\mathbf{a}_j$ as the code word that was transmitted. This is a special case of a decoding error where there is no way the decoder can know that there is an error in transmission since a valid code word was received. Thus the error is undetected. For error detecting codes the probability of undetected block error is an important measure of the performance of the code. It should be noted that the probability of undetected block error is also relevant for error correcting codes, but such codes are more likely to experience decoding errors which would otherwise be detectable.

EXAMPLE 5.19

Consider the Hamming code of Example 5.16. If the code word 0000000 is transmitted and the word 0000011 is received the decoder will decode this as code word 1000011; thus the actual 2-bit error has been treated as a more likely 1-bit error. There is a decoding error, but this is not an undetected block error since the received word is not a code word and an error is present, be it 1-bit or 2-bit. Indeed if the Hamming code were to operate purely as an error detecting code it would be able to detect the 2-bit error (since $d(\mathcal{K}_7) = 3$, all 2-bit errors can be detected). However if the code word 0000000 is transmitted and the word 1000011 is received then since 1000011 is a code word the decoder will assume there has been no error in transmission. $\Box$

The probability of undetected block error is calculated as:

$$P_u(\mathcal{K}_n) = \sum_{i=1}^{M} P(\mathbf{a}_i) \sum_{\substack{j=1 \\ j \neq i}}^{M} P(\mathbf{a}_j | \mathbf{a}_i)$$

$$= \sum_{i=1}^{M} P(\mathbf{a}_i) \sum_{\substack{j=1 \\ j \neq i}}^{M} (1-q)^{n-d(\mathbf{a}_j, \mathbf{a}_i)} \, q^{d(\mathbf{a}_j, \mathbf{a}_i)}$$

$$= \frac{1}{M} \sum_{i=1}^{M} \sum_{\substack{j=1 \\ j \neq i}}^{M} (1-q)^{n-d(\mathbf{a}_j, \mathbf{a}_i)} \, q^{d(\mathbf{a}_j, \mathbf{a}_i)} \tag{5.29}$$

where it is assumed that code words are equally likely to occur, that is, $P(\mathbf{a}_i) = \frac{1}{M}$. The main complication is the double summation and the need to consider all pairs of code words and their Hamming distance, that is, having to enumerate the Hamming distance between all pairs of distinct code words. For some codes, in particular the binary linear codes discussed in Chapter 6, the distribution of the Hamming distance, $d(\mathbf{a}_j, \mathbf{a}_i)$, for given $\mathbf{a}_i$ is independent of $\mathbf{a}_i$, and this simplifies the calculation as follows:

RESULT 5.6 Probability of Undetected Block Error for Binary Linear Codes

Let d be the Hamming distance between the transmitted code word, $\mathbf{a}_i$, and the received code word, $\mathbf{a}_j$, where $j \neq i$. Let A_d be the number of choices of $\mathbf{a}_j$ which yield the same d, where it is assumed this is independent of the transmitted code word $\mathbf{a}_i$. Then the probability of undetected block error is:

$$P_u(\mathcal{K}_n) = \sum_{d=d(\mathcal{K}_n)}^{n} A_d \, (1-q)^{n-d} \, q^d \tag{5.30}$$

EXAMPLE 5.20

Consider the Hamming code of Example 5.16. The Hamming code is an example of binary linear code. The reader can verify that for any choice of $\mathbf{a}_i$ then:

- there are 7 choices of $\mathbf{a}_j$ such that $d(\mathbf{a}_i, \mathbf{a}_j) = 3$
- there are 7 choices of $\mathbf{a}_j$ such that $d(\mathbf{a}_i, \mathbf{a}_j) = 4$
- there is 1 choice of $\mathbf{a}_j$ such that $d(\mathbf{a}_i, \mathbf{a}_j) = 7$

- there are no choices of $\mathbf{a}_j$ for $d = 5$ and $d = 6$

Thus the probability of undetected block error is:

$$P_u(\mathcal{K}_n) = 7(1-q)^4 q^3 + 7(1-q)^3 q^4 + q^7$$

For comparison the block error probability of the Hamming code, from Equation 5.28, is:

$$P_e(\mathcal{K}_n) = 1 - \sum_{i=0}^{1} \binom{7}{i} q^i (1-q)^{7-i}$$
$$= 1 - (1-q)^7 - 7q(1-q)^6$$

Assume a BSC with $q = 0.001$; then if the Hamming code is used as a 1-bit error correcting code it will suffer a block error decoding with probability $P_e(\mathcal{K}_n) = 2 \times 10^{-5}$, or an incorrect decoding once every 50,000 code words. On the other hand if the Hamming code is used as a 2-bit error detecting code it will suffer from an undetected block error with probability $P_u(\mathcal{K}_n) = 7 \times 10^{-9}$, an undetected block error once every 14,000,000 or so code words. The fact the error detection is so much more robust than error correction explains why practical communication systems are ARQ when a feedback channel is physically possible. ▯

5.7 Shannon's Fundamental Coding Theorem

In the previous section we saw, especially for repetition codes, that there is a trade-off between the code rate, R, and the block error probability, $P_e(\mathcal{K}_n)$. This is generally true, but we would like a formal statement on the achievable code rates and block error probabilities. Such a statement is provided by Shannon's Fundamental Coding Theorem [7].

THEOREM 5.3 *Shannon's Fundamental Coding Theorem for BSCs*

Every BSC of capacity $C > 0$ can be encoded with an arbitrary reliability and with code rate, $R(\mathcal{K}_n) < C$, arbitrarily close to C for increasing code word lengths n. That is, there exist codes $\mathcal{K}_1, \mathcal{K}_2, \mathcal{K}_3, \ldots$ such that $P_e(\mathcal{K}_n)$ tends to zero and $R(\mathcal{K}_n)$ tends to C with increasing n:

$$\lim_{n \to \infty} P_e(\mathcal{K}_n) = 0, \qquad \lim_{n \to \infty} R(\mathcal{K}_n) = C \qquad (5.31)$$

The proof of Shannon's Fundamental Coding Theorem can be quite long and tedious. However it is instructive to examine how the theorem is proved (by concentrating on the structure of the proof) since this may lead to some useful insight into achieving the limits posed by the theorem (achieving error-free coding with code rate as close to channel capacity as possible).

PROOF   Let $\epsilon_1 > 0$ be an arbitrary small number and suppose we have to design a code $\mathcal{K}_n$ of length n such that $R(\mathcal{K}_n) = C - \epsilon_1$. To do this we need messages of length $L = n(C - \epsilon_1)$ since:

$$R(\mathcal{K}_n) = \frac{L}{N} = \frac{n(C - \epsilon_1)}{n} = C - \epsilon_1 \tag{5.32}$$

This means we need $M = 2^{n(C-\epsilon_1)}$ code words.

We prove the theorem by making use of random codes. Given any number n we can pick M out of the 2^n binary words in a random way and obtain the random code $\mathcal{K}_n$. If $M = 2^{n(C-\epsilon_1)}$ then we know that $R(\mathcal{K}_n) = C - \epsilon_1$ but $P_e(\mathcal{K}_n)$ is a random variable denoted by:

$$\tilde{P}_e(n) = E[P_e(\mathcal{K}_n)]$$

which is the expected value of $P_e(\mathcal{K}_n)$ for a fixed value of n, but a completely random choice of M code words. The main, and difficult part of the proof (see [1, 8]) is to show that:

$$\lim_{n \to \infty} \tilde{P}_e(n) = 0$$

We can intuitively see this by noting that a smaller value of $P_e(\mathcal{K}_n)$ results for codes with a larger $d(\mathcal{K}_n)$ and a larger minimum distance is possible if $M \ll 2^n$. Since for a BSC, $C \leq 1$, $\epsilon_1 > 0$ and $M = 2^{n(C-\epsilon_1)}$ then we see that only for sufficiently large n can we get $M \ll 2^n$ and thus a smaller $P_e(\mathcal{K}_n)$.   ∐

The surprising part of the proof is that we can achieve a small $P_e(\mathcal{K}_n)$ and large $R(\mathcal{K}_n)$ with a *random* choice of $\mathcal{K}_n$. Thus the theorem is only of theoretical importance since no practical coding scheme has yet been devised that realises the promises of Shannon's Fundamental Coding Theorem for low error, but high code rate codes. Another stumbling block is that a large n is needed. However this is a *sufficient*, not *necessary*, condition of the proof. It may be possible, with clever coding schemes, to approach the limits of the theorem but without resorting to excessively large values of n. Indeed the family of *turbo codes* discovered in 1993 and discussed in Chapter 9 nearly achieves the limits of the theorem without being overly complex or requiring large values of n.

The converse of the theorem also exists:

THEOREM 5.4 *Converse of Shannon's Fundamental Coding Theorem for BSCs*

For every BSC of capacity C wherever codes $\mathcal{K}_n$ of length n have code rates $R(\mathcal{K}_n) > C$ then the codes tend to be unreliable, that is:

$$\lim_{n \to \infty} P_e(\mathcal{K}_n) = 1 \qquad (5.33)$$

EXAMPLE 5.21

Consider designing a channel coder for BSC with $q = 0.001$. The channel capacity of the BSC is:

$$C_{BSC} = 1 - \left((1 - q) \log \frac{1}{(1 - q)} + q \log \frac{1}{q} \right) = 0.9886$$

By Shannon's Fundamental Coding Theorem we should be able to design a code with $M = 2^L$ code words and code word length N such that if the code rate $R = \frac{L}{N}$ satisfies $R < C$ then for sufficiently large N we can achieve an arbitrary small block error probability $P_e(\mathcal{K}_n)$ and make R as close to C as we like. However neither the Theorem nor its proof provide any practical mechanism or coding scheme for constructing such codes.

Conversely if we choose $M = 2^L$ and N such that $R > C$ then we can never find a code with arbitrarily small block error probability and indeed the block error probability will tend to 1 (completely unreliable codes). However this does not mean that we can't find a specific code with $R > C$ which will have a reasonably small block error probability. $\quad\Box$

5.8 Exercises

1. The input to the channel encoder arrives at 4 bits per second (bps). The binary channel can transmit at 5 bps. In the following for different values of the block encoder length, L, determine the length of the code word and where and when any dummy bits are needed and how error protection can be encoded:

    (a) $L = 2$
    (b) $L = 3$
    (c) $L = 4$

(d) $L = 5$

2. Consider the following binary channel:

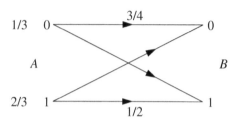

(a) Find the *maximum likelihood decoding* rule and consequent error probability.

(b) Find the *minimum error decoding* rule and consequent error probability.

(c) How useful is the above channel and source with a minimum error decoding rule?

3. A BSC has the following channel matrix and input probability distribution:

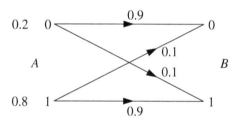

(a) Find the maximum likelihood decoding rule and consequent error probability.

(b) Find the minimum error decoding rule and consequent error probability.

(c) For practical BSCs $p \gg q$ and the inputs can be assumed to be close
to equiprobable (i.e., $P(0) \approx P(1)$). Show why this implies that the
maximum likelihood decoding rule is indeed minimum error.

*4. Consider the following BEC:

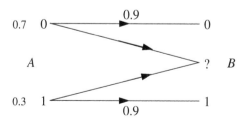

Design a decoding rule for each pair of outputs in terms of the corresponding pair of inputs (e.g., $D(?0) = 00$ means a ? followed by a 0 at the output decides in favour of a 0 followed by a 0 being transmitted at the input of the channel):

(a) Design the decoding rule assuming maximum likelihood decoding.

(b) Design the decoding rule assuming minimum error decoding.

Assume inputs are produced independently of one another (e.g., $P(00) = P(0)P(0) = 0.49$)

5. Consider the following $L = 3$ and $N = 6$ channel code:

Message	Code word
0 0 0	0 0 0 0 0 0
0 0 1	0 0 1 1 1 0
0 1 0	0 1 0 0 1 1
0 1 1	0 1 1 1 0 1
1 0 0	1 0 0 1 1 1
1 0 1	1 0 1 0 0 1
1 1 0	1 1 0 1 0 0
1 1 1	1 1 1 0 1 0

(a) Perform Hamming channel decoding for the following received words, indicating the Hamming distance between the received word and decoded word, and any error detection/correction that arises:

   i. 0 1 1 1 0 1
   ii. 1 1 1 0 0 1
   iii. 0 1 0 0 1 0
   iv. 0 0 1 1 1 0
   v. 1 0 0 0 1 0

(b) Calculate the minimum distance of the code and hence comment on the error detection/correction properties.

6. What is the minimum code word length you can use for designing a single-bit error correcting code for a binary stream that is block encoded on blocks of length 4?

7. Repeat Qu. 6 for the case of double-bit error correction.

8. Prove the following result for binary codes:

$$B(N, d) \leq 2B(N - 1, d)$$

by considering a code of length N with $M = B(N, d)$ code words and then extracting a code of length $N - 1$ from this code.

9. Prove the following result for binary codes:

$$B(N, 2t + 1) = B(N + 1, 2t + 2)$$

by considering a code of length $N + 1$ with $M = B(N + 1, d)$ code words and then removing one bit from each code word in such a way that a new code of length N with the same number of code words but distance $d - 1$ is created.

10. Consider a channel coding with $L = 16$ and $N = 21$. What is the best error detection and error correction that can be designed?

11. A channel coding system uses 4-byte code words and all triple-bit or less errors have to be detected. What is the L for maximum code rate?

*12. A channel coding system can only process message blocks and code blocks which occupy an integer number of bytes (i.e., 8, 16, 32, etc.). Specify a L and N for maximum code rate and best error detection / correction. What is the error detection / correction capabilities of your design?

13. Is it possible to design a *maximal code* for $t = 3$ bit error correction if code words are 16 bits in length and only 4 code words are used? How about if 5 code words are used?

*14. Write a program that attempts to generate up to M words of length N such that the minimum distance between the M words is $d(\mathcal{K}_N)$. One way to do this is as follows:

(a) Start with $m = 0$.

(b) Randomly generate a word of length N, $\mathbf{r}_N$.

(c) Calculate the Hamming distance $d(\mathbf{r}_N, \mathbf{a}_i)$ between the generated word, $\mathbf{r}_N$, and each of the m words found so far, $\{\mathbf{a}_i : i = 1, 2, \ldots, m\}$.

(d) If $d(\mathbf{r}_N, \mathbf{a}_i) < d(\mathcal{K}_N)$ for any i, reject $\mathbf{r}_N$; otherwise define $\mathbf{a}_{m+1} = \mathbf{r}_N$ and let $m = m + 1$.

(e) Repeat steps (b) to (d) and terminate when $m = M$.

Can a code with $N = 8$, $M = 25$ and $d(\mathcal{K}_N) = 3$ exist? Use your program to answer this question!

15. You are required to design a single-bit error correcting code for messages of length $L = 11$. What are the minimum number of check bits that you need? Is this is a maximal code? Is this a *perfect code*?

16. Prove that a perfect code for t-bit error correction is also a maximal code.

*17. Prove that a perfect code for t-bit error correction has $d(\mathcal{K}_n) = 2t + 1$.

18. Consider the block code of length n:

(a) If the *repetition code* is used, derive the expression for the block error probability $P_e(\mathcal{K}_n)$ as a function of n.

(b) A BSC has error probability $q = 0.1$. Find the smallest length of a repetition code such that $P_{eb}(\mathcal{K}_n) < 10^{-2}/2$. What is the code rate?

*19. Calculate the *block error probability*, $P_e(\mathcal{K}_6)$, for code $\mathcal{K}_6$ from Example 5.8 for a BSC with $q = 0.001$. Now provide a reasonable estimate for the *bit-error probability*, $P_{eb}(\mathcal{K}_6)$, and compare this value with $P_e(\mathcal{K}_6)$.

20. Consider the following design parameters for different channel codes:

	Code $\mathcal{K}_A$	Code $\mathcal{K}_B$	Code $\mathcal{K}_C$	Code $\mathcal{K}_D$
$d(\mathcal{K})$	3	2	4	5
L	26	11	11	21
N	31	12	16	31

(a) Indicate whether the code implies an FEC system, an ARQ system or both. In each case specify the error detection/correction properties of the system.

(b) Derive a simple worst-case expression for the block error probability, $P_e(\mathcal{K}_n)$, as a function of the BSC channel error probability, q, for both types of error-control systems.

(c) Will any of the FEC systems fail to operate (i.e., not be able to correct errors, only detect them)?

21. A channel encoder receives 3 bits per second and is connected to a BSC. If $q = 0.25$, how many bits per second can we send through the channel for the-oretical error-free transmission according to Shannon's Fundamental Coding Theorem?

22. Consider three single-bit error correcting codes of length $n = 7, 12$ and 16, respectively. Calculate the block-error probability $P_e(\mathcal{K}_n)$ for each code as-suming $q = 0.01$ and a code rate of $R = 0.5$. What happens to $P_e(\mathcal{K}_n)$ for increasing length n? Is this contrary to Shannon's Fundamental Coding Theorem?

*23. Let us explore Shannon's Fundamental Coding Theorem further. Consider error correcting codes of length $n = 6, 12, 24, 48, 96$ and a code rate of $R = 1/2$. What is the best t-bit error correction that can be achieved for each n and what is the corresponding block error probability $P_e(\mathcal{K}_n)$ for $q = 0.01$? Is this in line with the expectations of Shannon's Fundamental Coding Theorem? Explain! Repeat your analysis for different values of R and q.

24. A BSC has a probability of error of $q = 0.003$ and can transmit no more than 3 bits per second. Can an error-free coding be devised if:

(a) 165 bits per minute enter the channel encoder?

(b) 170 bits per minute enter the channel encoder?

(c) 175 bits per minute enter the channel encoder?

25. A BSC transmits bits with error probability, q. Letters of the alphabet arrive at 2,100 letters per second. The letters are encoded with a binary code of average length 4.5 bits per letter. The BSC is transmitting at 10,000 bps. Can an arbitrarily error-free coding be theoretically devised if:

(a) $q = 0.1$

(b) $q = 0.01$

(c) $q = 0.005$

(d) $q = 0.001$

5.9   References

[1] T.M. Cover, and J.A. Thomas, *Elements of Information Theory*, John Wiley & Sons, New York, 1991.

[2] E.N. Gilbert, A comparison of signalling alphabets, *Bell System Tech. J.*, 31, 504-522, 1952.

[3] R.W. Hamming, Error detecting and error correcting codes, *Bell System Tech. J.*, 29, 147-160, 1950.

[4] S. Haykin, *Communication Systems*, John Wiley & Sons, New York, 4th ed., 2001.

[5] J. C. A. van der Lubbe, *Information Theory*, Cambridge University Press, London, 1997.

[6] S. Roman, *Coding and Information Theory*, Springer-Verlag, New York, 1992.

[7] C.E. Shannon, A mathematical theory of communication, *Bell System Tech. J.*, 28, 379-423, 623-656, 1948.

[8] R. Wells, *Applied Coding and Information Theory for Engineers*, Prentice-Hall, New York, 1999.

Chapter 6

Error-Correcting Codes

6.1 Introduction

We have seen that the existence of redundancy makes it possible to compress data and so reduce the amount of space or time taken to transmit or store it. A consequence of removing redundancy is that the data become more susceptible to noise.

Conversely, it is possible to increase the redundancy of a set of data by adding to it. The added redundancy can be used to detect when the data have been corrupted by noise and even to undo the corruption. This idea is the the basis of the theory of *error-correcting codes*.

EXAMPLE 6.1

The simplest example of an error-correcting code is the *parity check*. (A special case of this was discussed in Example 5.5 of Chapter 5.) Suppose that we have a sequence of binary digits that is to be transmitted over a noisy channel that may have the effect of changing some of the 1's to 0's and some of the 0's to 1's. We can implement a parity check by breaking the sequence into blocks of, say, seven bits and adding a redundant bit to make a block of eight bits.

The added bit is a parity bit. It is 0 if there is an even number of 1s in the seven-bit block and 1 if there is an odd number of 1s in the seven-bit block. Alternatively, it is the result of XORing all the seven bits together.

When the eight bits are transmitted over a noisy channel, there may be one or more differences between the block that was transmitted and the block that is received. Suppose, first that the parity bit is different. It will then not be consistent with the other seven bits. If the parity bit is unchanged, but one of the other seven bits is changed, the parity bit will once again be inconsistent with the other seven bits. Checking to see if the parity bit is consistent with the other seven bits enables us to check whether there has been a single error in the transmission of the eight bit block. It does not tell us where the error has occurred, so we cannot correct it.

The parity check also cannot tell us whether more than one error has occurred. If two bits are changed, the parity bit will still be consistent with the other seven bits. The same is true if four or six bits are changed. If three bits are changed, the parity bit will not be consistent with the other seven bits, but this is no different from the situation where only one bit is changed. The same is true if five or seven bits are changed.

The parity check can be carried out easily by XORing all eight bits together. If the result is 0, then either no error has occurred, or two, four or six errors have occurred. If the result is 1, then one, three, five or seven errors have occurred.

The parity check is a simple error-detection technique with many limitations. More effective error-correcting capabilities require more sophisticated approaches.   ▢

EXAMPLE 6.2

Another simple way of introducing redundancy (discussed in Chapter 5, Example 5.6) is by repetition. If we have a sequence of bits, we can send each bit three times, so that the sequence 01100011 is transmitted as 000111111000000000111111. At the receiving end, the sequence is broken up into blocks of three bits. If there is an error in that block, one of the bits will be different, for example 001 instead of 000. In this case, we take the most frequently occurring bit as the correct one. This scheme allows us to detect and correct one error. If two errors occur within a block, they will be detected, but the correction procedure will give the wrong answer as the majority of bits will be erroneous. If all three bits are changed, this will not be detected.

If it is likely that errors will occur in bursts, this scheme can be modified by breaking the original sequence into blocks and transmitting the blocks three times each. The error correction procedure will then be applied to the corresponding bits in each of the three copies of the block at the receiving end.

This is an inefficient way of adding redundancy, as it triples the amount of data but only makes it possible to correct single errors.   ▢

The theory of error-correcting codes uses many concepts and results from the mathematical subject of *abstract algebra*. In particular, it uses the notions of *rings*, *fields* and *linear spaces*. The following sections give an introduction to these notions. A rigorous treatment of this material can be found in any textbook on abstract algebra, such as [3].

6.2 Groups

A group is a collection of objects that can be combined in pairs to produce another object of the collection according to certain rules that require that there be an object that does not change anything it combines with, and objects that undo the changes resulting from combining with other objects. The formal definition is as follows.

DEFINITION 6.1 Group   *A group, $(G, *)$, is a pair consisting of a set G and an operation $*$ on that set, that is, a function from the Cartesian product $G \times G$ to G, with the result of operating on a and b denoted by $a * b$, which satisfies*

1. associativity: $a * (b * c) = (a * b) * c$ *for all $a, b, c \in G$;*

2. existence of identity: *there exists $e \in G$ such that $e * a = a$ and $a * e = a$ for all $a \in G$;*

3. existence of inverses: *for each $a \in G$ there exists $a^{-1} \in G$ such that $a * a^{-1} = e$ and $a^{-1} * a = e$.*

If the operation also satisfies the condition that $a * b = b * a$ for all $a, b \in G$ (*commutativity*), the group is called a *commutative group* or an *Abelian group*. It is common to denote the operation of an Abelian group by $+$, its identity element by 0 and the inverse of $a \in G$ by $-a$.

EXAMPLE 6.3

The simplest group has just two elements. It is Abelian, so we can denote the elements by 0 and g and the operation by $+$. From the definition, we must have $0 + 0 = 0$, $0 + g = g$, and $g + 0 = g$. g must be its own inverse, so $g + g = 0$. We can describe the operation by the following table where the first operand is listed in the column on the left, the second operand is listed in the row at the top and the results of the operation appear in the body of the table.

$$
\begin{array}{c|cc}
+ & 0 & g \\
\hline
0 & 0 & g \\
g & g & 0
\end{array}
$$

This table will be symmetric about the main diagonal if and only if the group is Abelian. □

When we use an operation table to define a group operation, each element of the group will appear exactly once in each row and each column of the body of the table.

DEFINITION 6.2 Order of a Group   *The order of a group is the number of elements in it.*

EXAMPLE 6.4

It is easy to show that there is only one group of order 3. Let us denote the elements of the group by $0, 1, 2$ and the operation by $+$.

If we let 0 denote the identity element, we can start building the operation table as follows:

$$
\begin{array}{c|ccc}
+ & 0 & 1 & 2 \\
\hline
0 & 0 & 1 & 2 \\
1 & 1 & & \\
2 & 2 & &
\end{array}
$$

Now suppose that $1 + 2 = 1$. This would make 1 appear twice in the second row of the table. This breaks the rule that each element appears exactly once in each row of the body of the table, so we cannot have $1 + 2 = 1$. If $1 + 2 = 2$, 2 would appear twice in the third column of the table, so we cannot have this either. This leaves $1 + 2 = 0$, which does not break any rules. To complete the second row of the table we have to have $1 + 1 = 2$, and the table now looks like this:

$$
\begin{array}{c|ccc}
+ & 0 & 1 & 2 \\
\hline
0 & 0 & 1 & 2 \\
1 & 1 & 2 & 0 \\
2 & 2 & &
\end{array}
$$

We can now fill in the second and third columns with the missing elements to complete the table:

$$
\begin{array}{c|ccc}
+ & 0 & 1 & 2 \\
\hline
0 & 0 & 1 & 2 \\
1 & 1 & 2 & 0 \\
2 & 2 & 0 & 1
\end{array}
$$

This is the operation table for addition modulo 3.   ▯

In the example above, it is claimed that there is "only one group of order 3." This is true in the sense that any set of three elements with an operation that satisfies the group axioms must have the same structure as the group

$$
\begin{array}{c|ccc}
+ & 0 & 1 & 2 \\
\hline
0 & 0 & 1 & 2 \\
1 & 1 & 2 & 0 \\
2 & 2 & 0 & 1
\end{array}
$$

regardless of the names of the elements. We regard

$$
\begin{array}{c|ccc}
+ & \alpha & \beta & \gamma \\
\hline
\alpha & \alpha & \beta & \gamma \\
\beta & \beta & \gamma & \alpha \\
\gamma & \gamma & \alpha & \beta
\end{array}
$$

as the same group because, if we replace α by 0, β by 1 and γ by 2 everywhere in the table above, we will finish up with the previous table.

To make these ideas precise, we need to consider functions that preserve the structure of the group in the following way.

DEFINITION 6.3 Group Homomorphism   *Let $(G, *)$ and $(H, \circ)$ be two groups and let $h : G \rightarrow H$ be a function defined on G with values in H. h is a* group homomorphism *if it preserves the structure of G, that is, if*

$$
h(a * b) = h(a) \circ h(b) \tag{6.1}
$$

for all $a, b \in G$.

DEFINITION 6.4 Group Isomorphism   *A group homomorphism that is both one-one and onto is a* group isomorphism.

Two groups that are isomorphic have the same structure and the same number of elements.

If $h : G \rightarrow H$ is a homomorphism, the image of the identity of G is the identity of H.

EXAMPLE 6.5

There are two groups of order 4. We will denote the elements of the first by $0, 1, 2, 3$ and use $+$ for the operation. The operation table is:

$$
\begin{array}{c|cccc}
+ & 0 & 1 & 2 & 3 \\
\hline
0 & 0 & 1 & 2 & 3 \\
1 & 1 & 2 & 3 & 0 \\
2 & 2 & 3 & 0 & 1 \\
3 & 3 & 0 & 1 & 2
\end{array}
$$

For the other, we will use e, a, b, c for the elements and $*$ for the operation. The operation table is

$$
\begin{array}{c|cccc}
* & e & a & b & c \\
\hline
e & e & a & b & c \\
a & a & e & c & b \\
b & b & c & e & a \\
c & c & b & a & e
\end{array}
$$

These groups obviously do not have the same structure. In the first group $1 + 1 = 2$ and $3 + 3 = 2$, but in the second group combining any element with itself under $*$ gives the identity, e.   ☐

EXAMPLE 6.6

There is one group of order 5. Its operation table is

$$
\begin{array}{c|ccccc}
+ & 0 & 1 & 2 & 3 & 4 \\
\hline
0 & 0 & 1 & 2 & 3 & 4 \\
1 & 1 & 2 & 3 & 4 & 0 \\
2 & 2 & 3 & 4 & 0 & 1 \\
3 & 3 & 4 & 0 & 1 & 2 \\
4 & 4 & 0 & 1 & 2 & 3
\end{array}
$$

☐

EXAMPLE 6.7

There are two groups of order 6. The operation table of the first is

$$
\begin{array}{c|cccccc}
+ & 0 & 1 & 2 & 3 & 4 & 5 \\
\hline
0 & 0 & 1 & 2 & 3 & 4 & 5 \\
1 & 1 & 2 & 3 & 4 & 5 & 0 \\
2 & 2 & 3 & 4 & 5 & 0 & 1 \\
3 & 3 & 4 & 5 & 0 & 1 & 2 \\
4 & 4 & 5 & 0 & 1 & 2 & 3 \\
5 & 5 & 0 & 1 & 2 & 3 & 4
\end{array}
$$

For the second, we will use E, F, G, H, I, J for the elements and $*$ for the operation. The operation table is

*	E	F	G	H	I	J
E	E	F	G	H	I	J
F	F	E	I	J	G	H
G	G	H	E	F	J	I
H	H	G	J	I	E	F
I	I	J	F	E	H	G
J	J	I	H	G	F	E

The table is not symmetric, so this group is not Abelian. This is the smallest example of a non-commutative group. □

For any positive integer p, there is at least one group of order p. There may be more than one group of order p.

DEFINITION 6.5 The Cyclic Groups   *For each positive integer p, there is a group called the* cyclic group of order p, *with set of elements*

$$\mathbb{Z}_p = \{0, 1, \ldots, (p-1)\}$$

and operation $\oplus$ defined by

$$i \oplus j = i + j \qquad (6.2)$$

if $i + j < p$, where $+$ denotes the usual operation of addition of integers, and

$$i \oplus j = i + j - p \qquad (6.3)$$

if $i + j > p$, where $-$ denotes the usual operation of subtraction of integers. The operation in the cyclic group is addition modulo p. We shall use the sign $+$ instead of $\oplus$ to denote this operation in what follows and refer to "the cyclic group $(\mathbb{Z}_p, +)$," or simply "the cyclic group $\mathbb{Z}_p$."

There are also plenty of examples of infinite groups.

EXAMPLE 6.8

The set of integers, $\mathbb{Z} = \{\ldots - 3, -2, -1, 0, 1, 2, 3, \ldots\}$, with the operation of addition is a group. 0 is the identity and $-k$ is the inverse of k. This is an Abelian group. □

EXAMPLE 6.9

The set of real numbers $\mathbb{R}$ with the operation of addition is another Abelian group. Again, 0 is the identity and $-r$ is the inverse of $r \in \mathbb{R}$. □

EXAMPLE 6.10

The set of positive real numbers $\mathbb{R}^+$ with the operation of multiplication is also an Abelian group. This time, the identity is 1 and the inverse of $r \in \mathbb{R}^+$ is $1/r$. ⬜

EXAMPLE 6.11

The set of $n \times n$ real matrices with the operation of matrix addition is a group. The zero matrix is the identity and the inverse of a matrix M is the matrix $-M$, consisting of the same elements as M but with their signs reversed. ⬜

EXAMPLE 6.12

The set of non-singular $n \times n$ real matrices with the operation of matrix multiplication is a group for any positive integer n. The identity matrix is the group identity and the inverse of a matrix is its group inverse. These groups are not commutative. ⬜

Groups are not used directly in the construction of error-correcting codes. They form the basis for more complex algebraic structures which are used.

6.3 Rings and Fields

Having two operations that interact with each other makes things more interesting.

DEFINITION 6.6 Ring   A ring *is a triple* $(R, +, \times)$ *consisting of a set R, and two operations + and ×, referred to as* addition *and* multiplication, *respectively, which satisfy the following conditions:*

1. associativity of +: $a + (b + c) = (a + b) + c$ *for all* $a, b, c \in R$;

2. commutativity of +: $a + b = b + a$ *for all* $a, b \in R$;

3. existence of additive identity: *there exists* $0 \in R$ *such that* $0 + a = a$ *and* $a + 0 = a$ *for all* $a \in R$;

4. existence of additive inverses: *for each* $a \in R$ *there exists* $-a \in R$ *such that* $a + (-a) = 0$ *and* $(-a) + a = 0$;

5. associativity of ×: $a \times (b \times c) = (a \times b) \times c$ *for all* $a, b, c \in R$;

6. distributivity of × over +: $a \times (b + c) = (a \times b) + (a \times c)$ *for all* $a, b, c \in R$.

The additive part of a ring, $(R, +)$, is an Abelian group.

In any ring $(R, +, \times)$, $0 \times r = 0$ and $r \times 0 = 0$ for all $r \in R$.

In many cases, it is useful to consider rings with additional properties.

DEFINITION 6.7 Commutative Ring   *A ring $(R, +, \times)$ is a commutative ring if $\times$ is a commutative operation, that is, $a \times b = b \times a$ for all $a, b \in R$.*

DEFINITION 6.8 Ring with Unity   *A ring $(R, +, \times)$ is a ring with unity if it has an identity element of the multiplication operation, that is, if there exists $1 \in R$ such that $1 \times a = a$ and $a \times 1 = a$ for all $a \in R$.*

DEFINITION 6.9 Division Ring   *A ring $(R, +, \times)$ is a division ring if it is a ring with unity and every non-zero element has a multiplicative inverse, that is, for every $a \in R$, $a \neq 0$, there exists $a^{-1} \in R$ such that $a \times a^{-1} = 1$ and $a^{-1} \times a = 1$.*

DEFINITION 6.10 Field   *A commutative division ring in which $0 \neq 1$ is a* field.

EXAMPLE 6.13

There is one ring with two elements. The additive part of the ring is the cyclic group of order 2, with operation table

$$
\begin{array}{c|cc}
+ & 0 & 1 \\
\hline
0 & 0 & 1 \\
1 & 1 & 0
\end{array}
$$

and the multiplicative part has the operation table

$$
\begin{array}{c|cc}
\times & 0 & 1 \\
\hline
0 & 0 & 0 \\
1 & 0 & 1
\end{array}
$$

These tables represent the arithmetic operations of addition and multiplication modulo 2. The ring will be denoted by $(\mathbb{Z}_2, +, \times)$ or simply by $\mathbb{Z}_2$.

If we consider 0 and 1 to represent truth values, then these tables are the truth tables of the bit operations XOR and AND, respectively.

$\mathbb{Z}_2$ is a commutative ring, a ring with unity and a division ring, and hence a field. It is the smallest field and the one of most relevance to the construction of error-correcting codes. ⬜

We can add a multiplicative structure to the cyclic groups to form rings.

DEFINITION 6.11 The Cyclic Rings   *For every positive integer p, there is a ring $(\mathbb{Z}_p, +, \times)$, called the* cyclic ring *of order p, with set of elements*

$$Z_p = \{0, 1, \ldots, (p-1)\}$$

and operations $+$ denoting addition modulo p, and $\times$ denoting multiplication modulo p.

The previous example described the cyclic ring of order 2. The following examples show the operation tables of larger cyclic rings.

EXAMPLE 6.14

The operation tables of the cyclic ring of order 3 are

$$
\begin{array}{c|ccc}
+ & 0 & 1 & 2 \\
\hline
0 & 0 & 1 & 2 \\
1 & 1 & 2 & 0 \\
2 & 2 & 0 & 1
\end{array}
\qquad
\begin{array}{c|ccc}
\times & 0 & 1 & 2 \\
\hline
0 & 0 & 0 & 0 \\
1 & 0 & 1 & 2 \\
2 & 0 & 2 & 1
\end{array}
$$

$\mathbb{Z}_3$ is a field. ⬜

EXAMPLE 6.15

The operation tables of the cyclic ring of order 4 are

$$
\begin{array}{c|cccc}
+ & 0 & 1 & 2 & 3 \\
\hline
0 & 0 & 1 & 2 & 3 \\
1 & 1 & 2 & 3 & 0 \\
2 & 2 & 3 & 0 & 1 \\
3 & 3 & 0 & 1 & 2
\end{array}
\qquad
\begin{array}{c|cccc}
\times & 0 & 1 & 2 & 3 \\
\hline
0 & 0 & 0 & 0 & 0 \\
1 & 0 & 1 & 2 & 3 \\
2 & 0 & 2 & 0 & 2 \\
3 & 0 & 3 & 2 & 1
\end{array}
$$

$\mathbb{Z}_4$ is not a field; 2 does not have a multiplicative inverse. ⬜

EXAMPLE 6.16

The operation tables of the cyclic ring of order 5 are

$$
\begin{array}{c|ccccc}
+ & 0 & 1 & 2 & 3 & 4 \\
\hline
0 & 0 & 1 & 2 & 3 & 4 \\
1 & 1 & 2 & 3 & 4 & 0 \\
2 & 2 & 3 & 4 & 0 & 1 \\
3 & 3 & 4 & 0 & 1 & 2 \\
4 & 4 & 0 & 1 & 2 & 3
\end{array}
\qquad
\begin{array}{c|ccccc}
\times & 0 & 1 & 2 & 3 & 4 \\
\hline
0 & 0 & 0 & 0 & 0 & 0 \\
1 & 0 & 1 & 2 & 3 & 4 \\
2 & 0 & 2 & 4 & 1 & 3 \\
3 & 0 & 3 & 1 & 4 & 2 \\
4 & 0 & 4 & 3 & 2 & 1
\end{array}
$$

$\mathbb{Z}_5$ is a field. $\square$

The cyclic ring of order p is a field if and only if p is a prime number.

The following are examples of infinite rings.

EXAMPLE 6.17

The integers $(\mathbb{Z}, +, \times)$ with the usual operations of addition and multiplication form a commutative ring with unity. It is not a division ring. $\square$

EXAMPLE 6.18

The real numbers $(\mathbb{R}, +, \times)$ with the usual operations of addition and multiplication form a field. $\square$

EXAMPLE 6.19

The complex numbers $(\mathbb{C}, +, \times)$ with the operations of complex addition and complex multiplication form a field. $\square$

EXAMPLE 6.20

The set of $n \times n$ real matrices with operations matrix operation and matrix multiplication forms a ring with unity. The zero matrix is the additive identity and the identity matrix is the multiplicative identity. It is not commutative, and is not a division ring as only non-singular matrices have multiplicative inverses. $\square$

Just as in the case of groups, we can talk about rings having the same structure and about functions which preserve structure.

DEFINITION 6.12 Ring Homomorphism   *Let $(R, +, \times)$ and $(S, \oplus, \otimes)$ be two rings and let $h : R \to S$ be a function defined on R with values in S. h is a ring homomorphism if it preserves the structure of R, that is, if*

$$h(a + b) = h(a) \oplus h(b) \tag{6.4}$$

and

$$h(a \times b) = h(a) \otimes h(b) \tag{6.5}$$

for all $a, b \in G$.

DEFINITION 6.13 Ring Isomorphism   *A ring homomorphism that is both one-one and onto is a* ring isomorphism.

As with groups, two rings are isomorphic if they have the same structure and the same number of elements.

6.4   Linear Spaces

Linear spaces consist of things which can be added, subtracted and re-scaled.

DEFINITION 6.14 Linear Space   *A linear space over a field F is a 6-tuple $(V, F, \oplus, +, \times, \circ)$, where $(V, \oplus)$ is a commutative group, $(F, +, \times)$ is a field and $\circ : F \times V \to V$ is a function that satisfies the following conditions:*

1. *$a \circ (b \circ v) = (a \times b) \circ v$ for all $a, b \in F$ and all $v \in V$;*

2. *$(a + b) \circ v = (a \circ v) \oplus (b \circ v)$ for all $a, b \in F$ and all $v \in V$;*

3. *$a \circ (v \oplus w) = (a \circ v) \oplus (a \circ w)$ for all $a \in F$ and all $v, w \in V$;*

4. *$1 \circ v = v$ for all $v \in V$.*

Most of the elementary examples of linear spaces are spaces of geometric vectors; so an alternative name for a linear space is a *vector space*. The elements of V are called *vectors* and the operation $\oplus$ is called *vector addition*, while the elements of f are called *scalars* and the function $\circ : F \times V \to V$ is called *multiplication by scalars*.

While we have been careful to use different symbols for the various operations in the definition above, we shall hereafter abuse notation by using the same symbol, $+$, for addition in V as well as for addition in F, and by omitting the symbol $\circ$ and denoting multiplication by a scalar by juxtaposition of a scalar and a vector. We shall also use the same symbol, 0, for both the group identity in V and the additive identity in F.

The following properties are simple consequences of the definition above:

1. $0v = 0$ for all $v \in V$, where the 0 on the left hand side belongs to F and the 0 on the right hand side belongs to V;

2. $a0 = 0$ for all $a \in F$, where the 0 on both sides belongs to V;

3. $(-a)v = a(-v) = -(av)$ for all $a \in F$ and $v \in V$.

The archetypal linear spaces are the n-dimensional real vector spaces.

EXAMPLE 6.21

Let $\mathbb{R}^n$ denote the n-fold Cartesian product $\mathbb{R} \times \mathbb{R} \times \ldots \times \mathbb{R}$, consisting of n-tuples $(x^1, x^2, \ldots, x^n)$ for $x^i \in \mathbb{R}$, $1 \le i \le n$. If we define vector addition by

$$(x^1, x^2, \ldots, x^n) + (y^1, y^2, \ldots, y^n) = (x^1 + y^1, x^2 + y^2, \ldots, x^n + y^n)$$

and multiplication by a scalar by

$$a(x^1, x^2, \ldots, x^n) = (ax^1, ax^2, \ldots, ax^n)$$

for $(x^1, x^2, \ldots, x^n)$ and $(y^1, y^2, \ldots, y^n) \in \mathbb{R}^n$ and $a \in \mathbb{R}$, $\mathbb{R}^n$ becomes a linear space over $\mathbb{R}$. ▯

EXAMPLE 6.22

For an example of a linear space that is not $\mathbb{R}^n$, consider the set $\mathcal{C}$ of continuous functions on the real line. For f and $g \in \mathcal{C}$, define the sum $f + g$ by

$$(f + g)(x) = f(x) + g(x),$$

for all $x \in \mathbb{R}$, and the product of $r \in \mathbb{R}$ and $f \in \mathcal{C}$ by

$$(rf)(x) = rf(x),$$

for all $x \in \mathbb{R}$, where the right hand side of the equation denotes the product of r and $f(x)$ as real numbers.

If f and g are continuous, so are $f + g$ and rf. The other properties that addition in $\mathcal{C}$ and multiplication by scalars must satisfy follow from the properties of the real numbers. ▯

Parts of linear spaces can be linear spaces in their own right.

DEFINITION 6.15 Linear Subspace   *A subset S of a linear space V over a field F is a* linear subspace *of V if for all v, w ∈ S and all a ∈ F, v + w ∈ S and av ∈ S.*

EXAMPLE 6.23

If $x = (x^1, x^2, \ldots, x^n) \in \mathbb{R}^n$ then the set

$$S_1 = \{rx : r \in \mathbb{R}\}$$

is a linear subspace of $\mathbb{R}^n$. Geometrically, this is a line through the origin.

If $y = (y^1, y^2, \ldots, y^n)$ also belongs to $\mathbb{R}^n$ then the set

$$S_2 = \{rx + sy : r, s \in \mathbb{R}\}$$

is a linear subspace of $\mathbb{R}^n$. This is a plane through the origin.   ▯

EXAMPLE 6.24

The set $\{f \in \mathcal{C} : f(0) = 0\}$ is a linear subspace of $\mathcal{C}$. To check this, note that if $f(0) = 0$ and $g(0) = 0$, then $(f + g)(0) = f(0) + g(0) = 0$ and if $a \in \mathbb{R}$, then $(rf(0)) = rf(0) = 0$.   ▯

It is quite easy to create subspaces of a linear space.

DEFINITION 6.16 Linear Combination   *If V is a linear space over F, a* (finite) linear combination *of elements of V is a sum*

$$c = \sum_{i=1}^{k} a_i s_i \tag{6.6}$$

where k is a positive integer, the $a_i \in F$ and the $s_i \in V$.

This definition includes the case where $k = 1$.

DEFINITION 6.17 Linear Span   *Let S be a subset of the linear space V. The* linear span *of S is the set*

$$span[S] = \left\{ \sum_{i=1}^{k} a_i s_i : k \geq 1, a_i \in F, s_i \in S \right\} \tag{6.7}$$

consisting of all finite linear combinations of elements of S.

The linear span of S is also known as the *linear subspace generated by S.*

It is possible for different sets to generate the same linear subspace.

EXAMPLE 6.25

Consider the linear subspace of $\mathbb{R}^3$ generated by the vectors $(1, 0, 0)$ and $(0, 1, 0)$. This is the set $\{(x, y, 0) \in \mathbb{R}^3, x, y \in \mathbb{R}\}$. It is also generated by the vectors $(1, 1, 0)$, and $(1, -1, 0)$.

The same subspace is also generated by the vectors $(1, 0, 0)$, $(0, 1, 0)$ and $(3, -2, 0)$. There is a redundancy in this, since we can express $(3, -2, 0)$ as a linear combination of the other two vectors,

$$(3, -2, 0) = 3(1, 0, 0) + (-2)(0, 1, 0),$$

and so we can reduce any linear combination of the three vectors to a linear combination of the first two by putting

$$a(1, 0, 0) + b(0, 1, 0) + c(3, -2, 0) = (a + 3c)(1, 0, 0) + (b - 2c)(0, 1, 0),$$

for any $a, b, c \in \mathbb{R}$. ▯

It is important to distinguish sets that have the redundancy property illustrated in the example above from sets that do not possess this property.

DEFINITION 6.18 Linearly Independent   *A subset S of a linear space V over F is* linearly independent *if for any set of vectors $\{s_1, s_2, \ldots, s_n\}$ contained in S, the equation*

$$\sum_{i=1}^{n} a_i s_i = 0 \tag{6.8}$$

implies that all the $a_i = 0$.

A set that is not linearly independent is *linearly dependent.*

EXAMPLE 6.26

$\{(1, 1, 1), (1, -1, 1)\}$ is a linearly independent subset of $\mathbb{R}^3$. The equation

$$a_1(1, 1, 1) + a_2(1, -1, 1) = (0, 0, 0)$$

implies the three equations

$$a_1 + a_2 = 0$$
$$a_1 - a_2 = 0$$
$$a_1 + a_2 = 0.$$

The first and third equation are identical, but the only solution of the first and second equations is $a_1 = 0$, $a_2 = 0$.

$\{(1, 1, 1), (1, -1, 1), (0, 1, 0)\}$ is a linearly dependent subset of $\mathbb{R}^3$. The equation

$$a_1(1, 1, 1) + a_2(1, -1, 1) + a_3(0, -1, 0) = (0, 0, 0)$$

implies the three equations

$$a_1 + a_2 = 0$$
$$a_1 - a_2 - a_3 = 0$$
$$a_1 + a_2 = 0.$$

This set of equations has infinitely many solutions, for example, $a_1 = 1$, $a_2 = -1$, $a_3 = 2$. □

A linearly independent set that generates a vector space has the property that removing any vector from the set will produce a set that no longer generates the space while adding a vector to the set will produce a set that is no longer linearly independent. Such sets are important enough to be given a special name.

DEFINITION 6.19 Basis   *If V is a linear space over F, a basis for V is a linearly independent subset of V that generates the whole of V.*

EXAMPLE 6.27

The set of vectors $\{(1, 0, 0, \ldots, 0), (0, 1, 0, \ldots, 0), (0, 0, 1, \ldots, 0), \ldots, (0, 0, 0, \ldots, 1)\}$, forms a basis for $\mathbb{R}^n$, known as the standard basis. There are lots of others.

For $n = 3$, the standard basis is $\{(1, 0, 0), (0, 1, 0), (0, 0, 1)\}$. Other bases are $\{(1, 1, 0), (1, -1, 0), (0, 0, -1)\}$, and $\{(1, -2, 3), (1, 2, 1), (4, -1, -2)\}$. □

It can be shown that every basis of a linear space contains the same number of vectors. This makes the following definition unambiguous.

DEFINITION 6.20 Dimension   *The dimension of a linear space is the number of elements in any basis for the space.*

EXAMPLE 6.28

The existence of the standard basis for $\mathbb{R}^n$ shows that its dimension is n. □

The dimension of a linear space is also the maximum number of elements that can be contained in a linearly independent subset of that space. There are linear spaces in which it is possible to find arbitrarily large linearly independent subsets. These spaces do not have a finite basis.

EXAMPLE 6.29

Consider the space C of continuous functions on the real line. Define the polynomial functions by

$$p_0(x) = 1,$$
$$p_1(x) = x,$$
$$p_2(x) = x^2,$$
$$\vdots$$
$$p_n(x) = x^n,$$

for $x \in \mathbb{R}$.

Any collection of these functions is a linearly independent subset of C, but none of them generates it. C therefore does not have a finite basis. $\square$

6.5 Linear Spaces over the Binary Field

We have seen that the ring $\mathbb{Z}_2$, with addition and multiplication tables

+	0	1
0	0	1
1	1	0

×	0	1
0	0	0
1	0	1

is the smallest example of a field. We will be using this field almost exclusively in the development of the theory of error-correcting codes. From now on, we will refer to it as the *binary field* and denote it by $\mathbb{B}$.

For any n, there are exactly 2^n n-tuples of elements of $\mathbb{B}^n$. For convenience, we will denote these simply by concatenating the bits, without commas in between or round brackets before and after them. (This also makes the elements of $\mathbb{B}^n$ look like sequences of bits.) Using these conventions, we have

$$\mathbb{B} = \{0, 1\},$$

$$\mathbb{B}^2 = \{00, 01, 10, 11\},$$

$$\mathbb{B}^3 = \{000, 001, 010, 011, 100, 101, 110, 111\},$$

$$\mathbb{B}^4 = \{0000, 0001, 0010, 0011, 0100, 0101, 0110, 0111,$$
$$1000, 1001, 1010, 1011, 1100, 1101, 1110, 1111\},$$

and so on.

We can define addition on $\mathbb{B}^n$ component-wise, so that, for example, in $\mathbb{B}^4$,

$$0011 + 0101 = 0110,$$

while multiplication by elements of $\mathbb{B}$ is defined very simply by setting $0b = 0$ and $1b = b$ for all $b \in \mathbb{B}^n$. These operations make $\mathbb{B}^n$ into a linear space over $\mathbb{B}$.

We can represent the members of $\mathbb{B}^n$ as the vertices of the unit cube in n-dimensional space.

The definitions of linear combinations, linear independence, bases and dimension given above for general linear spaces all apply to linear spaces over $\mathbb{B}$. A linear subspace of $\mathbb{B}^n$ of dimension k has exactly 2^k elements.

EXAMPLE 6.30

The following are linear subspaces of $\mathbb{B}^4$:

$$\{0000\}$$

$$\{0000, 1111\}$$

$$\{0000, 1000, 0010, 1010\}$$

$$\{0000, 1100, 0110, 0011, 1010, 1111, 0101, 1001\}.$$

Their dimensions are zero, one, two and three, respectively.

The following are not linear subspaces of $\mathbb{B}^4$:

$$\{0101\}$$

$$\{0101, 1010\}$$

$$\{0101, 1010, 1111\}$$

$$\{1000, 0100, 0010, 0001\}$$

$$\{0000, 1000, 0010, 1010, 1111\}.$$

⬚

DEFINITION 6.21 Coset   *If L is a linear subspace of* $\mathbb{B}$, *and* $x \in \mathbb{B}$, *then the set of vectors*

$$x + L = \{x + y : y \in L\} \tag{6.9}$$

is a coset of L.

Cosets are also known as *affine subspaces.* $x + L = L$ if and only if $x \in L$, and $y + L = z + L$ if and only if $y \in z + L$. In particular, $0 + L = L$ as 0 always belongs to L.

EXAMPLE 6.31

$L = \{00, 11\}$ is a subspace of $\mathbb{B}^2$. The cosets of L are:

$$00 + L = 11 + L = L,$$

$$01 + L = 10 + L = \{01, 10\}.$$

EXAMPLE 6.32

$L = \{000, 111\}$ is a subspace of $\mathbb{B}^3$. The cosets of L are:

$$000 + L = 111 + L = L,$$

$$001 + L = 110 + L = \{001, 110\},$$
$$010 + L = 101 + L = \{010, 101\},$$
$$100 + L = 011 + L = \{100, 011\}.$$

EXAMPLE 6.33

$L = \{000, 101, 010, 111\}$ is a subspace of $\mathbb{B}^3$. The cosets of L are:

$$000 + L = 101 + L = 010 + L = 111 + L = L,$$

$$001 + L = 100 + L = 011 + L = 110 + L = \{001, 100, 011, 110\}.$$

DEFINITION 6.22 Weight   *The* weight *of an element of* $\mathbb{B}^n$ *is the number of ones in it.*

We repeat the following definition from Chapter 5.

DEFINITION 6.23 Hamming Distance　　　The Hamming distance *between x and y* $\in \mathbb{B}^n$ *is the number of places where they differ.*

The Hamming distance between x and y is equal to the weight of $x - y$.

EXAMPLE 6.34

The following table shows the weights of the elements of $\mathbb{B}^2$.

Element	Weight
00	0
01	1
10	1
11	2

The following table shows the Hamming distance between the pairs of elements of $\mathbb{B}^2$.

	00	01	10	11
00	0	1	1	2
01	1	0	2	1
10	1	2	0	1
11	2	1	1	0

▯

EXAMPLE 6.35

The following table shows the weights of the elements of $\mathbb{B}^3$.

Element	Weight
000	0
001	1
010	1
011	2
100	1
101	2
110	2
111	3

The following table shows the Hamming distance between the pairs of elements of $\mathbb{B}^3$.

	000	001	010	011	100	101	110	111
000	0	1	1	2	1	2	2	3
001	1	0	2	1	2	1	3	2
010	1	2	0	1	2	3	1	2
011	2	1	1	0	3	2	2	1
100	1	2	2	3	0	1	1	2
101	2	1	3	2	1	0	2	1
110	2	3	1	2	1	2	0	1
111	3	2	2	1	2	1	1	0

⬛

6.6   Linear Codes

We construct error-correcting codes by finding subsets of $\mathbb{B}^n$ with desirable properties.

DEFINITION 6.24 Binary Block Code   A binary block code *is a subset of* $\mathbb{B}^n$ *for some n. Elements of the code are called* code words.

EXAMPLE 6.36

We can encode the alphabet using a subset of $\mathbb{B}^5$. We let 0001 stand for A, let 0010 stand for B, and so on, until 11010 stands for Z. The remaining six elements of $\mathbb{B}^5$ are not included in the code.   ⬛

Such simple codes do not have error-correcting properties. We need to have codes with more structure.

DEFINITION 6.25 Linear Code   A linear code *is a linear subspace of* $\mathbb{B}^n$.

Linear codes are subsets of $\mathbb{B}^n$ with a linear structure added. Because multiplication is trivial in linear spaces over a binary field, a binary code is a linear code if and only if the sum of two code words is also a code word.

DEFINITION 6.26 Minimum Distance   *The* minimum distance *of a linear code is the minimum of the weights of the non-zero code words.*

The minimum distance of a linear code is the minimum of the Hamming distances between pairs of code words.

The relationships between the minimum distance of a code and its capabilities in respect of detecting and correcting and detecting errors that were discussed in Chapter 5 also hold for linear codes. The additional structure gives us systematic ways of constructing codes with good error detecting and error correcting properties. Details of such results can be found in Chapter 4 of [1], Chapter 7 of [4] and Chapter 4 of [5].

EXAMPLE 6.37

In $\mathbb{B}^2$, $\{00, 01\}$ and $\{00, 10\}$ are linear codes with minimum distance 1 while $\{00, 11\}$ is a linear code with minimum distance 2.   ⬚

EXAMPLE 6.38

The following two-dimensional codes in $\mathbb{B}^3$ have minimum distance 1:

$$\{000, 001, 010, 011\},$$

$$\{000, 001, 100, 101\},$$

$$\{000, 010, 100, 110\},$$

$$\{000, 001, 110, 111\},$$

$$\{000, 010, 101, 111\},$$

$$\{000, 100, 011, 111\}.$$

The code $\{000, 011, 101, 110\}$ has minimum distance 2.   ⬚

Let L be a linear code. If L is a k-dimensional subspace of $\mathbb{B}^n$, then we can find a basis for L consisting of k code words $b_1, b_2, \ldots, b_k$ in $\mathbb{B}^n$. Every code word in L is a linear combination of these basis code words. There is a convenient matrix notation for linear codes that uses this fact.

DEFINITION 6.27 Generator Matrix   *A* generator matrix *for a linear code is a binary matrix whose rows are the code words belonging to some basis for the code.*

A generator matrix for a k-dimensional linear code in $\mathbb{B}^n$ is a $k \times n$ matrix whose rank is k. Conversely, any $k \times n$ binary matrix with rank k is a generator matrix for some code.

We can compute the code words from the generator matrix by multiplying it on the left by all the row vectors of dimension k.

EXAMPLE 6.39

The code $\{0000, 0001, 1000, 1001\}$ is a two-dimensional linear code in $\mathbb{B}^4$. $\{0001, 1000\}$ is a basis for this code, which gives us the generator matrix

$$G = \begin{bmatrix} 0\,0\,0\,1 \\ 1\,0\,0\,0 \end{bmatrix}.$$

To find the code words from the generator matrix, we perform the following matrix multiplications:

$$\begin{bmatrix} 0\,0 \end{bmatrix} \begin{bmatrix} 0\,0\,0\,1 \\ 1\,0\,0\,0 \end{bmatrix} = \begin{bmatrix} 0\,0\,0\,0 \end{bmatrix},$$

$$\begin{bmatrix} 0\,1 \end{bmatrix} \begin{bmatrix} 0\,0\,0\,1 \\ 1\,0\,0\,0 \end{bmatrix} = \begin{bmatrix} 1\,0\,0\,0 \end{bmatrix},$$

$$\begin{bmatrix} 1\,0 \end{bmatrix} \begin{bmatrix} 0\,0\,0\,1 \\ 1\,0\,0\,0 \end{bmatrix} = \begin{bmatrix} 0\,0\,0\,1 \end{bmatrix},$$

$$\begin{bmatrix} 1\,1 \end{bmatrix} \begin{bmatrix} 0\,0\,0\,1 \\ 1\,0\,0\,0 \end{bmatrix} = \begin{bmatrix} 1\,0\,0\,1 \end{bmatrix}.$$

This gives us the four code words, 0000, 1000, 0001 and 1001, with which we started.

☐

EXAMPLE 6.40

$\{0000, 0011, 0110, 1100, 0101, 1111, 1010, 1001\}$ is a three-dimensional linear code in $\mathbb{B}^4$.

$\{0011, 0110, 1100\}$ is a basis for this code, which gives us the generator matrix

$$G = \begin{bmatrix} 0\,0\,1\,1 \\ 0\,1\,1\,0 \\ 1\,1\,0\,0 \end{bmatrix}.$$

To recover the code words from the generator matrix, we perform the following matrix multiplications:

$$[0\ 0\ 0]\,G = [0\ 0\ 0\ 0]\,,$$

$$[0\ 0\ 1]\,G = [1\ 1\ 0\ 0]\,,$$

$$[0\ 1\ 0]\,G = [0\ 1\ 1\ 0]\,,$$

$$[0\ 1\ 1]\,G = [1\ 0\ 1\ 0]\,,$$

$$[1\ 0\ 0]\,G = [0\ 0\ 1\ 1]\,,$$

$$[1\ 0\ 1]\,G = [1\ 1\ 1\ 1]\,,$$

$$[1\ 1\ 0]\,G = [0\ 1\ 0\ 1]\,,$$

$$[1\ 1\ 1]\,G = [1\ 0\ 0\ 1]\,.$$

$\{0101, 1001, 1010\}$ is also a basis for the code. It gives us the generator matrix

$$G' = \begin{bmatrix} 0\ 1\ 0\ 1 \\ 1\ 0\ 0\ 1 \\ 1\ 0\ 1\ 0 \end{bmatrix}.$$

The code words can also be recovered from this matrix, but in a different order.

$$[0\ 0\ 0]\,G' = [0\ 0\ 0\ 0]\,,$$

$$[0\ 0\ 1]\,G' = [1\ 0\ 1\ 0]\,,$$

$$[0\ 1\ 0]\,G' = [1\ 0\ 0\ 1]\,,$$

$$\begin{bmatrix} 0\ 1\ 1 \end{bmatrix} G' = \begin{bmatrix} 0\ 0\ 1\ 1 \end{bmatrix},$$

$$\begin{bmatrix} 1\ 0\ 0 \end{bmatrix} G' = \begin{bmatrix} 0\ 1\ 0\ 1 \end{bmatrix},$$

$$\begin{bmatrix} 1\ 0\ 1 \end{bmatrix} G' = \begin{bmatrix} 1\ 1\ 1\ 1 \end{bmatrix},$$

$$\begin{bmatrix} 1\ 1\ 0 \end{bmatrix} G' = \begin{bmatrix} 1\ 1\ 0\ 0 \end{bmatrix},$$

$$\begin{bmatrix} 1\ 1\ 1 \end{bmatrix} G' = \begin{bmatrix} 0\ 1\ 1\ 0 \end{bmatrix}.$$

⬚

EXAMPLE 6.41

$$G = \begin{bmatrix} 0\ 0\ 0\ 0\ 1 \\ 0\ 0\ 1\ 1\ 1 \\ 1\ 1\ 1\ 1\ 1 \end{bmatrix}$$

is a 3×5 binary matrix of rank 3. The code words of the three-dimensional linear code for which G is a generator matrix can be found by the following matrix multiplications.

$$\begin{bmatrix} 0\ 0\ 0 \end{bmatrix} G = \begin{bmatrix} 0\ 0\ 0\ 0\ 0 \end{bmatrix},$$

$$\begin{bmatrix} 0\ 0\ 1 \end{bmatrix} G = \begin{bmatrix} 1\ 1\ 1\ 1\ 1 \end{bmatrix},$$

$$\begin{bmatrix} 0\ 1\ 0 \end{bmatrix} G = \begin{bmatrix} 0\ 0\ 1\ 1\ 1 \end{bmatrix},$$

$$\begin{bmatrix} 0\ 1\ 1 \end{bmatrix} G = \begin{bmatrix} 1\ 1\ 0\ 0\ 0 \end{bmatrix},$$

$$\begin{bmatrix} 1\ 0\ 0 \end{bmatrix} G = \begin{bmatrix} 0\ 0\ 0\ 0\ 1 \end{bmatrix},$$

$$\left[1\ 0\ 1\right] G = \left[1\ 1\ 1\ 1\ 0\right],$$

$$\left[1\ 1\ 0\right] G = \left[0\ 0\ 1\ 1\ 0\right],$$

$$\left[1\ 1\ 1\right] G = \left[1\ 1\ 0\ 0\ 1\right].$$

The code is

$$\{00000, 00001, 00110, 00111, 11000, 11001, 11110, 11111\}.$$

☐

Because a linear space can have many bases, a linear code can have many generator matrices. This raises the question of when two generator matrices generate the same code.

DEFINITION 6.28 Elementary Row Operation   *An elementary row operation on a binary matrix consists of replacing a row of the matrix with the sum of that row and any other row.*

If we have a generator matrix G for a linear code L, all other generator matrices for L can be obtained by applying a sequence of elementary row operations to G.

EXAMPLE 6.42

In a previous example, we saw that the linear code

$$\{0000, 0011, 0110, 1100, 0101, 1111, 1010, 1001\}$$

has the following generator matrices:

$$G = \begin{bmatrix} 0\ 0\ 1\ 1 \\ 0\ 1\ 1\ 0 \\ 1\ 1\ 0\ 0 \end{bmatrix},$$

and

$$G' = \begin{bmatrix} 0\ 1\ 0\ 1 \\ 1\ 0\ 0\ 1 \\ 1\ 0\ 1\ 0 \end{bmatrix}.$$

We can change G to G' by the following sequence of elementary row operations.

1. Replace the third row with the sum of the second and third rows. This gives

$$G_1 = \begin{bmatrix} 0 & 0 & 1 & 1 \\ 0 & 1 & 1 & 0 \\ 1 & 0 & 1 & 0 \end{bmatrix}.$$

2. Replace the first row with the sum of the first and second rows. This gives

$$G_2 = \begin{bmatrix} 0 & 1 & 0 & 1 \\ 0 & 1 & 1 & 0 \\ 1 & 0 & 1 & 0 \end{bmatrix}.$$

3. Replace the second row with the sum of the first and second rows. This gives

$$G_3 = \begin{bmatrix} 0 & 1 & 0 & 1 \\ 0 & 0 & 1 & 1 \\ 1 & 0 & 1 & 0 \end{bmatrix}.$$

4. Finally, replace the second row with the sum of the second and third rows. This gives

$$G' = \begin{bmatrix} 0 & 1 & 0 & 1 \\ 1 & 0 & 0 & 1 \\ 1 & 0 & 1 & 0 \end{bmatrix}.$$

$\Box$

For error-correcting purposes, two codes that have the same minimum distance have the same properties. One way in which we can change a linear code without changing its minimum distance is to change the order of the bits in all the code words.

DEFINITION 6.29 Equivalent Codes   *Two codes are* equivalent *if each can be constructed from the other by reordering the bits of each code word in the same way.*

The generator matrices of equivalent codes can be obtained from each other by interchanging columns.

DEFINITION 6.30 Canonical Form   *The generator matrix G of a k-dimensional linear code in $\mathbb{B}^n$ is in* canonical form *if it is of the form*

$$G = [I : A],$$

where I is a $k \times k$ identity matrix and A is an arbitrary $k \times (n-k)$ binary matrix.

It is possible to convert the generator matrix for a code using elementary row operations (which do not change the set of code words) and column interchanges into the generator matrix of an equivalent code in the canonical form. The code words derived from a generator matrix in canonical form are also said to be in canonical form or in *systematic form*.

If the generator matrix G is in canonical form, and $\mathbf{w}$ is any k-bit word, the code word $\mathbf{s} = \mathbf{w}G$ is in systematic form and the first k bits of $\mathbf{s}$ are the same as the bits of $\mathbf{w}$.

EXAMPLE 6.43

We have seen in a previous example that

$$G = \begin{bmatrix} 0\,0\,0\,1 \\ 1\,0\,0\,0 \end{bmatrix}$$

is a generator matrix for the code $\{0000, 0001, 1000, 1001\}$.

To reduce this to canonical form, we first interchange the first and second columns to get

$$G_1 = \begin{bmatrix} 0\,0\,0\,1 \\ 0\,1\,0\,0 \end{bmatrix}$$

and then interchange the first and last columns to get

$$G_2 = \begin{bmatrix} 1\,0\,0\,0 \\ 0\,1\,0\,0 \end{bmatrix}.$$

This is now in canonical form, with

$$A = \begin{bmatrix} 0\,0 \\ 0\,0 \end{bmatrix}.$$

The code generated by the canonical form of the generator matrix is

$$\{0000, 1000, 0100, 1100\}.$$

Note that both codes have minimum distance 1.

⬚

EXAMPLE 6.44

$$G = \begin{bmatrix} 0\,0\,1\,1 \\ 0\,1\,1\,0 \\ 1\,1\,0\,0 \end{bmatrix}$$

is a generator matrix of the linear code $\{0000, 0011, 0110, 1100, 0101, 1111, 1010, 1001\}$.

To reduce G to canonical form, we begin by applying the elementary row operation of replacing the second row of G by the sum of the first and second rows to give

$$G_1 = \begin{bmatrix} 0 & 0 & 1 & 1 \\ 0 & 1 & 0 & 1 \\ 1 & 1 & 0 & 0 \end{bmatrix}.$$

Next we replace the third row by the sum of the second and third rows to give

$$G_2 = \begin{bmatrix} 0 & 0 & 1 & 1 \\ 0 & 1 & 0 & 1 \\ 1 & 0 & 0 & 1 \end{bmatrix}.$$

Finally we interchange the first and third columns to give

$$G_3 = \begin{bmatrix} 1 & 0 & 0 & 1 \\ 0 & 1 & 0 & 1 \\ 0 & 0 & 1 & 1 \end{bmatrix}.$$

This is in canonical form with

$$A = \begin{bmatrix} 1 \\ 1 \\ 1 \end{bmatrix}.$$

The code generated by the canonical form of G is the same as the code generated by G. ☐

EXAMPLE 6.45

$$G = \begin{bmatrix} 0 & 0 & 0 & 0 & 1 \\ 0 & 0 & 1 & 1 & 1 \\ 1 & 1 & 1 & 1 & 1 \end{bmatrix}$$

is a generator matrix of the linear code

$$\{00000, 00001, 00110, 00111, 11000, 11001, 11110, 11111\}.$$

To reduce G to canonical form we start by replacing the third row with the sum of the second and third rows to give

$$G_1 = \begin{bmatrix} 0 & 0 & 0 & 0 & 1 \\ 0 & 0 & 1 & 1 & 1 \\ 1 & 1 & 0 & 0 & 0 \end{bmatrix}.$$

Next, we replace the second row with the sum of the first and second rows to give

$$G_2 = \begin{bmatrix} 0\,0\,0\,0\,1 \\ 0\,0\,1\,1\,0 \\ 1\,1\,0\,0\,0 \end{bmatrix}.$$

We interchange the first and last columns, obtaining

$$G_3 = \begin{bmatrix} 1\,0\,0\,0\,0 \\ 0\,0\,1\,1\,0 \\ 0\,1\,0\,0\,1 \end{bmatrix},$$

and finally interchange the second and third columns, to get

$$G_4 = \begin{bmatrix} 1\,0\,0\,0\,0 \\ 0\,1\,0\,1\,0 \\ 0\,0\,1\,0\,1 \end{bmatrix}.$$

G_4 is now in canonical form, with

$$A = \begin{bmatrix} 0\,0 \\ 1\,0 \\ 0\,1 \end{bmatrix}.$$

It generates the code

$$\{00000, 10000, 01010, 00101, 11010, 10101, 01111, 11111\}.$$

⬚

DEFINITION 6.31 Parity Check Matrix   *The* parity check matrix *of a linear code with $k \times n$ generator matrix G is the $k \times n$ matrix H satisfying*

$$GH^T = 0, \tag{6.10}$$

where H^T is the transpose of H and 0 denotes the $k \times (n - k)$ zero matrix.

If G is in canonical form, $G = [I : A]$, with I the $k \times k$ identity matrix, then $H = [A^T : I]$, where I is the $(n - k) \times (n - k)$ identity matrix. (This is true if G and H are binary matrices. In general, we should have $H = [-A^T : I]$. In the binary case, $A = -A$.) If G is not in canonical form, we can find H by reducing G to canonical form, finding the canonical form of H using the equation above, and then reversing the column operations used to convert G to canonical form to convert the canonical form of H to the parity check matrix of G.

EXAMPLE 6.46

$$G = \begin{bmatrix} 1 & 0 & 0 & 0 & 1 \\ 0 & 1 & 0 & 1 & 1 \\ 0 & 0 & 1 & 1 & 0 \end{bmatrix}$$

is in canonical form with

$$A = \begin{bmatrix} 0 & 1 \\ 1 & 1 \\ 1 & 0 \end{bmatrix}.$$

H is obtained by transposing A and adjoining a 2×2 identity matrix to get

$$H = \begin{bmatrix} 0 & 1 & 1 & 1 & 0 \\ 1 & 1 & 0 & 0 & 1 \end{bmatrix}.$$

As expected, we have

$$GH^T = \begin{bmatrix} 1 & 0 & 0 & 0 & 1 \\ 0 & 1 & 0 & 1 & 1 \\ 0 & 0 & 1 & 1 & 0 \end{bmatrix} \begin{bmatrix} 0 & 1 \\ 1 & 1 \\ 1 & 0 \\ 1 & 0 \\ 0 & 1 \end{bmatrix} = \begin{bmatrix} 0 & 0 \\ 0 & 0 \\ 0 & 0 \end{bmatrix}.$$

▯

EXAMPLE 6.47

We have seen in a previous example that

$$G = \begin{bmatrix} 0 & 0 & 0 & 0 & 1 \\ 0 & 0 & 1 & 1 & 1 \\ 1 & 1 & 1 & 1 & 1 \end{bmatrix}$$

is a generator matrix of the linear code

$$\{00000, 00001, 00110, 00111, 11000, 11001, 11110, 11111\},$$

which can be reduced to the canonical form

$$G_c = \begin{bmatrix} 1 & 0 & 0 & 0 & 0 \\ 0 & 1 & 0 & 1 & 0 \\ 0 & 0 & 1 & 0 & 1 \end{bmatrix}$$

by the following operations:

1. Replace the third row by the sum of the second and third rows.

2. Replace the second row by the sum of the first and second rows.

3. Interchange the first and fifth columns.

4. Interchange the second and third columns.

The canonical form of the parity check matrix is

$$H_c = \begin{bmatrix} 0 & 1 & 0 & 1 & 0 \\ 0 & 0 & 1 & 0 & 1 \end{bmatrix}.$$

We have

$$G_c H_c^T = \begin{bmatrix} 0 & 0 \\ 0 & 0 \\ 0 & 0 \end{bmatrix}.$$

To find the parity check matrix of G, we apply the column operations used to reduce G to canonical form to H_c in reverse order. We start by interchanging the second and third columns to get

$$H_1 = \begin{bmatrix} 0 & 0 & 1 & 1 & 0 \\ 0 & 1 & 0 & 0 & 1 \end{bmatrix}.$$

We interchange the first and fifth columns to get

$$H = \begin{bmatrix} 0 & 0 & 1 & 1 & 0 \\ 1 & 1 & 0 & 0 & 0 \end{bmatrix}.$$

We now check that

$$GH^T = \begin{bmatrix} 0 & 0 \\ 0 & 0 \\ 0 & 0 \end{bmatrix}.$$

☐

EXAMPLE 6.48

$$G = \begin{bmatrix} 1 & 0 & 1 \\ 0 & 1 & 1 \end{bmatrix}$$

is a generator matrix of the linear code $\{000, 101, 011, 110\}$. It is in canonical form; so the parity check matrix is

$$H = \begin{bmatrix} 1 & 1 & 1 \end{bmatrix}.$$

Let us see what happens to elements of $\mathbb{B}^3$ when they are multiplied by H^T:

$$000H^T = 0$$
$$001H^T = 1$$
$$010H^T = 1$$
$$011H^T = 0$$
$$100H^T = 1$$
$$101H^T = 0$$
$$110H^T = 0$$
$$111H^T = 1.$$

The product of H and any of the code words in the code generated by G is zero, while the remaining products are non-zero. $\square$

The example above should not be surprising, for we have the following result.

RESULT 6.1

If L is a k-dimensional linear code in $\mathbb{B}^n$, and G is a generator matrix for L, every code word in L can be obtained by taking some $b \in \mathbb{B}^k$ and multiplying it by G, to get the code word bG. If we now multiply this by the transpose of the parity check matrix H, we get

$$(bG)H^T = b(GH^T) = b0 = 0. \tag{6.11}$$

The parity check matrix gives us an easy way of determining if a word $w \in \mathbb{B}^n$ belongs to the linear code L: we compute wH^t. If the result is the zero matrix, w is a code word. If not, then w is not a code word. As we shall see, the parity check matrix enables us to do more than just check if a word is a code word.

EXAMPLE 6.49

$$G = \begin{bmatrix} 0 & 0 & 1 & 1 \\ 1 & 1 & 0 & 0 \end{bmatrix}$$

is a generator matrix of the linear code $\{0000, 0011, 1100, 1111\}$. We reduce it to canonical form by interchanging the first and last columns to get

$$G_c = \begin{bmatrix} 1 & 0 & 1 & 0 \\ 0 & 1 & 0 & 1 \end{bmatrix},$$

which generates the code $\{0000, 1010, 0101, 1111\}$.

The canonical form of the parity check matrix is

$$H_c = \begin{bmatrix} 1 & 0 & 1 & 0 \\ 0 & 1 & 0 & 1 \end{bmatrix}.$$

If we look at the products

$$0000H_c^T = 00$$
$$0001H_c^T = 01$$
$$0010H_c^T = 10$$
$$0011H_c^T = 11$$
$$0100H_c^T = 01$$
$$0101H_c^T = 00$$
$$0110H_c^T = 11$$
$$0111H_c^T = 10$$
$$1000H_c^T = 10$$
$$1001H_c^T = 11$$
$$1010H_c^T = 00$$
$$1011H_c^T = 01$$
$$1100H_c^T = 11$$
$$1101H_c^T = 10$$
$$1110H_c^T = 01$$
$$1111H_c^T = 00$$

we see that only the code words in the code generated by G_c have products equal to 00.

If we now interchange the first and last columns of H_c to get the parity check matrix of G, we get

$$H = \begin{bmatrix} 0 & 0 & 1 & 1 \\ 1 & 1 & 0 & 0 \end{bmatrix},$$

and the products are

$$0000H^T = 00$$
$$0001H^T = 10$$
$$0010H^T = 10$$
$$0011H^T = 00$$
$$0100H^T = 01$$
$$0101H^T = 11$$
$$0110H^T = 11$$

$$0111H^T = 01$$
$$1000H^T = 01$$
$$1001H^T = 11$$
$$1010H^T = 11$$
$$1011H^T = 01$$
$$1100H^T = 00$$
$$1101H^T = 10$$
$$1110H^T = 10$$
$$1111H^T = 00$$

and again only the code words in the code generated by G have products equal to 00.

In this example, we have $G = H$ and $G_c = H_c$. A generator matrix can be the same as its parity check matrix. □

(Although we have not introduced any terminology regarding linear transformations, readers who are familiar with this topic will realise that a linear code L is the range of the linear transformation determined by the generator matrix, and the kernel of the linear transformation determined by the transpose of the associated parity check matrix.)

6.7 Encoding and Decoding

The use of a linear code for error correction involves an encoding step to add redundancy and a later decoding step which attempts to correct errors before it removes the redundancy. This is essentially the same process as was described in Chapter 5, but the additional structure of linear codes enables us to find new algorithms for the encoding and decoding processes.

In the simplest case, the encoding step uses a generator matrix G of a linear code, while the decoding step uses the corresponding parity check matrix.

The encoding process takes a string, $\mathbf{b}$, k bits in length and produces a code word, $\mathbf{w}$, n bits in length, $\mathbf{w} = \mathbf{b}G$. The first k bits of $\mathbf{w}$ are the *message bits*, and the remaining bits are the *check bits*. The latter represent the redundancy which has been added to the message.

If G is in canonical form, the first k bits of $\mathbf{w}$ are the bits of $\mathbf{b}$ and the code is said to be in *systematic form*. In this case, code words can be decoded by simply removing the last $(n - k)$ bits.

Suppose that a code word is corrupted by noise between encoding and decoding. One or more bits will be changed, zeros becoming ones and ones becoming zeros. It is possible that the result will be another code word, in which case it will not be possible to detect that corruption has occurred. Otherwise, the corrupted string will not be a code word, and attempts may be made to restore the uncorrupted code word as part of the decoding process.

The decoding process therefore has two stages. If the received string is not a code word, the uncorrupted code word has to be restored. The code word is then decoded to recover the original string.

If we assume that the corruption occurred while transmission through a *binary symmetric channel*, and use the *maximum likelihood decoding strategy*, it follows that a corrupted string should be restored to the code word which is closest to it (in terms of the Hamming distance). (Other assumptions about the characteristics of the noise process may lead to other procedures.)

We suppose that the generator matrix G generates the linear code L, and that the code word $\mathbf{w}$ is corrupted by a noise vector $\mathbf{e}$, and the result is the vector $\mathbf{x} = \mathbf{w} + \mathbf{e}$. If $\mathbf{x}$ is not a code word, the decoder must find the closest code word $\mathbf{y}$, or equivalently, the error vector of smallest weight.

If $\mathbf{x}$ is not a code word, it belongs to the coset $\mathbf{x} + L$; so we assume that $\mathbf{x}$ is of the form $\mathbf{e} + \mathbf{u}$ where $\mathbf{u} \in L$ and $\mathbf{e}$ is the vector of least weight in the coset.

DEFINITION 6.32 Syndrome   *If L is a linear code with generator matrix G and parity check matrix H, the* syndrome *of $\mathbf{x}$ is given by $s = \mathbf{x}H^T$.*

Members of the same coset of L have the same syndrome, so there is a one-to-one correspondence between cosets and syndromes.

We now have a procedure for removing noise. Draw up the *syndrome table*, which is a list of the cosets, showing the vector of minimum weight and the syndrome. When a string that is not a code word is received, compute its syndrome. Add the vector of minimum weight from the corresponding coset to the corrupted string to recover the uncorrupted code word.

EXAMPLE 6.50

Suppose that

$$H = \begin{bmatrix} 0 & 0 & 1 & 1 \\ 1 & 1 & 0 & 0 \end{bmatrix}$$

is the parity check matrix of the linear code L.

To draw up the syndrome table, we start with the elements of L itself, which have syndrome 00. The first row of the syndrome table looks like

$$00|0000\ 0011\ 1100\ 1111|0000$$

where the syndrome appears at the left, the code words in L appear in the middle and the vector of minimum weight appears on the right.

We now choose a vector that does not belong to L and compute its syndrome. Let us choose 0001, whose syndrome is 10. We add this vector to all the code words in L to get the next row of the syndrome table:

$$00|0000\ 0011\ 1100\ 1111|0000$$
$$10|0001\ 0010\ 1101\ 1110|0001$$

Note that we have two vectors of minimum weight in this coset. We have chosen one of them arbitrarily.

We now choose another vector and compute its syndrome. If we choose 0110, the syndrome is 11 and we can add another row to the syndrome table:

$$00|0000\ 0011\ 1100\ 1111|0000$$
$$10|0001\ 0010\ 1101\ 1110|0001$$
$$11|0110\ 0101\ 1010\ 1001|0101$$

Again, we have chosen the vector of minimum weight arbitrarily.

Again we choose a vector and compute its syndrome. The syndrome of 0100 is 01. We use this to complete the table:

$$00|0000\ 0011\ 1100\ 1111|0000$$
$$01|0100\ 0111\ 1000\ 1011|0100$$
$$10|0001\ 0010\ 1101\ 1110|0001$$
$$11|0110\ 0101\ 1010\ 1001|0101$$

We can now use the syndrome table for error correction. Suppose the vector 1010 is to be corrected. We compute its syndrome, which is 11. We add the vector of minimum weight in the coset with syndrome 11 to 1010,

$$1010 + 0101 = 1111.$$

So the corrected code word is 1111.

If the vector 0111 is to be corrected, its syndrome is 01, so we add

$$0111 + 0100 = 0011$$

to obtain the corrected code word 0011.

If the vector 1110 is to be corrected, its syndrome is 10, so we add

$$1110 + 0001 = 1111$$

to obtain the corrected code word 1111.

If the vector 1100 is to be corrected, its syndrome is 00, which means that it is a code word and no correction is necessary.

Note that the error correction process may not give the code word that was corrupted by noise. Suppose that the original code word is 0000 and that the leftmost bit gets corrupted, giving 1000. The correction procedure will produce 1100, not 0000. This kind of error is unavoidable if a coset has more than one vector of minimum weight.

☐

6.8   Codes Derived from Hadamard Matrices

> **DEFINITION 6.33 Hadamard Matrix**   A Hadamard matrix *is an orthogonal matrix whose elements are either* 1 *or* −1.

The simplest Hadamard matrix is the 2×2 matrix

$$H_2 = \begin{bmatrix} 1 & 1 \\ 1 & -1 \end{bmatrix}.$$

If H is an $n \times n$ Hadamard matrix, then the $2n \times 2n$ matrix

$$H' = \begin{bmatrix} H & H \\ H & -H \end{bmatrix}$$

is also a Hadamard matrix. This gives us a means of constructing $2^k \times 2^k$ Hadamard matrices for all $k \geq 1$.

EXAMPLE 6.51

We construct a 4×4 Hadamard matrix by combining the matrices H_2 above:

$$H_4 = \begin{bmatrix} 1 & 1 & 1 & 1 \\ 1 & -1 & 1 & -1 \\ 1 & 1 & -1 & -1 \\ 1 & -1 & -1 & 1 \end{bmatrix}.$$

We construct a 8×8 Hadamard matrix by combining the matrices H_4:

$$H_8 = \begin{bmatrix}
1 & 1 & 1 & 1 & 1 & 1 & 1 & 1 \\
1 & -1 & 1 & -1 & 1 & -1 & 1 & -1 \\
1 & 1 & -1 & -1 & 1 & 1 & -1 & -1 \\
1 & -1 & -1 & 1 & 1 & -1 & -1 & 1 \\
1 & 1 & 1 & 1 & -1 & -1 & -1 & -1 \\
1 & -1 & 1 & -1 & -1 & 1 & -1 & 1 \\
1 & 1 & -1 & -1 & -1 & -1 & 1 & 1 \\
1 & -1 & -1 & 1 & -1 & 1 & 1 & -1
\end{bmatrix}.$$

We can continue in this way to construct 16×16 Hadamard matrices, 32×32 Hadamard matrices, and so on. □

$n \times n$ Hadamard matrices exist for n other than 2^m, but n must be a multiple of 4. Some results about the existence of Hadamard matrices for various values of n are cited in Section 5.7 of [5]. Finding Hadamard matrices is still a research topic.

If H is a $n \times n$ Hadamard matrix, we can construct a code with $2n$ code words, each n bits long, and distance $n/2$, in the following way. Let $v_1, v_2, \ldots, v_n$ denote the rows of H. For each v_i, we construct two code words. For the first, we change all the 1's to 0's and the -1's to 1; for the second, we do not change the 1's, but change the -1's to 0's.

If n is not a power of 2, the resulting code is not a linear code. If n is a power of 2 and the Hadamard matrix is constructed recursively as described above, the resulting code is a first-order Reed-Muller code. (Reed-Muller codes are discussed in the next chapter.)

EXAMPLE 6.52

We apply the construction described above to

$$H_2 = \begin{bmatrix} 1 & 1 \\ 1 & -1 \end{bmatrix}.$$

The first row of H_2 is $[1\ 1]$, which gives us the code words 00 and 11.

The second row is $[1\ -1]$, which gives us the code words 01 and 10.

The code is $\{00, 01, 10, 11\}$, the whole of $\mathbb{B}^2$. □

EXAMPLE 6.53

We get a more interesting example if we use H_4.

The first row of H_4 is $[1\ 1\ 1\ 1]$, which gives us the code words 0000 and 1111.

The second row of H_4 is $[1\ -1\ 1\ -1]$, which gives us the code words 0101 and 1010.

The third row of H_4 is $[1\ 1\ -1\ -1]$, which gives us the code words 0011 and 1100.

The fourth row of H_4 is $[1\ -1\ -1\ 1]$, which gives us the code words 0110 and 1001.

The code is $\{0000, 0011, 0101, 0110, 1001, 1010, 1100, 1111\}$.   ☐

The use of Hadamard matrices gives good separation between code words, which results in good error-correcting capabilities.

6.9   Exercises

1. Find all the homomorphisms from $\mathbb{Z}_2$ into the two groups of order 4.

*2. Find all the continuous homomorphisms from the group of real numbers with the operation of addition to the group of positive real numbers with the operation of multiplication.

3. Find the values of the following expressions when the numbers and operations belong to the specified Cyclic Ring:

   (a) $(1+2) \times (2+3)$ in $\mathbb{Z}_4$;

   (b) $(1+2) \times (2+3)$ in $\mathbb{Z}_5$;

   (c) $(1+2) \times (2+3)$ in $\mathbb{Z}_6$;

   (d) $(1+3) \times (2+4)$ in $\mathbb{Z}_5$;

   (e) $(1+3) \times (2+4)$ in $\mathbb{Z}_6$;

   (f) $(1+3) \times (2+4)$ in $\mathbb{Z}_7$.

4. Which of the following subsets of $\mathbb{R}^4$ are linearly independent?

   (a) $\{(1, -1, 1, -1)\}$;

   (b) $\{(1, -1, 1, -1), (-1, 1, -1, 1)\}$;

   (c) $\{(1, -1, 1, -1), (1, 1, 1, 1)\}$;

   (d) $\{(1, 1, 1, 1), (-1, 1, -1, 1), (0, 2, 0, 2)\}$;

   (e) $\{(1, 1, 1, 1), (-1, 1, -1, 1), (0, 2, 0, 3)\}$;

   (f) $\{(1, 1, 1, 1), (-1, 1, -1, 1), (0, 0, 0, 0)\}$.

5. Which of the following sets are bases for $\mathbb{R}^3$?

    (a) $\{(1,0,0),(0,0,1)\}$;

    (b) $\{(2,-3,4),(-4,5,-6)\}$;

    (c) $\{(1,1,0),(1,-1,0),(0,0,1)\}$;

    (d) $\{(1,1,1),(1,-1,0),(0,0,1)\}$;

    (e) $\{(1,0,0),(0,1,0),(0,0,1),(1,1,1)\}$.

6. Which of the following sets are linear subspaces of $\mathbb{B}^5$?

    (a) $\{00000\}$;

    (b) $\{11111\}$;

    (c) $\{00000,11111\}$;

    (d) $\{01010,10101\}$;

    (e) $\{00000,01010,10101\}$;

    (f) $\{00000,01010,10101,11111\}$;

    (g) $\{10000,01000,00100,00010,00001\}$;

    (h) $\{00000,10000,01000,00100,00010,00001\}$;

    (i) $\{00000,00100,01010,10001,01110,10101,11011,11111\}$.

7. Determine which of the following subsets of $\mathbb{B}^4$ are linear subspaces and write down all the cosets of those which are.

    (a) $\{0000,1100,0011,1001\}$;

    (b) $\{0000,1100,0011,1111\}$;

    (c) $\{0000,1000,0010,1010\}$;

    (d) $\{1100,0011,1001,1111\}$.

8. Construct a linear subspace of $\mathbb{B}^6$ of dimension 3 and find the weight of each of its elements.

9. Construct a table showing the Hamming distance between each pair of elements in $\mathbb{B}^4$.

10. Find the dimension and the minimum distance of each of the following linear codes in $\mathbb{B}^4$ or $\mathbb{B}^5$.

    (a) $\{0000,1000,0001,1001\}$;

    (b) $\{0000,1010,0101,1111\}$;

    (c) $\{0000,1000,0110,1110\}$;

    (d) $\{0000,1100,0110,0011,1001,1010,0101,1111\}$;

    (e) $\{00000,10101,01010,11111\}$;

(f) $\{00000, 01000, 00010, 01010\}$;

(g) $\{00000, 11000, 01110, 00011, 10110, 11011, 01101, 10101\}$.

11. Write down a generator matrix for each of the linear codes in the previous exercise.

12. Find the code words of the linear codes defined by the following generator matrices:

   (a)
   $$G = \begin{bmatrix} 0\,0\,1 \\ 1\,1\,0 \end{bmatrix};$$

   (b)
   $$G = \begin{bmatrix} 0\,1\,1\,0 \\ 1\,1\,0\,1 \end{bmatrix};$$

   (c)
   $$G = \begin{bmatrix} 1\,1\,0\,0 \\ 0\,1\,1\,0 \\ 0\,0\,1\,1 \end{bmatrix};$$

   (d)
   $$G = \begin{bmatrix} 1\,1\,1\,0\,0 \\ 0\,0\,0\,1\,1 \end{bmatrix};$$

   (e)
   $$G = \begin{bmatrix} 1\,1\,0\,0\,0 \\ 1\,0\,1\,0\,1 \\ 0\,0\,0\,1\,1 \end{bmatrix};$$

   (f)
   $$G = \begin{bmatrix} 1\,1\,1\,1\,0 \\ 1\,1\,1\,0\,1 \\ 1\,1\,0\,1\,1 \\ 1\,0\,1\,1\,1 \end{bmatrix}.$$

13. Reduce the generator matrices in the previous exercise to canonical form and find their parity check matrices.

14. Find the linear codes defined by the following parity check matrices:

(a)

$$H = \begin{bmatrix} 1 & 1 & 0 \\ 0 & 1 & 1 \end{bmatrix};$$

(b)

$$H = \begin{bmatrix} 1 & 0 & 0 & 1 \\ 0 & 1 & 0 & 1 \\ 0 & 0 & 1 & 1 \end{bmatrix};$$

(c)

$$H = \begin{bmatrix} 1 & 1 & 0 & 0 \\ 0 & 1 & 1 & 0 \\ 0 & 0 & 1 & 1 \end{bmatrix};$$

(d)

$$H = \begin{bmatrix} 1 & 1 & 1 & 0 & 0 \\ 0 & 0 & 1 & 1 & 1 \end{bmatrix};$$

(e)

$$H = \begin{bmatrix} 1 & 1 & 1 & 0 & 0 \\ 1 & 1 & 1 & 1 & 1 \\ 0 & 0 & 1 & 1 & 1 \end{bmatrix}.$$

15. For each of the parity check matrices in the previous exercise, construct the syndrome table and use it to decode the following:

    (a) $100, 010, 001, 111$;
    (b) $1000, 0100, 0010, 0001$;
    (c) $1110, 1101, 1011, 0111$;
    (d) $10101, 01010, 11111$;
    (e) $10101, 01010, 11111$.

*16. Suppose that the code words of a k-dimensional linear code of length n are arranged as the rows of a matrix with 2^k rows and n columns, with no column consisting only of 0s. Show that each column consists of 2^{k-1} 0s and 2^{k-1} 1s. Use this to show that the sum of the weights of the code words is $n2^{k-1}$.

*17. Use the result of the previous Exercise to show that the minimum distance of a k-dimensional linear code of length n is no more than $n2^{k-1}/(2^k - 1)$.

*18. Show that a k-dimensional linear code of length n whose minimum distance is more than $2m$ must have at least $\log_2(1 + C_1^n + C_2^n + \ldots + C_m^n)$ parity check bits, where C_p^n denotes the number of combinations of p objects taken from a collection of n objects.

*19. Suppose we have the codes

$$\{000, 011, 101, 110\}$$

in which the third bit is the parity of the first two, and

$$\{0000, 0011, 0101, 0110, 1001, 1010, 1100, 1111\}$$

where the fourth bit is the parity of the first three. We can use these to construct a code of length 12 with 6 information bits and 6 parity check bits by arranging the 6 information bits in an array with 2 rows and 3 columns, adding the parity check bits of the rows and the columns and completing the array with the parity of the parity row (or column). For example, the word 000111 is arranged as

$$\begin{matrix} 0 & 0 & 0 \\ 1 & 1 & 1 \end{matrix}$$

the parity check bits are added to the rows and columns to give

$$\begin{matrix} 0 & 0 & 0 & 0 \\ 1 & 1 & 1 & 1 \\ 1 & 1 & 1 & \end{matrix}$$

and adding the final parity check bit

$$\begin{matrix} 0 & 0 & 0 & 0 \\ 1 & 1 & 1 & 1 \\ 1 & 1 & 1 & 1 \end{matrix}$$

completes the array. The rows are then concatenated to give the code word 000011111111. Write down the generator matrix of this code.

*20. The construction of the previous Exercise, which is due to Elias [2], can be applied to any pair of codes of length m and n to produce a code of length mn. Show that the minimum distance of the resulting code is equal to the product of the minimum distances of the two codes used in the construction.

6.10 References

[1] R. Ash, *Information Theory*, John Wiley & Sons, New York, 1965.

[2] P. Elias, Error-free coding, *Transactions of the I.R.E. Professional Group on Information Theory*, PGIT-4, 29–37, September 1954.

[3] J. B. Fraleigh, *A First Course in Abstract Algebra,* 5th ed., Addison-Wesley, Reading, MA, 1994.

[4] R. J. McEliece, *The Theory of Information and Coding,* 2nd ed., Cambridge University Press, Cambridge, 2002.

[5] W. W. Peterson, *Error-correcting Codes,* MIT Press, Cambridge, MA, 1961.

Chapter 7

Cyclic Codes

7.1 Introduction

The use of the linear space structure of $\mathbb{B}^n$ enabled us to define binary codes whose encoding and decoding operations depend on this structure. The development of more powerful codes requires the use of additional algebraic structure, in particular, a ring structure on $\mathbb{B}^n$. In this chapter, we will develop the algebraic concepts that enable us to define a ring structure on $\mathbb{B}^n$ and apply them to the design of *cyclic codes*. As in the previous chapter, the presentation of the algebraic concepts will be informal; a rigorous treatment can be found in textbooks such as [1].

7.2 Rings of Polynomials

In the previous sections on linear codes we have made use of the facts that $\mathbb{B}$ is a field and $\mathbb{B}^n$ can be made into a linear space over $\mathbb{B}$ by defining the operations of addition and multiplication by a scalar component by component.

There are other algebraic structures that can also be introduced on $\mathbb{B}^n$.

DEFINITION 7.1 Polynomial of degree n   A polynomial of degree n with coefficients in a ring R is an element of R^{n+1}.

We may also refer to polynomials with coefficients in a ring R as polynomials over R.

This does not look like the more familiar definition of a polynomial function, say, $p(x) = ax^2 + bx + c$, but it is equivalent. The essential information about a polynomial is contained in its coefficients, and that is what the components of an element of R^{n+1} give us.

There is additional structure associated with polynomials, and to emphasize this we will not denote polynomials by $(n+1)$-tuples such as $(r_0, r_1, r_2, \ldots, r_n)$, but by the notation $r_0 + r_1 X + r_2 X^2 + \ldots + r_n X^n$.

Polynomials over R may be added and multiplied by elements of R. Addition is carried out by adding the coefficients of like powers of X, while multiplication is carried out by multiplying every coefficient by the multiplier.

EXAMPLE 7.1

The following are polynomials over the ring of integers, $\mathbb{Z}$:

$$p_1(X) = 1 + 2X + 3X^2 + 4X^3 + 5X^4$$

$$p_2(X) = 1 - 2X + 3X^2 - 4X^3 + 5X^4$$

$$p_3(X) = 1 - 3X^2 + 3X^4 - X^6$$

$$p_4(X) = 4X^7 - 8X^8 + 16X^9$$

$$p_5(X) = X^{10} - X^{100}.$$

Adding these polynomials means adding the coefficients:

$$(p_1 + p_2)(X) = (1+1) + (2-2)X + (3+3)X^2 + (4-4)X^3 + (5+5)X^4$$
$$= 2 + 6X^2 + 10X^4$$

$$(p_2 + p_3)(X) = (1+1) + 2X + (3-3)X^2 - 4X^3 + (5+3)X^4 - X^6$$
$$= 2 + 2X - 4X^3 + 8X^4 - X^6$$

$$(p_4 + p_5)(X) = 4X^7 - 8X^8 + 16X^9 + x^{10} - X^{100}.$$

Multiplying a polynomial by an integer means multiplying all the coefficients:

$$8p_1(X) = 8 + 16X + 24X^2 + 32X^3 + 40X^4$$

$$(-5)p_3(X) = -5 + 15X^2 - 15X^4 + 5X^6$$

$$100p_5(X) = 100X^{10} - 100X^{100}.$$

❑

EXAMPLE 7.2

We also have polynomials over the binary field $\mathbb{B}$:

$$p_1(X) = 1 + X + X^2 + X^3 + X^4$$

$$p_2(X) = 1 + X^2 + X^4$$
$$p_3(X) = 1 + X^4 + X^6$$
$$p_4(X) = X^7 + X^8 + X^9$$
$$p_5(X) = X^{10} + X^{100}.$$

Adding these polynomials uses the rules for addition in $\mathbb{B}$:

$$(p_1 + p_2)(X) = (1+1) + X + (X^2 + X^2) + X^3 + (X^4 + X^4)$$
$$= X + X^3$$

$$(p_2 + p_3)(X) = X^2 + X^6$$
$$(p_3 + p_4)(X) = 1 + X^4 + X^6 + X^7 + X^8 + X^9.$$

Multiplying polynomials by elements of $\mathbb{B}$ is trivial: we either multiply by 0 to get the zero polynomial, or multiply by 1, which gives the same polynomial. $\square$

If R is a field, then these operations make the set of polynomials of degree n over R into a linear space. Since $\mathbb{B}$ is a field, the set of polynomials of degree n over $\mathbb{B}$ is a linear space; it is clearly the space $\mathbb{B}^{n+1}$. We identify elements of the linear space $\mathbb{B}^{n+1}$ with polynomials by matching the components of the vector with coefficients of the polynomial, matching the leftmost bit with the constant term. So, for $n+1 = 8$, 11000111 is matched with $1 + X + X^5 + X^6 + X^7$.

EXAMPLE 7.3

For $n = 1$, we identify 00 with (the polynomial) 0, 10 with (the polynomial) 1, 01 with X and 11 with $1 + X$.

For $n = 2$, we identify 010 with X, 011 with $X + X^2$, 111 with $1 + X + X^2$, and so on.

For $n = 9$, we identify 101010101 with $1 + X^3 + X^5 + X^7 + X^9$, 000111000 with $X^3 + X^4 + X^5$, 0110001111 with $X + X^2 + X^6 + X^7 + X^8 + X^9$, and so on. $\square$

We can also define a multiplication operation on polynomials. To multiply polynomials, we proceed term by term, multiplying the coefficients and adding the exponents of X, so that the product of aX^m and bX^n is $(ab)X^{m+n}$.

EXAMPLE 7.4

Here are some examples of multiplying polynomials with coefficients in $\mathbb{Z}$:

$$(1 + 2X + 3X^2) \times (4 + 5X) = 4 + 5X + 8X + 10X^2 + 12X^2 + 15X^3$$
$$= 4 + 13X + 22X^2 + 15X^3$$

$$(4 - 3X^2) \times (2X - X^3) = 8X - 4X^3 - 6X^3 + 3X^5$$
$$= 8X - 10X^3 + 3X^5$$

$$(X + X^3) \times (X^5 - X^9) = X^6 + X^8 - X^{10} - X^{12}.$$

⬚

EXAMPLE 7.5

Here are some examples of multiplying polynomials over $\mathbb{B}$:

$$(1 + X + X^2) \times (1 + X) = 1 + X + X + X^2 + X^2 + X^3$$
$$= 1 + X^3$$

$$(1 + X^2) \times (X + X^3) = X + X^3 + X^3 + X^5$$
$$= X + X^5$$

$$(X + X^3) \times (X^5 + X^9) = X^6 + X^8 + X^{10} + X^{12}.$$

If we represent polynomials as elements of $\mathbb{B}^{n+1}$, we can also write these products as

$$111 \times 11 = 1001$$

$$101 \times 0101 = 010001$$

and

$$0101 \times 0000010001 = 0000001010101.$$

⬚

We shall denote the ring of all polynomials with coefficients in $\mathbb{B}$ with operations of addition and multiplication defined above by $\mathbb{B}[X]$.

We can also define a division operation for polynomials. If the ring of coefficients is a field, we can use the *synthetic division* algorithm that is taught in high school algebra courses to carry out the division. In most cases, the division will not be exact, and there will be a remainder.

EXAMPLE 7.6

Consider $X + 4$ and $X^3 + 2X^2 - 5X + 15$ to be polynomials with coefficients in the field of real numbers. We start the synthetic division by setting them out as follows:

$$X + 4 \overline{\smash{\big)}\, X^3 + 2X^2 - 5X + 15}$$

We multiply $X + 4$ by X^2, and subtract the result from $X^3 + 2X^2 - 5X + 15$:

$$
\begin{array}{r}
X^2 \\
X + 4 \,\overline{\big)\, X^3 + 2X^2 - 5X + 15} \\
\underline{X^3 + 4X^2 } \\
- 2X^2
\end{array}
$$

We bring down the $-5X$, multiply $X + 4$ by $-2X$ and subtract the result from $-2X^2 - 5X$:

$$
\begin{array}{r}
X^2 - 2X \\
X + 4 \,\overline{\big)\, X^3 + 2X^2 - 5X + 15} \\
\underline{X^3 + 4X^2 } \\
- 2X^2 - 5X \\
\underline{- 2X^2 - 8X } \\
3X
\end{array}
$$

We bring down the 15, multiply $X + 4$ by 3 and subtract the result from $3X + 15$:

$$
\begin{array}{r}
X^2 - 2X + 3 \\
X + 4 \,\overline{\big)\, X^3 + 2X^2 - 5X + 15} \\
\underline{X^3 + 4X^2 } \\
- 2X^2 - 5X \\
\underline{- 2X^2 - 8X } \\
3X + 15 \\
\underline{3X + 12} \\
3
\end{array}
$$

This shows that

$$X^3 + 2X^2 - 5X + 15 = (X + 4)(X^2 - 2X + 3) + 3.$$

⧠

EXAMPLE 7.7

We can perform synthetic division in the same way when the coefficients of the polynomials come from the binary field $\mathbb{B}$. In this case, addition and subtraction are the same operation, so the procedure uses addition.

To divide $X^3 + X^2 + X + 1$ by $X + 1$ we start by setting the polynomials out in the usual way:

$$
X + 1 \,\overline{\big)\, X^3 + X^2 + X + 1}
$$

We multiply $X + 1$ by X^2, and add the result to $X^3 + X^2 + X + 1$:

$$
\begin{array}{r}
X^2 \\
X + 1 \,\overline{\big)\, X^3 + X^2 + X + 1} \\
\underline{X^3 + X^2 } \\
0
\end{array}
$$

We bring down the $X + 1$, multiply $X + 1$ by 1 and add the result to $X + 1$:

$$
\begin{array}{r}
X^2 \quad + \quad 1 \\
X + 1 \,\overline{\big)\, X^3 + X^2 + X + 1} \\
\underline{X^3 + X^2} \\
X + 1 \\
\underline{X + 1} \\
0
\end{array}
$$

There is no remainder. This shows that

$$X^3 + X^2 + X + 1 = (X + 1)(X^2 + 1).$$

☐

EXAMPLE 7.8

To divide $X^5 + 1$ by $X^2 + X$ in the ring of polynomials over $\mathbb{B}$, we set out the polynomials in the usual way:

$$X^2 + X \,\overline{\big)\, X^5 \qquad + 1}$$

leaving space for the terms that are not present in $X^5 + 1$.

We multiply $X^2 + X$ by X^3 and add the product to $X^5 + 1$:

$$
\begin{array}{r}
X^3 \\
X^2 + X \,\overline{\big)\, X^5 \qquad\qquad\qquad + 1} \\
\underline{X^5 + X^4} \\
X^4
\end{array}
$$

We multiply $X^2 + X$ by X^2 and add the product to X^4:

$$
\begin{array}{r}
X^3 + X^2 \\
X^2 + X \,\overline{\big)\, X^5 \qquad\qquad\qquad + 1} \\
\underline{X^5 + X^4} \\
X^4 \\
X^4 + X^3 \\
\underline{X^3}
\end{array}
$$

We multiply $X^2 + X$ by X and add the product to X^3:

$$
\begin{array}{r}
X^3 + X^2 + X \\
X^2 + X \,\overline{\big)\, X^5 \qquad\qquad\qquad + 1} \\
\underline{X^5 + X^4} \\
X^4 \\
X^4 + X^3 \\
\underline{X^3} \\
X^3 + X^2 \\
\underline{X^2}
\end{array}
$$

We bring down the 1 and add $X^2 + X$ to $X^2 + 1$:

$$
\begin{array}{r}
X^3 + X^2 + X + 1 \\
X^2 + X \overline{\smash{\big)}\ X^5 \hspace{4.5cm} + 1} \\
\underline{X^5 + X^4} \hspace{3.2cm} \\
\overline{X^4} \hspace{3.0cm} \\
X^4 + X^3 \hspace{2.0cm} \\
\underline{\overline{X^3}} \hspace{1.8cm} \\
X^3 + X^2 \hspace{0.8cm} \\
\overline{X^2}\quad + 1 \\
X^2 + X \\
\overline{X + 1}
\end{array}
$$

The remainder is $X + 1$. This shows that

$$X^5 + 1 = (X^2 + X)(X^3 + X^2 + X^1 + X) + (X + 1).$$

⬚

If we multiply two polynomials of degree n together, the result is usually a polynomial of degree $2n$. This means that we cannot make the set of polynomials of degree n into a ring under the operations of polynomial addition and multiplication, as multiplying two polynomials in this set will give us a polynomial that is not in this set.

There is another way that we can define a multiplication operation on polynomials that will give a useful ring structure.

DEFINITION 7.2 Multiplication modulo a polynomial   *Let p, q and r be polynomials with coefficients in some ring R. The* product of p and q modulo r is *the remainder when pq is divided by r.*

EXAMPLE 7.9

Let $p(X) = X^2 - 3$, $q(X) = X^2 + 5$ and $r(X) = X^3 - X^2$, where these polynomials have coefficients in the field of real numbers, $\mathbb{R}$.

The product of p and q is

$$p(X)q(X) = (X^2 - 3)(X^2 + 5) = X^4 + 2X^2 - 15.$$

Using synthetic division as in the examples above, we find that

$$X^4 + 2X^2 - 15 = (X + 1)(X^3 - X^2) + (-X^2 - 15).$$

So the product of $(X^2 - 3)$ and $(X^2 + 5)$ modulo $(X^3 - X^2)$ is $(-X^2 - 15)$.

⬜

EXAMPLE 7.10

Let $p(X) = X^3 + X$, $q(X) = X^2 + X$ and $r(X) = X^4 + X^2$ be polynomials with coefficients in $\mathbb{B}$.

The product of p and q is

$$p(X)q(X) = (X^3 + X)(X^2 + X) = X^5 + X^4 + X^3 + X^2.$$

Using synthetic division as in the examples above, we find that

$$X^5 + X^4 + X^3 + X^2 = (X + 1)(X^4 + X^2).$$

There is no remainder, so the product of $(X^3 + X)$ and $(X^2 + X)$ modulo $(X^4 + X^2)$ is 0.

⬜

EXAMPLE 7.11

Let $p(X) = X^4 + X^2$, $q(X) = X^4 + X$ and $r(X) = X^5 + X$ be polynomials with coefficients in $\mathbb{B}$.

The product of p and q is

$$p(X)q(X) = (X^4 + X^2)(X^4 + X) = X^8 + X^6 + X^5 + X^3.$$

Using synthetic division as in the examples above, we find that

$$X^8 + X^6 + X^5 + X^3 = (X^3 + X)(X^5 + X) + (X^4 + X^3 + X^2 + X).$$

So the product of $(X^4 + X^2)$ and $(X^4 + X)$ modulo $(X^5 + X)$ is $(X^4 + X^3 + X^2 + X)$.

⬜

The operation of multiplication modulo the polynomial $X^n + 1$ gives us products which are polynomials of degree $n - 1$ or less. The set of polynomials of degree less than n forms a ring with respect to addition and multiplication modulo $X^n + 1$; this ring will be denoted $\mathbb{B}_n[X]/(X^n + 1)$. It can be identified with $\mathbb{B}^n$ and the identification can be used to give $\mathbb{B}^n$ the structure of a ring.

The *Remainder Theorem* of high school algebra states that the remainder that is obtained when a polynomial $p(X)$ with real coefficients is divided by $(X - a)$ can be calculated by replacing a for X in $p(X)$. In particular we have the following:

RESULT 7.1

The remainder that is obtained when a polynomial in $\mathbb{B}[X]$ is divided by $X^n + 1$ can be calculated by replacing X^n with 1 in the polynomial. This operation "wraps" the powers of X around from X^n to $X^0 = 1$.

EXAMPLE 7.12

Using the rule above, we can quickly calculate some remainders.

The remainder of $X^3 + X^2 + X$ when divided by $X^3 + 1$ is $1 + X^2 + X = X^2 + X + 1$.

The remainder of $X^4 + X^3 + X^2$ when divided by $X^3 + 1$ is $1X + 1 + X^2 = X^2 + X + 1$.

The remainder of $X^7 + X^6 + X^5$ when divided by $X^4 + 1$ is $1X^3 + 1X^2 + 1X = X^3 + X^2 + X$.

The product of $X^3 + X^2$ and $X^2 + X$ modulo $X^4 + 1$ is the remainder when $(X^3 + X^2)(X^2 + X) = X^5 + X^3$ is divided by $X^4 + 1$, namely $1X + X^3 = X^3 + X$.

$\square$

We can now compute the multiplication tables of the rings $\mathbb{B}_n[X]/(X^n + 1)$.

EXAMPLE 7.13

The elements of $\mathbb{B}_2[X]/(X^2 + 1)$ are the polynomials 0, 1, X and $1 + X$. Multiplication by 0 and 1 is trivial. The other products are

$(X)(X)$ modulo $(X^2 + 1) = X^2$ modulo $(X^2 + 1) = 1$,

$(X)(1 + X)$ modulo $(X^2 + 1) = X + X^2$ modulo $(X^2 + 1) = 1 + X$,

$(1 + X)(1 + X)$ modulo $(X^2 + 1) = 1 + X^2$ modulo $(X^2 + 1) = 0$.

So the multiplication table is

	0	1	X	$1 + X$
0	0	0	0	0
1	0	1	X	$1 + X$
X	0	X	1	$1 + X$
$1 + X$	0	$1 + X$	$1 + X$	0

Identifying the polynomial 0 with 00, 1 with 10, X with 01 and $1 + X$ with 11, the multiplication table becomes

	00	10	01	11
00	00	00	00	00
10	00	10	01	11
01	00	01	10	11
11	00	11	11	00

EXAMPLE 7.14

The elements of $\mathbb{B}_3[X]/(X^3 + 1)$ are the polynomials $0, 1, X, 1 + X, X^2, 1 + X^2,$ $X + X^2, 1 + X + X^2$. The multiplication table is too large to fit on the page, so we display it in two parts:

	0	1	X	$1 + X$
0	0	0	0	0
1	0	1	X	$1 + X$
X	0	X	X^2	$X + X^2$
$1 + X$	0	$1 + X$	$X + X^2$	$1 + X^2$
X^2	0	X^2	1	$1 + X^2$
$1 + X^2$	0	$1 + X^2$	$1 + X$	$X + X^2$
$X + X^2$	0	$X + X^2$	$1 + X^2$	$1 + X$
$1 + X + X^2$	0	$1 + X + X^2$	$1 + X + X^2$	0

	X^2	$1 + X^2$	$X + X^2$	$1 + X + X^2$
0	0	0	0	0
1	X^2	$1 + X^2$	$X + X^2$	$1 + X + X^2$
X	1	$1 + X$	$1 + X^2$	$1 + X + X^2$
$1 + X$	$1 + X$	$X + X^2$	$1 + X$	0
X^2	X	$X + X^2$	$1 + X$	$1 + X + X^2$
$1 + X^2$	$1 + X^2$	$1 + X$	$1 + X^2$	0
$X + X^2$	$1 + X$	$1 + X^2$	$X + X^2$	0
$1 + X + X^2$	$1 + X + X^2$	0	0	$1 + X + X^2$

Identifying the polynomial 0 with 000, 1 with 100, X with 010, $1 + X$ with 110, X^2 with 001, and so on, the multiplication table becomes

	000	100	010	110	001	101	011	111
000	000	000	000	000	000	000	000	000
100	000	100	010	110	001	101	011	111
010	000	010	001	011	100	010	101	111
110	000	110	011	101	110	011	110	000
001	000	001	100	101	010	011	110	111
101	000	101	110	011	101	110	101	000
011	000	011	101	110	110	101	011	000
111	000	111	111	000	111	000	000	111

The following result is very important.

RESULT 7.2
Multiplication by X in $\mathbb{B}_n[X]/(X^n + 1)$ is equivalent to shifting the components of the corresponding vector in $\mathbb{B}^n$ cyclically one place to the right.

PROOF   Let $p(X)$ be a polynomial in $\mathbb{B}_n[X]/(X^n + 1)$,

$$p(X) = b_0 + b_1 X + b_2 X^2 + \ldots + b_{n-1}X^{n-1}. \tag{7.1}$$

The corresponding vector in $\mathbb{B}^n$ is $b_0 b_1 b_2 \ldots b_{n-1}$.

$$\begin{aligned}
Xp(X) &= b_0 X + b_1 X^2 + b_2 X^3 + \ldots + b_{n-1}X^n \\
&= b_{n-1} + b_0 X + b_1 X^2 + \ldots + b_{n-1}(X^n + 1)
\end{aligned} \tag{7.2}$$

so

$$Xp(X) \bmod (X^n + 1) = b_{n-1} + b_0 X + \ldots b_{n-2}X^{n-1}. \tag{7.3}$$

The vector corresponding to $Xp(X) \bmod (X^n + 1)$ is $b_{n-1}b_0 b_1 \ldots b_{n-2}$, which is the result of cyclically shifting $b_0 b_1 b_2 \ldots b_{n-1}$ one place to the right.

□

7.3   Cyclic Codes

We now look at codes whose construction uses both the additive and multiplicative structures of the ring $\mathbb{B}_n[X]/(X^n + 1)$. Since the elements of this ring can be represented both as code words in $\mathbb{B}^n$ and as polynomials, we will use the terms interchangeably as convenient.

DEFINITION 7.3 Cyclic Code   *A cyclic code is a linear code with the property that any cyclic shift of a code word is also a code word.*

EXAMPLE 7.15

Consider the code whose generator matrix is

$$G = \begin{bmatrix} 1 & 0 & 1 & 0 \\ 0 & 1 & 0 & 1 \end{bmatrix}.$$

The linear code generated by G is $\{0000, 1010, 0101, 1111\}$. Note that shifting 0000 cyclically gives 0000, shifting 1010 one place cyclically gives 0101, shifting 0101 one place cyclically gives 1010 and shifting 1111 cyclically gives 1111. So this is a cyclic code.

In polynomial notation, the code is $\{0, 1 + X^2, X + X^3, 1 + X + X^2 + X^3\}$. A cyclic shift to the right can be accomplished by multiplying by X modulo $(X^4 + 1)$. Multiplying 0 by X gives 0, multiplying $1 + X^2$ by X gives $X + X^3$, multiplying $X + X^3$ by X (modulo $X^4 + 1$) gives $1 + X^2$ and multiplying $1 + X + X^2 + X^3$ by X (also modulo $X^4 + 1$) gives $1 + X + X^2 + X^3$.

⬚

RESULT 7.3

A cyclic code contains a unique non-zero polynomial of minimal degree.

PROOF   There are only a finite number of non-zero polynomials in the code, so at least one of them is of minimal degree.

Suppose that there are two polynomials of minimum degree r, say

$$g(X) = g_0 + g_1 X + \dots X^r \tag{7.4}$$

and

$$h(X) = h_0 + h_1 X + \dots X^r. \tag{7.5}$$

(The X^r terms must be present if the degree of the polynomials is r.)

The sum of these polynomials is

$$(g + h)(X) = (g_0 + h_0) + (g_1 + h_1)X + \dots (g_{r-1} + h_{r-1})X^{r-1} \tag{7.6}$$

$(g + h)$ must belong to the code, but it is a polynomial of degree $(r - 1)$, contradicting the assumption that the polynomials of minimum degree are of degree r. Hence the polynomial of minimum degree must be unique.

⬚

DEFINITION 7.4 Generator Polynomial   *The unique non-zero polynomial of minimal degree in a cyclic code is the* generator polynomial *of the code.*

RESULT 7.4
If $g \in \mathbb{B}_n[X]/(X^n + 1)$ is the generator polynomial for some cyclic code, then every polynomial in the code can be generated by multiplying g by some polynomial in $\mathbb{B}_n[X]/(X^n + 1)$.

PROOF Since multiplication by X modulo $(X^n + 1)$ has the effect of shifting a code word cyclically one place to the right, multiplying by X^k modulo $(X^n + 1)$ has the effect of shifting a code word cyclically k places to the right. It follows that the product g by X^k modulo $(X^n + 1)$ will be a code word for any $k \geq 0$.

Multiplying g by any polynomial is equivalent to multiplying g by various powers of X modulo $(X^n + 1)$ and adding the products together. Since the products are all code words, so is their sum.

To show that every code word can be generated in this way, note that if the degree of g is r, then the polynomials $g, gX, gX^2 \ldots, gX^{n-r-1}$ form a basis for the code and hence that every code word can be generated by taking a linear combination of these basis elements.

$\square$

If the generator of a cyclic code is $g(X) = g_0 + g_1 X + \ldots + g_r X^r$, the fact that the polynomials $g, gX, gX^2 \ldots, gX^{n-r-1}$ form a basis for the code means that the generator matrix of the code can be written in the form

$$
G = \begin{bmatrix}
g_0 & g_1 & \cdots & g_r & 0 & 0 & \cdots & 0 \\
0 & g_0 & \cdots & g_{r-1} & g_r & 0 & \cdots & 0 \\
& & & \vdots & & & & \\
0 & \cdots & 0 & g_0 & g_1 & \cdots & g_r & 0 \\
0 & \cdots & 0 & 0 & g_0 & \cdots & g_{r-1} & g_r
\end{bmatrix}.
\tag{7.7}
$$

G is a cyclic matrix (each row is obtained by shifting the previous row one column to the right).

EXAMPLE 7.16

$\{0000000, 1011100, 0101110, 0010111, 1001011, 1100101, 1110010, 0111001\}$ is a cyclic code. The code word 1011100 corresponds to the polynomial $1 + X^2 + X^3 + X^4$, which is the polynomial of minimal degree and hence the generator polynomial.

The generator matrix is
$$
G = \begin{bmatrix}
1 & 0 & 1 & 1 & 1 & 0 & 0 \\
0 & 1 & 0 & 1 & 1 & 1 & 0 \\
0 & 0 & 1 & 0 & 1 & 1 & 1
\end{bmatrix}.
$$

We interchange the third and seventh columns to reduce the generator matrix to canonical form:
$$
G_c = \begin{bmatrix}
1 & 0 & 0 & 1 & 1 & 0 & 1 \\
0 & 1 & 0 & 1 & 1 & 1 & 0 \\
0 & 0 & 1 & 0 & 1 & 1 & 1
\end{bmatrix}.
$$

The canonical form of the parity check matrix is:

$$H_c = \begin{bmatrix} 1 & 1 & 0 & 1 & 0 & 0 & 0 \\ 1 & 1 & 1 & 0 & 1 & 0 & 0 \\ 0 & 1 & 1 & 0 & 0 & 1 & 0 \\ 1 & 0 & 1 & 0 & 0 & 0 & 1 \end{bmatrix}.$$

Interchanging the third and seventh columns gives us the parity check matrix of G:

$$H = \begin{bmatrix} 1 & 1 & 0 & 1 & 0 & 0 & 0 \\ 1 & 1 & 0 & 0 & 1 & 0 & 1 \\ 0 & 1 & 0 & 0 & 0 & 1 & 1 \\ 1 & 0 & 1 & 0 & 0 & 0 & 1 \end{bmatrix}.$$

☐

The next result gives us a way of finding generator polynomials for cyclic codes.

RESULT 7.5
g is the generator polynomial of a cyclic code in $\mathbb{B}_n[X]/(X^n + 1)$ if and only if it is a factor of $(X^n + 1)$.

PROOF   Let g by the generator polynomial of a cyclic code. If the degree of g is r, then $X^{n-r}g$ is a polynomial of degree n. Let h be the remainder when $X^{n-r}g$ is divided by $(X^n + 1)$. The quotient is 1, so we have

$$X^{n-r}g = (X^n + 1) + h. \tag{7.8}$$

h is the result of multiplying g by X^{n-r} modulo $(X^n + 1)$, so it belongs to the code. Therefore, there is a polynomial $p \in \mathbb{B}_n[X]/(X^n + 1)$ such that

$$h = pg. \tag{7.9}$$

Adding these equations, we get

$$(X^n + 1) + h + h = X^{n-r}g + pg, \tag{7.10}$$

or

$$(X^n + 1) = (X^{n-r} + p)g, \tag{7.11}$$

which shows that g is a factor of $(X^n + 1)$.

Conversely, if g divides $(X^n + 1)$, there is a polynomial h such that

$$g(X)h(X) = (X^n + 1), \tag{7.12}$$

and the polynomials g, Xg, X^2g, ..., $X^{n-r-1}g$ represent linearly independent code words obtained by cyclically shifting the bits of g. These form the basis of the linear code generated by g.

Let $p(X) = p_0g(X) + p_1Xg(X) + \ldots + p_{n-r-1}X^{n-r-1}g(X)$ be an arbitrary element of the code generated by g. Then

$$Xp(X) = p_0Xg(X) + p_1X^2g(X) + \ldots + p_{n-r-1}X^{n-r}g(X). \tag{7.13}$$

If $p_{n-r-1} = 0$, then $Xp(X)$ is a linear combination of the basis vectors $Xg(X)$, $X^2g(X)\ldots X^{n-r-1}g(X)$, and so it belongs to the code.

If $p_{n-r-1} = 1$, $Xp(X)$ is a polynomial of degree n. The remainder when it is divided by $(X^n + 1)$ is

$$r(X) = p_{n-r-1} + p_0Xg(X) + \ldots + p_{n-r-2}X^{n-r-1}g(X), \tag{7.14}$$

so

$$r(X) = Xp(X) + p_{n-r-1}(X^n + 1). \tag{7.15}$$

By construction,

$$Xp(X) = (p_0X + p_1X^2 + \ldots + p_{n-r-1}X^{n-r})g(X), \tag{7.16}$$

and by assumption

$$g(X)h(X) = (X^n + 1), \tag{7.17}$$

so

$$r(X) = (p_0X + p_1X^2 + \ldots + p_{n-r-1}X^{n-r} + h(X))g(X). \tag{7.18}$$

This shows that $r(X)$ is a linear combination of the basis polynomials g, Xg, X^2g, ..., $X^{n-r-1}g$ and so $r(X)$ belongs to the code. Hence the code generated by $g(X)$ is cyclic. ∎

$(X^n + 1)$ always has at least two factors, because

$$(X^n + 1) = (X + 1)(X^{n-1} + X^{n-2} + \ldots + X^2 + X + 1). \tag{7.19}$$

The factorizations of $(X^n + 1)$ determine the cyclic codes of length n. There is a cyclic code for each factor.

EXAMPLE 7.17

The factorization of $(X^9 + 1)$ is

$$(X^9 + 1) = (1 + X)(1 + X + X^2)(1 + X^3 + X^6).$$

This gives six factors and six cyclic codes.

The cyclic code with generator polynomial $(1 + X)$ has a basis of eight code words:

$$\{110000000, 011000000, 001100000, 000110000,$$
$$000011000, 000001100, 000000110, 000000011\}.$$

The cyclic code with generator polynomial $(1 + X + X^2)$ has a basis of seven code words:

$$\{111000000, 011100000, 001110000,$$
$$000111000, 000011100, 000001110, 000000111\}.$$

The cyclic code with generator polynomial $(1 + X)(1 + X + X^2) = (1 + X^3)$ has a basis of six code words:

$$\{100100000, 010010000, 001001000, 000100100, 000010010, 000001001\}.$$

The cyclic code with generator polynomial $(1 + X^3 + X^6)$ has a basis of three code words:
$$\{100100100, 010010010, 001001001\}.$$

The cyclic code with generator polynomial $(1 + X)(1 + X^3 + X^6) = (1 + X + X^3 + X^4 + X^6 + X^7)$ has a basis of two code words:

$$\{110110110, 011011011\}.$$

The cyclic code with generator polynomial $(1 + X + X^2)(1 + X^3 + X^6) = (1 + X + X^2 + X^3 + X^4 + X^5 + X^6 + X^7 + X^8)$ has a basis with one code word: $\{111111111\}.$

⬜

7.4 Encoding and Decoding of Cyclic Codes

Cyclic codes are linear codes; so the techniques described in the previous chapter can be used to encode and decode messages using them. However, the additional structure they possess allows us to devise other encoding and decoding techniques, some of which can be carried out efficiently in hardware.

We can define polynomial equivalents of the parity check matrix and the syndrome. We have seen that g is the generator polynomial of a cyclic code if and only if it is a factor of $(X^n + 1)$. This means that there is a polynomial h such that

$$(X^n + 1) = g(X)h(X). \tag{7.20}$$

h is the *parity check polynomial*; we shall see that it has properties similar to those of the parity check matrix.

The encoding process produces a code polynomial c from a message polynomial m by multiplication by the generator polynomial:

$$c(X) = m(X)g(X). \tag{7.21}$$

If we multiply c by the parity check polynomial, we get

$$c(X)h(X) = m(X)g(X)h(X) = m(X)(X^n + 1) = 0 \bmod (X^n + 1), \quad (7.22)$$

which is analogous to the result that the matrix product of a code word and the parity check matrix is the zero vector.

If the code polynomial is corrupted in some way, the result can be represented as the sum of the code polynomial and an error polynomial, e, that is,

$$r(X) = c(X) + e(X). \tag{7.23}$$

The *syndrome polynomial*, s, is the product of the result and the parity check polynomial,

$$s(X) = r(X)h(X) = c(x)h(X) + e(X)h(X) = e(X)h(X). \tag{7.24}$$

If the code polynomial is not corrupted, the error polynomial and its syndrome will be zero.

We have seen that we can reduce cyclic codes to systematic form using the canonical form of the generator matrix. We now consider how to describe the systematic form of the code in terms of polynomials. Suppose the code has k message bits and $n - k$ check bits. If it is in systematic form, the first k terms of the code polynomial will be the same as the message polynomial and there will be a polynomial d of degree $n - k - 1$ such that

$$c(X) = m(X) + d(X)X^k. \tag{7.25}$$

The code polynomials will still be multiples of the generator polynomial, but they will not be equal to the product of the message polynomial and the generator polynomial. Instead there will be some other polynomial q such that

$$c(X) = q(X)g(X) = m(X) + d(X)X^k. \tag{7.26}$$

We do not need to know what q is, but we do need to be able to compute d. If we take the remainder after dividing by g, we get

$$(m(X) + d(X)X^k) \bmod g(X) = 0. \qquad (7.27)$$

If we multiply through by X^{n-k}, and use the fact that the multiplication is carried out in $\mathbb{B}_n[X]/(X^n + 1)$, we get

$$(m(X)X^{n-k} + d(X)) \bmod g(X) = 0, \qquad (7.28)$$

or

$$d(X) \bmod g(X) = m(X)X^{n-k} \bmod g(X). \qquad (7.29)$$

Since the degree of d is no greater than $n - k - 1$ and the degree of g is $n - k$, $d \bmod g = d$ and so

$$d(X) = m(X)X^{n-k} \bmod g(X). \qquad (7.30)$$

EXAMPLE 7.18

Since

$$X^6 + 1 = (X^2 + X + 1)(X^4 + X^3 + X + 1)$$

there is a cyclic code whose code words are 6 bits long with generator polynomial

$$g(X) = 1 + X + X^2$$

and parity check polynomial

$$h(X) = 1 + X + X^3 + X^4.$$

The generator matrix of this code is

$$G = \begin{bmatrix} 1\,1\,1\,0\,0\,0 \\ 0\,1\,1\,1\,0\,0 \\ 0\,0\,1\,1\,1\,0 \\ 0\,0\,0\,1\,1\,1 \end{bmatrix};$$

and its canonical form is

$$G_c = \begin{bmatrix} 1\,0\,0\,0\,1\,1 \\ 0\,1\,0\,0\,1\,0 \\ 0\,0\,1\,0\,0\,1 \\ 0\,0\,0\,1\,1\,1 \end{bmatrix}.$$

We can verify that the same encoding is given when we use the procedure involving polynomials given above. We only need to verify it for the message polynomials 1, X, X^2 and X^3; all the others are linear combinations of these.

We have $n = 6$, and $k = 4$, so

$$d(X) = m(X)X^2 \bmod g(X)$$

and

$$c(X) = m(X) + d(X)X^4.$$

If $m(X) = 1$, we can use synthetic division to show that

$$d(X) = X^2 \bmod g(X) = 1 + X$$

and so

$$c(X) = 1 + (1 + X)X^4 = 1 + X^4 + X^5.$$

This corresponds to the code word 100011, which is the first row of the canonical form of the generator matrix.

If $m(X) = X$, we get

$$d(X) = X^3 \bmod g(X) = 1,$$

$$c(X) = X + X^4.$$

The corresponding code word is 010010, the second row of the canonical form of the generator matrix.

If $m(X) = X^2$,

$$d(X) = X^4 \bmod g(X) = X,$$

$$c(X) = X^2 + X^5.$$

The corresponding code word is 001001, the third row of the canonical form of the generator matrix.

Finally, if $m(X) = X^3$,

$$d(X) = X^5 \bmod g(X) = 1 + X,$$

$$c(X) = X^3 + (1 + X)X^4 = X^3 + X^4 + X^5.$$

The corresponding code word is 000111, the fourth row of the canonical form of the generator matrix.

□

We define the syndrome polynomial to be the remainder when the received polynomial is divided by g. Since the code polynomials are precisely the multiples of g, the syndrome polynomial is zero if and only if the received polynomial is a code polynomial. Errors can be corrected by adding the syndrome polynomial to the first k terms of the received polynomial.

EXAMPLE 7.19

From the previous example, if we use the canonical form of the cyclic code of length 6 with generator polynomial $g(X) = 1 + X + X^2$, the message polynomial $1 + X^2$ is encoded as $1 + X^2 + X^4$, or 101010.

If we receive the code word 111010, which corresponds to $1 + X + X^2 + X^4$, the syndrome is X. Adding this to $1 + X + X^2$, we get $1 + X^2$, which is the polynomial that was encoded.

$\square$

7.5   Encoding and Decoding Circuits for Cyclic Codes

A number of circuits for encoding and decoding cyclic codes efficiently have been devised. The simplest ones use only binary adders and registers that can store single bits.

We can carry out the encoding process for the canonical form of a cyclic code by finding remainder polynomials that represent the parity check bits. We can carry out the decoding process by finding remainder polynomials that represent the syndrome of the received code word. Both the encoding and decoding can be carried out by a circuit that computes the remainder polynomial, an example of which is shown in Figure 7.1

In Figure 7.1, the squares represent registers that store a single bit, the circles labelled with the $+$ sign represent binary adders and the circles labelled with the $\times$ sign represent either binary multipliers or switches that are open if the control input is 0 and closed if it is 1. The symbols $g_0, g_1, \ldots, g_{n-k-1}$ represent either one of the inputs to the binary multipliers or the control inputs of the switches. The g_i are the coefficients of the divisor polynomial.

The one-bit registers act as delays, producing an output which is equal to the value stored in during the previous cycle. The circuit is controlled by a clocking mechanism that is not shown in the figure. There are two switches that control the movement of bits into and out of the circuit. The first switch either connects the input to, or disconnects it from, the first binary adder. The second switch connects the output from the last shift register to either the output of the circuit or feeds it back to the first binary adder via the first multiplier.

Before the encoding begins, the values in the registers are set to zero and the switches are set so that the input to the circuit is connected to the first binary adder, and the output of the last register is fed back to the first multiplier.

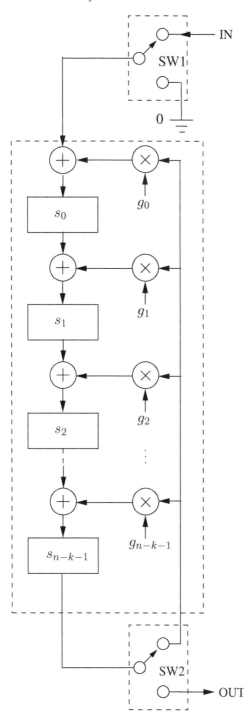

FIGURE 7.1
A circuit for computing remainder polynomials.

The synthetic division algorithm works by subtracting X^p times the divisor from the dividend for decreasing values of p. The feedback circuit accomplishes this in the following way.

The input to the circuit is the polynomial whose remainder is to be computed, represented as a string of bits with the coefficient of the highest power of X first. The bits are fed into the shift registers over the first $n - k$ clock cycles. While this is happening, the output of the last register will be 0.

At the next input step, the output of the last register will be 1, and this will be fed back to the adders after being multiplied by the appropriate values of the g_i. This has the effect of subtracting $g(X)X^p$ for some value of p from the input polynomial and shifting the polynomial coefficients up so that the coefficient of the second highest power of X is held in the last register. This process is repeated until there are no more bits in the input polynomial. The bits in the registers now represent the coefficients of the remainder. The input to the circuit is switched off and the output of the last register is now switched to the output of the circuit and the bits in the registers are shifted out.

EXAMPLE 7.20

Figure 7.2 shows the particular case of a circuit for computing remainders modulo $(1 + X^2 + X^4)$.

◻

When this circuit is used for encoding, the g_i are the coefficients of the generator polynomial and the input polynomial is $m(X)X^{n-k}$. The output of the circuit is $d(X)$. Additional circuitry is required to concatenate the bits of the message word and the check bits of the code word.

When this circuit is used for decoding, the g_i are again the coefficients of the generator polynomial. The received word is the input and the output is the syndrome. Additional circuitry is required to add the syndrome to the received word to correct it if necessary.

Other circuits for encoding and decoding cyclic codes have been devised. Details of various algorithms and their hardware implementations can be found in the following references: [2], Sections 4.4, 4.5, 5.1, 5.2, 5.3 and 5.5; [3], Chapter 7, Section 8 and Chapter 9; [4], Sections 8.4 and 8.5; and [5], Sections 5.4 and 5.5.

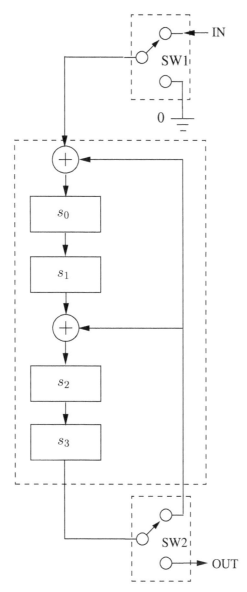

FIGURE 7.2
A circuit for computing remainder polynomials modulo $(1 + X^2 + X^4)$.

7.6   The Golay Code

An important example of a cyclic code that has had a number of practical applications is the *Golay code*.

The Golay code is a cyclic code with code words that are twenty-three bits long. It can be constructed from either of two generator polynomials, namely

$$g_1(X) = 1 + X^2 + X^4 + X^5 + X^6 + X^{10} + X^{11}, \qquad (7.31)$$

for which the generator matrix is

$$G_1 = \begin{bmatrix} 1 & 0 & 1 0 1 1 1 & 0 & 0 0 1 1 0 & \dots & 0 \\ 0 & 1 & 0 1 0 1 1 & 1 & 0 0 0 1 1 & \dots & 0 \\ \vdots & & & \ddots & & & \vdots \\ 0 & \dots & 0 1 0 1 0 & 1 & 1 1 0 0 0 & 1 & 1 \end{bmatrix},$$

and

$$g_2(X) = 1 + X + X^5 + X^6 + X^7 + X^9 + X^{11}, \qquad (7.32)$$

for which the generator matrix is

$$G_2 = \begin{bmatrix} 1 & 1 & 0 0 0 1 1 & 1 & 0 1 0 1 0 & \dots & 0 \\ 0 & 1 & 1 0 0 0 1 & 1 & 1 0 1 0 1 & \dots & 0 \\ \vdots & & & \ddots & & & \vdots \\ 0 & \dots & 0 1 1 0 0 & 0 & 1 1 1 0 1 & 0 & 1 \end{bmatrix}.$$

Note that both g_1 and g_2 divide $(X^{23} + 1)$, since

$$X^{23} + 1 = (1 + X)g_1(X)g_2(X). \qquad (7.33)$$

A Golay code can be used to correct up to three errors in a 23-bit code word. It is an example of a perfect code.

7.7   Hamming Codes

DEFINITION 7.5 Hamming Code   *Let $m \geq 3$. A* Hamming code *is a cyclic code with code words that are $2^m - 1$ bits long, having $2^m - m - 1$ information bits, m parity check bits and a minimum distance of 3.*

There is a simple way to construct the parity check matrix of a Hamming code. The columns of the parity check matrix are the binary representations of all the positive integers less than 2^m. If the columns are in the order of the numbers they represent, the syndrome will be the binary representation of the place where the error occurred.

EXAMPLE 7.21

For $m = 3$, the parity check matrix is

$$H = \begin{bmatrix} 0 & 0 & 0 & 1 & 1 & 1 & 1 \\ 0 & 1 & 1 & 0 & 0 & 1 & 1 \\ 1 & 0 & 1 & 0 & 1 & 0 & 1 \end{bmatrix}.$$

We interchange columns 1 and 7, columns 2 and 6 and columns 4 and 5 to reduce the parity check matrix to canonical form:

$$H_c = \begin{bmatrix} 1 & 1 & 0 & 1 & 1 & 0 & 0 \\ 1 & 1 & 1 & 0 & 0 & 1 & 0 \\ 1 & 0 & 1 & 1 & 0 & 0 & 1 \end{bmatrix}.$$

The canonical form of the generator matrix is

$$G_c = \begin{bmatrix} 1 & 0 & 0 & 0 & 1 & 1 & 1 \\ 0 & 1 & 0 & 0 & 1 & 1 & 0 \\ 0 & 0 & 1 & 0 & 0 & 1 & 1 \\ 0 & 0 & 0 & 1 & 1 & 0 & 1 \end{bmatrix}.$$

Interchanging columns 4 and 5, columns 2 and 6 and columns 1 and 7, we get

$$G = \begin{bmatrix} 1 & 1 & 0 & 1 & 0 & 0 & 1 \\ 0 & 1 & 0 & 1 & 0 & 1 & 0 \\ 1 & 1 & 1 & 0 & 0 & 0 & 0 \\ 1 & 0 & 0 & 1 & 1 & 0 & 0 \end{bmatrix}.$$

This gives us the code

$$\{0000000, 0001111, 0010110, 0011001, 0101010, 0101101, 0110011, 0111100,$$
$$1001011, 1001100, 1010101, 1011010, 1100110, 1101001, 1110000, 1111111\}.$$

To illustrate the error-correction procedure, suppose the third bit of the code word 1100110 gets corrupted, giving 1110110. The syndrome is $1110110H^T = 011$, indicating the third bit, as expected.

Similarly, the syndrome of 1110010 is $1110010H^T = 110$, indicating that the sixth bit has been corrupted, and that the correct code word is 1110000.

To show that the code generated by G is equivalent to a cyclic code, we replace the first row of G with the sum of the first and second rows to get

$$G_2 = \begin{bmatrix} 1\,0\,0\,0\,0\,1\,1 \\ 0\,1\,0\,1\,0\,1\,0 \\ 1\,1\,1\,0\,0\,0\,0 \\ 1\,0\,0\,1\,1\,0\,0 \end{bmatrix}.$$

We make the following columns interchanges: 1 and 7, 2 and 6, 3 and 6, 4 and 5, and 4 and 7 to get

$$G_c = \begin{bmatrix} 1\,1\,0\,1\,0\,0\,0 \\ 0\,1\,1\,0\,1\,0\,0 \\ 0\,0\,1\,1\,0\,1\,0 \\ 0\,0\,0\,1\,1\,0\,1 \end{bmatrix}.$$

This is the generator matrix of the cyclic code with generator polynomial $1 + X + X^3$.

□

DEFINITION 7.6 Primitive Polynomial   *A primitive polynomial is a polynomial of degree p which divides $(X^{2^p - 1} - 1)$ but does not divide $(X^k - 1)$ for any positive integer $k < p$.*

The generator polynomials of the Hamming codes are all primitive polynomials. The following table lists the generator polynomials for the Hamming codes for $3 \leq m \leq 10$.

Degree	Generator polynomial
3	$1 + X + X^3$
4	$1 + X + X^4$
5	$1 + X^2 + X^5$
6	$1 + X + X^6$
7	$1 + X^3 + X^7$
8	$1 + X^2 + X^3 + X^4 + X^8$
9	$1 + X^4 + X^9$
10	$1 + X^3 + X^{10}$

Hamming codes are perfect codes.

7.8 Cyclic Redundancy Check Codes

Cyclic Redundancy Check (CRC) codes are error-detecting codes that do not have any error-correcting capabilities. They are usually used in conjunction with an error-correcting code to improve performance of the system and to detect failures of the error-correction procedure.

When used for error detection, the syndrome of each received code word is computed. If it is non-zero, an error has been detected. The syndrome is not used to correct any error that may be detected.

CRC codes are commonly constructed from primitive polynomials, with a generator polynomial of the form

$$g(X) = (1 + X)p(X) \tag{7.34}$$

for some primitive polynomial p.

If the degree of the generator polynomial is r, the maximum block length is $n = 2^{r-1} - 1$ and the number of message bits is $n - r$. The usual practice is to select a large value of r and choose a block length slightly less than the maximum.

If $c(X)$ is a polynomial in the code generated by $g(X)$ then it is a multiple of $g(X)$, and so

$$c(X) \bmod G(X) = 0. \tag{7.35}$$

If the code word is corrupted, there is an error polynomial $e(X)$ such that the corrupted code word is $c(X) + e(X)$. Then

$$(c(X) + e(X)) \bmod g(X) = c(X) \bmod g(X) + e(X) \bmod g(X)$$
$$= e(X) \bmod g(X). \tag{7.36}$$

The error will be undetectable if $e(X) \bmod g(X) = 0$. We can use this fact to determine which generator polynomials will detect various classes of errors.

If a single bit is corrupted, we will have $e(X) = X^k$ for some k. If $g(X)$ has two or more terms, it will not divide X^k and single bit errors will be detected.

If two bits are corrupted, then $e(X) = X^k + X^m = X^k(1 + X^{m-k})$ for some k and m with $k < m$. If $g(X)$ has two or more terms and does not divide $(1 + X^j)$ for $j = 1, 2, \ldots, n$, then it will not divide any such error polynomial and two bit errors will be detected.

If there is a burst of r errors in succession, then $e(X) = X^j(1 + X + \ldots + X^{r-1})$ for some j. If $g(X)$ is polynomial of degree r, it will not divide $e(X)$. In fact, if $g(X)$ is of degree r all bursts of not more than r errors will be detected.

EXAMPLE 7.22

$1 + X + X^4$ is a primitive polynomial. We construct a CRC by taking

$$g(X) = (1 + X)(1 + X + X^4) = 1 + X^2 + X^4 + X^5$$

as our generator polynomial. The degree of $g(X)$ is 5; so we can have up to 15 bits in our code words. If we choose to have code words that are 11 bits long, the generator matrix is

$$G = \begin{bmatrix} 1\,0\,1\,0\,1\,1\,0\,0\,0\,0\,0 \\ 0\,1\,0\,1\,0\,1\,1\,0\,0\,0\,0 \\ 0\,0\,1\,0\,1\,0\,1\,1\,0\,0\,0 \\ 0\,0\,0\,1\,0\,1\,0\,1\,1\,0\,0 \\ 0\,0\,0\,0\,1\,0\,1\,0\,1\,1\,0 \\ 0\,0\,0\,0\,0\,1\,0\,1\,0\,1\,1 \end{bmatrix}.$$

Replacing the first row with the sum of the first and third rows and replacing the second row with the sum of the second and fourth rows, we get

$$G_2 = \begin{bmatrix} 1\,0\,0\,0\,0\,1\,1\,1\,0\,0\,0 \\ 0\,1\,0\,0\,0\,0\,1\,1\,1\,0\,0 \\ 0\,0\,1\,0\,1\,0\,1\,1\,0\,0\,0 \\ 0\,0\,0\,1\,0\,1\,0\,1\,1\,0\,0 \\ 0\,0\,0\,0\,1\,0\,1\,0\,1\,1\,0 \\ 0\,0\,0\,0\,0\,1\,0\,1\,0\,1\,1 \end{bmatrix}.$$

Replacing the first row with the sum of the first and sixth rows, replacing the third row with the sum of the third and fifth rows, and replacing the fourth row with the sum of the fourth and sixth rows, we get the canonical form of the generator matrix:

$$G_c = \begin{bmatrix} 1\,0\,0\,0\,0\,0\,1\,0\,0\,1\,1 \\ 0\,1\,0\,0\,0\,0\,1\,1\,1\,0\,0 \\ 0\,0\,1\,0\,0\,0\,0\,1\,1\,1\,0 \\ 0\,0\,0\,1\,0\,0\,0\,0\,1\,1\,1 \\ 0\,0\,0\,0\,1\,0\,1\,0\,1\,1\,0 \\ 0\,0\,0\,0\,0\,1\,0\,1\,0\,1\,1 \end{bmatrix}.$$

The parity check matrix for the code is therefore

$$H = \begin{bmatrix} 1\,1\,0\,0\,1\,0\,1\,0\,0\,0\,0 \\ 0\,1\,1\,0\,0\,1\,0\,1\,0\,0\,0 \\ 0\,1\,1\,1\,1\,0\,0\,0\,1\,0\,0 \\ 1\,0\,1\,1\,1\,1\,0\,0\,0\,1\,0 \\ 1\,0\,0\,1\,0\,1\,0\,0\,0\,0\,0 \end{bmatrix}.$$

11001011001 is a code word. Its syndrome is

$$11001011001 H^T = 00000,$$

as expected.

The CRC should be able to detect any one bit error. Suppose the fifth bit of the code word 11001011001 is corrupted, giving 11000011001. Its syndrome is

$$11000011001 H^T = 10110,$$

indicating an error has occurred.

Suppose the third and eight bits of the code word 11001011001 are corrupted, giving 11101010001. Its syndrome is

$$11101010001 H^T = 00110,$$

again indicating an error has occurred.

Since the degree of the generator polynomial is 5, the CRC should be able to detect a burst of five errors. Such a burst will change the code word 11001011001 to 11000100101. The syndrome of the corrupted code word is

$$11000100101 H^T = 00001,$$

indicating that an error has occurred.

However, the CRC should not be able to detect a burst of six or more errors. A burst of six errors will change the code word 11001011001 to 11110101101. The syndrome of the corrupted code word is

$$11110101101 H^T = 00000,$$

so this burst of errors will not be detected. ▯

A number of standard CRC codes have been defined. These include the CRC-12 code, with generator polynomial

$$g_{12}(X) = 1 + X^2 + X^9 + X^{10} + X^{11} + X^{12} \tag{7.37}$$

and the CRC-ANSI code with generator polynomial

$$g_{ANSI}(X) = X + X^2 + X^{15} + X^{17}. \tag{7.38}$$

CRC codes can detect bursts of errors, but cannot be used to correct them. Codes that can be used to correct bursts of errors will be discussed in the next chapter.

7.9 Reed-Muller Codes

Let $n = 2^m$ for some positive integer m, and let $v_0 \in \mathbb{B}^n$ be the word consisting of 2^m 1's. Let $v_1, v_2, \ldots, v_m$ be the rows of the matrix that has all possible m-tuples as columns.

EXAMPLE 7.23

If $m = 2$, $n = 4$ and the matrix whose columns are all possible 2-tuples is

$$V_2 = \begin{bmatrix} 0 & 0 & 1 & 1 \\ 0 & 1 & 0 & 1 \end{bmatrix}.$$

So for $m = 2$, $v_0 = 1111$, $v_1 = 0011$, $v_2 = 0101$. ▯

EXAMPLE 7.24

If $m = 3$, $n = 8$ and the matrix whose columns are all possible 3-tuples is

$$V_3 = \begin{bmatrix} 0 & 0 & 0 & 0 & 1 & 1 & 1 & 1 \\ 0 & 0 & 1 & 1 & 0 & 0 & 1 & 1 \\ 0 & 1 & 0 & 1 & 0 & 1 & 0 & 1 \end{bmatrix}.$$

So for $m = 3$, $v_0 = 11111111$, $v_1 = 00001111$, $v_2 = 00110011$ and $v_3 = 01010101$.

▯

Define $v \otimes w$ to be the component-wise product of elements in $\mathbb{B}^n$, that is, the kth component of $v \otimes w$ is the product of the kth component of the v and the kth component of w. Since the product in $\mathbb{B}$ is commutative, it follows that $v \otimes w = w \otimes v$.

EXAMPLE 7.25

For $m = 2$, we have $v_0 = 1111$, $v_1 = 0011$, $v_2 = 0101$. The products are

$$v_0 \otimes v_0 = 1111$$
$$v_0 \otimes v_1 = 0011$$
$$v_0 \otimes v_2 = 0101$$
$$v_1 \otimes v_1 = 0011$$
$$v_1 \otimes v_2 = 0001$$
$$v_2 \otimes v_2 = 0101.$$

▯

EXAMPLE 7.26

For $m = 3$, $v_0 = 11111111$, $v_1 = 00001111$, $v_2 = 00110011$ and $v_3 = 01010101$. The products are

$$v_0 \otimes v_0 = 11111111$$

$$v_0 \otimes v_1 = 00001111$$
$$v_0 \otimes v_2 = 00110011$$
$$v_0 \otimes v_3 = 01010101$$
$$v_1 \otimes v_1 = 00001111$$
$$v_1 \otimes v_2 = 00000011$$
$$v_1 \otimes v_3 = 00000101$$
$$v_2 \otimes v_2 = 00110011$$
$$v_2 \otimes v_3 = 00010001$$
$$v_3 \otimes v_3 = 01010101.$$

☐

We always have $v_i \otimes v_i = v_i$ and $v_0 \otimes v_i = v_i$ for all i.

DEFINITION 7.7 *rth-order Reed-Muller Code* *The rth-order Reed-Muller code with word length $n = 2^m$ is the linear code in $\mathbb{B}^n$ whose basis is the set of vectors $v_0, v_1, \ldots, v_m$, together with all products of r or fewer of these vectors.*

The rth-order Reed-Muller code has

$$k = 1 + \binom{m}{1} \cdots + \binom{m}{r} \tag{7.39}$$

information bits and a distance of 2^{m-r}.

EXAMPLE 7.27

For $m = 3$ and $r = 2$, the basis is $v_0, v_1, v_2, v_3, v_1 \otimes v_2 = 00000011$, $v_1 \otimes v_3 = 00000101$, $v_2 \otimes v_3 = 00010001$.

The generator matrix is

$$G = \begin{bmatrix} 1 & 1 & 1 & 1 & 1 & 1 & 1 & 1 \\ 0 & 0 & 0 & 0 & 1 & 1 & 1 & 1 \\ 0 & 0 & 1 & 1 & 0 & 0 & 1 & 1 \\ 0 & 1 & 0 & 1 & 0 & 1 & 0 & 1 \\ 0 & 0 & 0 & 0 & 0 & 0 & 1 & 1 \\ 0 & 0 & 0 & 0 & 0 & 1 & 0 & 1 \\ 0 & 0 & 0 & 1 & 0 & 0 & 0 & 1 \end{bmatrix}.$$

☐

The parity check matrix of the code is the matrix whose rows are the vectors v_0, $v_1, \ldots, v_m$, together with the vectors that are the products of no more than $m - r - 1$ of these vectors.

EXAMPLE 7.28

For $m = 3$ and $r = 2$, the parity check matrix is

$$H = \begin{bmatrix} 1 & 1 & 1 & 1 & 1 & 1 & 1 & 1 \\ 0 & 0 & 0 & 0 & 1 & 1 & 1 & 1 \\ 0 & 0 & 1 & 1 & 0 & 0 & 1 & 1 \\ 0 & 1 & 0 & 1 & 0 & 1 & 0 & 1 \end{bmatrix}.$$

In this case $m - r - 1 = 0$, so there are no products in the parity check matrix.

Note that the parity check matrix for the 7-bit Hamming code is a sub-matrix of this matrix. ▢

Reed-Muller codes are equivalent to cyclic codes with an added overall check bit.

To detect and correct errors, it is possible to determine 2^{m-r} independent values for each information bit from the bits of the code word. If there are no errors, all these determinations will give the same value. If there are errors, some of these determinations will give 0 and some will give 1. The more frequently occurring value is taken as the correct value of the bit.

EXAMPLE 7.29

To illustrate the decoding process, consider the case $m = 3$ and $r = 1$, for which the generator matrix is

$$G = \begin{bmatrix} 1 & 1 & 1 & 1 & 1 & 1 & 1 & 1 \\ 0 & 0 & 0 & 0 & 1 & 1 & 1 & 1 \\ 0 & 0 & 1 & 1 & 0 & 0 & 1 & 1 \\ 0 & 1 & 0 & 1 & 0 & 1 & 0 & 1 \end{bmatrix}.$$

The decoding process depends on the fact that all arithmetic is done in base 2, so that $1 + 1 = 0$. If the word to be encoded is $a_0 a_1 a_2 a_3$, and the resulting code word is $b_0 b_1 b_2 b_3 b_4 b_5 b_6 b_7$, where each a_i and b_j denote a single bit, then the generator matrix gives us the following equations:

$$a_0 = b_0$$
$$a_0 + a_3 = b_1$$
$$a_0 + a_2 = b_2$$
$$a_0 + a_2 + a_3 = b_3$$
$$a_0 + a_1 = b_4$$
$$a_0 + a_1 + a_3 = b_5$$
$$a_0 + a_1 + a_2 = b_6$$
$$a_0 + a_1 + a_2 + a_3 = b_7$$

Adding the equations in pairs gives us the following independent expressions for a_1, a_2 and a_3:

$$a_1 = b_0 + b_4$$
$$a_1 = b_2 + b_6$$
$$a_1 = b_3 + b_7$$
$$a_1 = b_1 + b_5;$$

$$a_2 = b_0 + b_2$$
$$a_2 = b_1 + b_3$$
$$a_2 = b_4 + b_6$$
$$a_2 = b_5 + b_7;$$

$$a_3 = b_0 + b_1$$
$$a_3 = b_2 + b_3$$
$$a_3 = b_4 + b_5$$
$$a_3 = b_6 + b_7.$$

We also have the following for a_0:

$$a_0 = b_0$$
$$a_0 = b_1 + b_6 + b_7$$
$$a_0 = b_2 + b_5 + b_7$$
$$a_0 = b_3 + b_4 + b_7.$$

To decode a code word, we simply compute these values from the b_j and choose the a_i from them.

Suppose 0101 is coded to give 01011010, but one bit is corrupted to give 00011010, so that $b_0 = 0$, $b_1 = 0$, $b_2 = 0$, $b_3 = 1$, $b_4 = 1$, $b_5 = 0$, $b_6 = 1$, and $b_7 = 0$.

We compute four values for each of the a_1 using the equations above. For a_0 we get

$$b_0 = 0$$
$$b_1 + b_6 + b_7 = 1$$
$$b_2 + b_5 + b_7 = 0$$
$$b_3 + b_4 + b_7 = 0.$$

Three of the four values are 0, so we take $a_0 = 0$.

For a_1 we get

$$b_0 + b_4 = 1$$
$$b_2 + b_6 = 1$$
$$b_3 + b_7 = 1$$
$$b_1 + b_5 = 0;$$

Three of the four values are 1, so we take $a_1 = 1$.

For a_2 we get

$$b_0 + b_2 = 0$$
$$b_1 + b_3 = 1$$
$$b_4 + b_6 = 0$$
$$b_5 + b_7 = 0;$$

Three of the four values are 0, so we take $a_2 = 0$.

For a_3 we get

$$b_0 + b_1 = 0$$
$$b_2 + b_3 = 1$$
$$b_4 + b_5 = 1$$
$$b_6 + b_7 = 1.$$

Three of the four values are 1, so we take $a_3 = 0$. This makes the decoded word 0101, which is correct.   ⬜

7.10   Exercises

1. Compute the following sums of polynomials with coefficients in the specified ring:

   (a) $(3 + 4X^2 + 9X^4) + (X - 2X^3 + X^5)$, coefficients in $\mathbb{Z}$;
   (b) $(3 - 4X^2 + 9X^4) + (X + 2X^2 - X^4)$, coefficients in $\mathbb{Z}$;
   (c) $(1 + X + X^2) + (X + X^3)$, coefficients in $\mathbb{B}$;
   (d) $(1 + X + X^4) + (X^2 + X^4 + X^6)$, coefficients in $\mathbb{B}$;
   (e) $(1 + X + X^4) + (X^2 + X^4 + X^6)$, coefficients in $\mathbb{Z}_3$;
   (f) $(1 + X + 2X^4) + (X^2 + 2X^4 + 3X^6)$, coefficients in $\mathbb{Z}_4$;

(g) $(4 + 3X + 2X^2) + (2X + X^3)$, coefficients in $\mathbb{Z}_5$;

(h) $(\pi + X) + (1 - \pi X)$, coefficients in $\mathbb{R}$.

2. Compute the following products of polynomials with coefficients in the specified ring:

(a) $(3 + 4X^2 + 9X^4) \times (X - 2X^3 + X^5)$, coefficients in $\mathbb{Z}$;

(b) $(3 - 4X^2 + 9X^4) \times (X + 2X^2 - X^4)$, coefficients in $\mathbb{Z}$;

(c) $(1 + X + X^2) \times (X + X^3)$, coefficients in $\mathbb{B}$;

(d) $(1 + X + X^4) \times (X^2 + X^4 + X^6)$, coefficients in $\mathbb{B}$;

(e) $(1 + X + X^4) \times (X^2 + X^4 + X^6)$, coefficients in $\mathbb{Z}_3$;

(f) $(1 + X + 2X^4) \times (X^2 + 2X^4 + 3X^6)$, coefficients in $\mathbb{Z}_4$;

(g) $(4 + 3X + 2X^2) \times (2X + X^3)$, coefficients in $\mathbb{Z}_5$;

(h) $(\pi + X) \times (1 - \pi X)$, coefficients in $\mathbb{R}$.

3. Show that if p is a prime number, $(1 + X)^p = (1 + X^p)$ if the addition and multiplication operations are performed in $\mathbb{Z}_p$.

4. Compute the following sums of polynomials with coefficients in $\mathbb{B}$, where the polynomials have been represented as elements of $\mathbb{B}^n$ for some n:

(a) $1001 + 1010$;

(b) $11001 + 10010$;

(c) $101001 + 101010$;

(d) $1010001 + 1010110$;

(e) $01010001 + 10101110$.

5. Compute the following products of polynomials with coefficients in $\mathbb{B}$, where the polynomials have been represented as elements of $\mathbb{B}^n$ for some n:

(a) 1001×1010;

(b) 11001×10010;

(c) 101001×101010;

(d) 1010001×1010110;

(e) 01010001×10101110.

6. Perform the following synthetic division operations for the polynomials below, with coefficients in the specified field:

(a) divide $(X^3 + 2X^2 + 3X + 4)$ by $(X^2 + X + 1)$, with coefficients in $\mathbb{R}$;

(b) divide $(X^5 + X^3 + X + 1)$ by $(X^3 + X^2 + 1)$, with coefficients in $\mathbb{B}$;

(c) divide $(X^7 + 1)$ by $(X^3 + X + 1)$, with coefficients in $\mathbb{B}$;

(d) divide $(2X^4 + X^2 + 2)$ by $(X^2 + 2)$, with coefficients in $\mathbb{Z}_3$;

(e) divide $(X^5 + 4X^3 + 3X^2 + 2X + 1)$ by $(X^2 + 3)$, with coefficients in $\mathbb{Z}_5$.

7. Compute the product of the polynomials p and q modulo r, when p, q and r are as given below, with coefficients from the specified field:

(a) $p(X) = (X^2 + 5X + 7)$, $q(x) = (X^3 + 11)$, $r(X) = (X^3 + 2X)$, with coefficients in $\mathbb{R}$;

(b) $p(X) = (X^2 + X + 1)$, $q(x) = (X^3 + 1)$, $r(X) = (X^3 + X)$, with coefficients in $\mathbb{B}$;

(c) $p(X) = (X^3 + X + 1)$, $q(x) = (X^5 + X)$, $r(X) = (X^4 + X^2 + 1)$, with coefficients in $\mathbb{B}$;

(d) $p(X) = (X^3 + 2X + 1)$, $q(x) = (2X^4 + X^3 + 2)$, $r(X) = (X^3 + 2X + 1)$, with coefficients in $\mathbb{Z}_3$;

(e) $p(X) = (X^3 + 4X + 2)$, $q(x) = (3X^4 + 2X^3 + 4)$, $r(X) = (X^3 + 3X + 2)$, with coefficients in $\mathbb{Z}_5$.

8. Use the Remainder Theorem to compute the following:

(a) the remainder of $(X^5 + 7X^3 + 9X)$ when divided by $(X^2 + 13)$, with coefficients in $\mathbb{R}$;

(b) the remainder of $(X^5 + X^3 + X)$ when divided by $(X^2 + 1)$, with coefficients in $\mathbb{B}$;

(c) the remainder of $(X^9 + X^6 + X^3)$ when divided by $(X^4 + 1)$, with coefficients in $\mathbb{B}$;

(d) the remainder of $(X^9 + X^6 + X^3)$ when divided by $(X^4 + 1)$, with coefficients in $\mathbb{Z}_3$;

(e) the remainder of $(X^8 + 4X^5 + 2X^2)$ when divided by $(X^3 + 3)$, with coefficients in $\mathbb{Z}_5$.

9. Draw up the multiplication table of $\mathbb{B}_4[X]/(X^4 + 1)$.

10. Write down the generator polynomials of the cyclic codes whose generator matrices are given below:

(a)

$$G = \begin{bmatrix} 1\,0\,1\,1\,0\,0\,0 \\ 0\,1\,0\,1\,1\,0\,0 \\ 0\,0\,1\,0\,1\,1\,0 \\ 0\,0\,0\,1\,0\,1\,1 \end{bmatrix} ;$$

(b)

$$G = \begin{bmatrix} 1\,1\,0\,0\,1\,1\,0\,0 \\ 0\,1\,1\,0\,0\,1\,1\,0 \\ 0\,0\,1\,1\,0\,0\,1\,1 \end{bmatrix} ;$$

(c)

$$G = \begin{bmatrix} 1\,0\,0\,1\,0\,0\,1\,0\,0 \\ 0\,1\,0\,0\,1\,0\,0\,1\,0 \\ 0\,0\,1\,0\,0\,1\,0\,0\,1 \end{bmatrix} ;$$

(d)

$$G = \begin{bmatrix} 1\,1\,0\,0\,0\,1\,1\,0\,0\,0 \\ 0\,1\,1\,0\,0\,0\,1\,1\,0\,0 \\ 0\,0\,1\,1\,0\,0\,0\,1\,1\,0 \\ 0\,0\,0\,1\,1\,0\,0\,0\,1\,1 \end{bmatrix} .$$

11. Write down the generator matrices of the cyclic codes whose generator polynomials and lengths are given below:

(a) $g(X) = (1 + X^3)$, $n = 6$;

(b) $g(X) = (1 + X)$, $n = 7$;

(c) $g(X) = (1 + X + X^4 + X^5)$, $n = 8$;

(d) $g(X) = (1 + X^3 + X^6)$, $n = 9$;

(e) $g(X) = (1 + X + X^5 + X^6)$, $n = 10$.

12. Find the parity check polynomials of the cyclic codes in the previous exercise.

13. Find all the cyclic codes of length 4.

14. Find all the cyclic codes of length 5.

15. Find all the cyclic codes of length 6.

16. Is

$$G = \begin{bmatrix} 1\,0\,0\,1\,0\,1\,0\,0 \\ 0\,1\,0\,0\,1\,0\,1\,0 \\ 0\,0\,1\,0\,0\,1\,0\,1 \end{bmatrix}$$

the generator matrix of a cyclic code?

17. Write down the synthetic division of $X^5 + X^4 + 1$ by $X^2 + 1$. Draw the circuit in Figure 7.1 with $g_0 = 1$, $g_1 = 0$ and $g_2 = 1$. Trace the operation of the circuit when the input is 110001 and match the states of the registers with the stages of the synthetic division.

18. Construct the parity check matrix of the Hamming code for $m = 4$, and show that it is equivalent to the cyclic code with $n = 15$ and generator polynomial $(1 + X + X^4)$.

*19. Derive a decoding procedure for the Reed-Muller code with $m = 3$ and $r = 2$, and apply it to the received code words 01010101 and 10000011.

*20. Let $g(X)$ be a polynomial with coefficients in $\mathbb{B}$ and let n be the smallest integer such that $g(X)$ is a factor of $X^n + 1$. Show that the minimum distance of the cyclic code generated by $g(X)$ is at least 3. Construct two examples, one where the condition is satisfied and the minimum distance is 3, and one where the condition is not satisfied and there is a code word of weight 2.

7.11   References

[1] J. B. Fraleigh, *A First Course in Abstract Algebra,* 5th ed., Addison-Wesley, Reading, MA, 1994.

[2] S. Lin, *An Introduction of Error-Correcting Codes,* Prentice-Hall, Englewood Cliffs, NJ, 1970.

[3] F. J. MacWilliams and N. J. A. Sloane, *The Theory of Error-Correcting Codes Part I,* North-Holland, Amsterdam, 1977.

[4] W. W. Peterson, *Error-correcting Codes,* MIT Press, Cambridge, MA, 1961.

[5] R. B. Wells, *Applied Coding and Information Theory for Engineers,* Prentice-Hall, Upper Saddle River, NJ, 1999.

Chapter 8

Burst-Correcting Codes

8.1   Introduction

Most of the error-correcting codes that were described in the previous chapter are designed to detect and correct errors that occur independently of each other. In many cases, however, disturbances last long enough to cause several errors in succession, so that this condition of independence is not met. In this chapter we will look at codes that are designed to deal with these cases, that is, codes that are designed to detect and correct *bursts* of errors. We begin by describing some more concepts relating to rings and fields. A more rigorous exposition of these concepts can be found in [1].

8.2   Finite Fields

In the previous chapter, it was stated that the ring $\mathbb{Z}_p$ is a field whenever p is a prime number. If p is a prime number and n is any positive integer, there is a unique (up to isomorphism) field with p^n elements.

DEFINITION 8.1 Galois Field   *For any prime number p and positive integer n, the unique field with p^n elements is called the* Galois Field *of order p^n and is denoted by $GF(p^n)$.*

EXAMPLE 8.1

In Chapter 6, we presented the addition and multiplication tables for $\mathbb{Z}_2$, $\mathbb{Z}_3$ and $\mathbb{Z}_5$, which are the Galois fields $GF(2)$, $GF(3)$ and $GF(5)$, respectively. The next Galois field is $GF(7)$ or $\mathbb{Z}_7$, whose addition and multiplication tables are shown below.

```
+ 0 1 2 3 4 5 6        × 0 1 2 3 4 5 6
0 0 1 2 3 4 5 6        0 0 0 0 0 0 0 0
1 1 2 3 4 5 6 0        1 0 1 2 3 4 5 6
2 2 3 4 5 6 0 1        2 0 2 4 6 1 3 5
3 3 4 5 6 0 1 2        3 0 3 6 2 5 1 4
4 4 5 6 0 1 2 3        4 0 4 1 5 2 6 3
5 5 6 0 1 2 3 4        5 0 5 3 1 6 4 2
6 6 0 1 2 3 4 5        6 0 6 5 4 3 2 1
```

☐

EXAMPLE 8.2

Taking $p = 2$ and $n = 2$, we see that there is a Galois field with four elements. Its addition and multiplication tables are:

```
+ 0 1 2 3        × 0 1 2 3
0 0 1 2 3        0 0 0 0 0
1 1 0 3 2        1 0 1 2 3
2 2 3 0 1        2 0 2 3 1
3 3 2 1 0        3 0 3 1 2
```

$GF(4)$ is not isomorphic to $\mathbb{Z}_4$.   ☐

EXAMPLE 8.3

The next Galois field after $GF(7)$ is $GF(2^3)$, or $GF(8)$. It has eight elements and its addition and multiplication tables are shown below.

```
+ 0 1 2 3 4 5 6 7        × 0 1 2 3 4 5 6 7
0 0 1 2 3 4 5 6 7        0 0 0 0 0 0 0 0 0
1 1 0 6 4 3 7 2 5        1 0 1 2 3 4 5 6 7
2 2 6 0 7 5 4 1 3        2 0 2 3 4 5 6 7 1
3 3 4 7 0 1 6 5 2        3 0 3 4 5 6 7 1 2
4 4 3 5 1 0 2 7 6        4 0 4 5 6 7 1 2 3
5 5 7 4 6 2 0 3 1        5 0 5 6 7 1 2 3 4
6 6 2 1 5 7 3 0 4        6 0 6 7 1 2 3 4 5
7 7 5 3 2 6 1 4 0        7 0 7 1 2 3 4 5 6
```

☐

EXAMPLE 8.4

The next Galois field is $GF(3^2)$, or $GF(9)$. It has nine elements and its addition and multiplication tables are shown below.

+	0	1	2	3	4	5	6	7	8
0	0	1	2	3	4	5	6	7	8
1	1	5	8	4	6	0	3	2	7
2	2	8	6	1	5	7	0	4	3
3	3	4	1	7	2	6	8	0	5
4	4	6	5	2	8	3	7	1	0
5	5	0	7	6	3	1	4	8	2
6	6	3	0	8	7	4	2	5	1
7	7	2	4	0	1	8	5	3	6
8	8	7	3	5	0	2	1	6	4

×	0	1	2	3	4	5	6	7	8
0	0	0	0	0	0	0	0	0	0
1	0	1	2	3	4	5	6	7	8
2	0	2	3	4	5	6	7	8	1
3	0	3	4	5	6	7	8	1	2
4	0	4	5	6	7	8	1	2	3
5	0	5	6	7	8	1	2	3	4
6	0	6	7	8	1	2	3	4	5
7	0	7	8	1	2	3	4	5	6
8	0	8	1	2	3	4	5	6	7

□

In the examples above, we have simply presented addition and multiplication tables and claimed that the structures they define are fields. We will now consider how these tables may be constructed.

In Chapter 7, we constructed rings of polynomials where the multiplication operation was polynomial multiplication modulo the polynomials $(X^n + 1)$. The Galois fields can also be constructed as rings of polynomials where the multiplication operation is multiplication modulo some polynomial. To see which polynomials we can use for this purpose, we need to look at factorization of polynomials.

8.3 Irreducible Polynomials

Polynomials that cannot be factored have different properties from those which can be factored. Whether factorization is possible depends on the ring to which the co-efficients of the polynomial belong. For example, the polynomial $(X^2 - 2)$ has no factors if the coefficients belong to the integers, but it can be factored if the coeffi-cients are real numbers, since $(X^2 - 2) = (X + \sqrt{2})(X - \sqrt{2})$.

In the rest of this chapter, $\mathbb{F}$ will stand for a field.

> **DEFINITION 8.2 Irreducible over $\mathbb{F}$**   *A polynomial $p(X) \in \mathbb{F}[X]$ of degree $d \geq 1$ is irreducible over $\mathbb{F}$ or is an irreducible polynomial in $\mathbb{F}[X]$, if $p(X)$ cannot be expressed as the product $q(X)r(X)$ of two polynomials $q(X)$ and $r(X)$ in $\mathbb{F}[X]$, where the degree of both $q(X)$ and $r(X)$ is greater than or equal to 1 but less than d.*

EXAMPLE 8.5

Any first-degree polynomial in $\mathbb{F}[X]$ is irreducible over $\mathbb{F}$. ▯

EXAMPLE 8.6

$(X^n + 1)$ is never irreducible over $\mathbb{B}[X]$ for any $n \geq 1$.

As we saw in Chapter 7, $(X + 1)$ is always a factor of $(X^n + 1)$ in $\mathbb{B}[X]$, since

$$X^n + 1 = (X + 1)(X^{n-1} + X^{n-2} + \ldots + X^2 + X + 1).$$

▯

EXAMPLE 8.7

The only polynomials of degree 1 in $\mathbb{B}[X]$ are X and $(X + 1)$. It is easy to determine whether these polynomials are factors of other polynomials in $\mathbb{B}[X]$. If $p(X) \in \mathbb{B}[X]$ has no constant term, then X is a factor of $p(X)$. If substituting 1 for X in $p(X)$ gives 0, then $(X + 1)$ divides $p(X)$.

Using these tests, we can determine which polynomials of degree 2 and 3 in $\mathbb{B}[X]$ are irreducible over $\mathbb{B}$.

X^2: This has no constant term, so X is a factor; $X^2 = X \times X$.

$(X^2 + 1)$: Substituting 1 for X gives $1 + 1 = 0$, so $(X + 1)$ is a factor; $(X^2 + 1) = (X + 1)(X + 1)$.

$(X^2 + X)$: This has no constant term, and substituting 1 for X gives 0, so both X and $(X + 1)$ are factors; $(X^2 + X) = X(X + 1)$.

$(X^2 + X + 1)$: This has a constant term, and substituting 1 for X gives $1 + 1 + 1 = 1$, so neither of X and $(X + 1)$ is a factor; it is irreducible over $\mathbb{B}$.

X^3: This has no constant term, so X is a factor; $X^3 = X \times X^2$.

$(X^3 + 1)$: Substituting 1 for X gives $1 + 1 = 0$, so $(X + 1)$ is a factor; $(X^3 + 1) = (X + 1)(X^2 + X + 1)$.

$(X^3 + X)$: This has no constant term, $(X^3 + X) = X(X^2 + 1)$.

$(X^3 + X + 1)$: This has a constant term, and substituting 1 for X gives $1 + 1 + 1 = 1$, so neither of X and $(X + 1)$ is a factor. If it had a factor of degree 2, it would also have to have a factor of degree 1, which is impossible. It is irreducible over $\mathbb{B}$.

$(X^3 + X^2)$: This has no constant term, $(X^3 + X^2) = X^2(X + 1)$.

$(X^3 + X^2 + 1)$: This has a constant term, and substituting 1 for X gives $1 + 1 + 1 = 1$, so neither of X and $(X + 1)$ is a factor. It is irreducible over $\mathbb{B}$.

$(X^3 + X^2 + X)$: This has no constant term, $(X^3 + X^2 + X) = X(X^2 + X + 1)$.

$(X^3 + X^2 + X + 1)$: $(X + 1)$ is a factor; $(X^3 + X^2 + X + 1) = (X + 1)(X^2 + 1)$.

$\Box$

EXAMPLE 8.8

$(X^4 + X^2 + 1) \in \mathbb{B}[X]$ has a constant term and substituting 1 for X gives $1 + 1 + 1 = 1$, so neither X or $(X + 1)$ is a factor. It is not irreducible over $\mathbb{B}$, however, as

$$(X^4 + X^2 + 1) = (X^2 + X + 1)(X^2 + X + 1).$$

$\Box$

EXAMPLE 8.9

$(X^4 + X + 1) \in \mathbb{B}[X]$ has a constant term and substituting 1 for X gives $1 + 1 + 1 = 1$, so neither X or $(X + 1)$ is a factor. This means that it cannot have a factor that is a third degree polynomial, so the only possible factorizations are of the form

$$(X^4 + X + 1) = (X^2 + aX + 1)(X^2 + bX + 1)$$

for some $a, b \in \mathbb{B}$.

Since

$$(X^2 + aX + 1)(X^2 + bX + 1) = (X^4 + (a+b)X^3 + (1+ab+1)X^+(a+b)X + 1),$$

equating coefficients of terms of equal degree gives us the equations

$$a + b = 0$$
$$ab = 0$$
$$a + b = 1.$$

The first and third of these equations contradict each other; so they have no solution. This shows that $(X^4 + X + 1)$ is irreducible over $\mathbb{B}$.

$\Box$

EXAMPLE 8.10

Consider $(X^7 + X + 1) \in \mathbb{B}[X]$. It has no linear factors, which means that it cannot have factors of degree six. There are two possible factorizations of this polynomial: either

$$(X^7 + X + 1) = (X^5 + a_4 X^4 + a_3 X_3 + a_2 X^2 + a_1 X + 1)(X^2 + b_1 X + 1)$$

or

$$(X^7 + X + 1) = (X^4 + a_3 X^3 + a_2 X^2 + a_1 X + 1)(X^3 + b_2 X^2 + b_1 X + 1).$$

Expanding the first factorization and rearranging terms gives

$$a_4 + b_1 = 0, \quad a_3 + a_4 b_1 = 1, \quad a_2 + a_3 b_1 + a_4 = 0,$$

$$a_1 + a_2 b_1 + a_3 = 0, \quad a_1 b_1 + a_2 = 1, \quad a_1 + b_1 = 1.$$

To find a solution of these equations, we can try all possible combinations of values for the coefficients. It is convenient to do this in a table.

a_1	a_2	a_3	a_4	b_1	$a_4 + b_1$	$a_3 + a_4 b_1$ $+a_4$	$a_2 + a_3 b_1$ $+a_3$	$a_1 + a_2 b_1$	$a_1 b_2 + a_2$	$a_1 + b_1$
0	0	0	0	0	0	0	0	0	0	0
0	0	0	0	1	1	0	0	0	0	1
0	0	0	1	0	1	0	1	0	0	0
0	0	0	1	1	0	1	1	0	0	1
0	0	1	0	0	0	1	0	1	0	0
0	0	1	0	1	1	1	1	1	0	1
0	0	1	1	0	1	1	1	1	0	0
0	0	1	1	1	0	0	0	1	0	1
0	1	0	0	0	0	0	1	0	1	0
0	1	0	0	1	1	0	1	1	1	1
0	1	0	1	0	1	0	0	0	1	0
0	1	0	1	1	0	1	0	1	1	1
0	1	1	0	0	0	1	1	1	1	0
0	1	1	0	1	1	1	0	0	1	1
0	1	1	1	0	1	1	0	1	1	0
0	1	1	1	1	0	0	1	0	1	1
1	0	0	0	0	0	0	0	1	0	1
1	0	0	0	1	1	0	0	1	1	0
1	0	0	1	0	1	0	1	1	0	1
1	0	0	1	1	0	1	1	1	1	0
1	0	1	0	0	0	1	0	0	0	1
1	0	1	0	1	1	1	1	0	1	0
1	0	1	1	0	1	1	1	0	0	1
1	0	1	1	1	0	0	0	0	1	0
1	1	0	0	0	0	0	1	1	1	1
1	1	0	0	1	1	0	1	0	0	0
1	1	0	1	0	1	0	0	1	1	1
1	1	0	1	1	0	1	0	0	0	0
1	1	1	0	0	0	1	1	0	1	1
1	1	1	0	1	1	1	0	1	0	0
1	1	1	1	0	1	1	0	0	1	1
1	1	1	1	1	0	0	1	1	0	0

The table shows that there is no solution to the equations, and hence there is no factorization of $(X^7 + X + 1)$ into the product of a fifth degree polynomial and a quadratic polynomial.

Expanding the second factorization and rearranging terms gives

$$a_3 + b_2 = 0, \quad a_2 + a_3 b_2 + b_1 = 0, \quad a_1 + a_2 b_2 + a_3 b_1 = 1,$$

$$a_1 b_2 + a_2 b_1 + a_3 = 1, \quad a_1 b_1 + a_2 + b_2 = 0, \quad a_1 + b_1 = 1.$$

a_1	a_2	a_3	b_1	b_2	$a_3 + b_2$	$a_2 + a_3 b_2$ $+ b_1$	$a_1 + a_2 b_2$ $+ a_3 b_1$	$a_1 b_2 + a_2 b_1$ $+ a_3$	$a_1 b_1 + a_2$ $+ b_2$	$a_1 + b_1$
0	0	0	0	0	0	0	0	0	0	0
0	0	0	0	1	1	0	0	0	0	0
0	0	0	1	0	0	1	0	0	0	1
0	0	0	1	1	1	1	0	0	0	1
0	0	1	0	0	1	0	0	1	0	0
0	0	1	0	1	0	1	0	1	0	0
0	0	1	1	0	1	1	1	1	0	1
0	0	1	1	1	0	1	1	1	0	1
0	1	0	0	0	0	1	0	0	0	0
0	1	0	0	1	1	1	1	0	1	0
0	1	0	1	0	0	0	0	1	0	1
0	1	0	1	1	1	0	1	1	1	1
0	1	1	0	0	1	1	0	1	0	0
0	1	1	0	1	0	0	1	1	1	0
0	1	1	1	0	1	0	1	0	0	1
0	1	1	1	1	0	0	0	0	1	1
1	0	0	0	0	0	0	1	0	0	1
1	0	0	0	1	1	0	1	1	0	1
1	0	0	1	0	0	1	1	0	1	0
1	0	0	1	1	1	1	1	1	1	0
1	0	1	0	0	1	0	1	1	0	1
1	0	1	0	1	0	1	1	0	0	1
1	0	1	1	0	1	1	0	1	1	0
1	0	1	1	1	0	1	0	0	1	0
1	1	0	0	0	0	1	1	0	0	1
1	1	0	0	1	1	1	0	1	1	1
1	1	0	1	0	0	0	1	1	1	0
1	1	0	1	1	1	0	0	0	0	0
1	1	1	0	0	1	1	1	1	0	1
1	1	1	0	1	0	0	0	0	1	1
1	1	1	1	0	1	0	0	0	1	0
1	1	1	1	1	0	0	1	1	0	0

The table above lists all possible values of the coefficients for the second set of equations. Again, there is no solution to the equations, and hence there is no factorization of $(X^7 + X + 1)$ into the product of a fourth degree polynomial and a third degree polynomial.

This exhausts all the possible factorizations, so $(X^7 + X + 1)$ is irreducible over $\mathbb{B}$. We will use this result in examples later. ◻

EXAMPLE 8.11

The polynomials of degree 1 in $GF(3)[X] = \mathbb{Z}_3[X]$ are X, $(X + 1)$, $(X + 2)$, $2X$, $(2X + 1)$ and $(2X + 2)$. X is a factor of any polynomial without a constant term. If substituting 1 for X in $p(X) \in GF(3)[X]$ gives 0, $(X + 2)$ is a factor of $p(X)$, and if substituting 2 for X in $p(X)$ gives 0, $(X + 1)$ is a factor of $p(X)$. (These conditions arise from the fact that $2 = -1$ in $GF(3)$.)

We can use these conditions to find the irreducible polynomials of degree 2 over $GF(3)$:

X^2: This has no constant term; $X^2 = X \times X$.

$(X^2 + 1)$: This has a constant term so X is not a factor. Substituting 1 for X gives $1 + 1 = 2$, so $X + 2$ is not a factor; substituting 2 for X gives $2^2 + 1 = 1 + 1 = 2$, so $X + 1$ is not a factor. $(X^2 + 1)$ is irreducible over $GF(3)$.

$(X^2 + 2)$: This has a constant term so X is not a factor. Substituting 1 for X gives $1 + 2 = 0$, so $X + 2$ is a factor; substituting 2 for X gives $2^2 + 2 = 1 + 2 = 0$, so $X + 1$ is a factor. $(X^2 + 1) = (X + 1)(X + 2)$.

$(X^2 + X)$: This has no constant term; $(X^2 + X) = X(X + 1)$.

$(X^2 + X + 1)$: This is not irreducible; it is the square of $(X + 2)$.

$(X^2 + X + 2)$: This is irreducible over $GF(3)$.

$(X^2 + 2X)$: This has no constant term; $(X^2 + 2X) = X(X + 2)$.

$(X^2 + 2X + 1)$: This is not irreducible; it is the square of $(X + 1)$.

$(X^2 + 2X + 2)$: This is irreducible over $GF(3)$.

$2X^2$: This has no constant term; $X^2 = 2X \times X$.

$(2X^2 + 1)$: Substituting 1 for X gives $2 + 1 = 0$, so $(X + 2)$ is a factor; $(2X^2 + 1) = (2X + 2)(X + 2)$.

$(2X^2 + 2)$: This is $2(X^2 + 1)$, so it is irreducible over $GF(3)$.

$(2X^2 + X)$: This has no constant term; $(2X^2 + X) = X(2X + 1)$.

$(2X^2 + X + 1)$: This is irreducible over $GF(3)$.

$(2X^2 + X + 2)$: This is not irreducible; it is $2(X + 1)^2$.

$(2X^2 + 2X)$: This has no constant term; $(2X^2 + 2X) = 2X(X + 1)$.

$(2X^2 + 2X + 1)$: This is irreducible over $GF(3)$.

$(2X^2 + 2X + 2)$: This is not irreducible; it is $2(X + 2)^2$.

⬜

8.4   Construction of Finite Fields

In Chapter 7, we constructed the rings $\mathbb{B}[X]/(X^n + 1)$ and used them to study the cyclic codes. If $\mathbb{F}$ is any field and $p(X)$ is any polynomial in $\mathbb{F}[X]$, we can construct the quotient ring $\mathbb{F}[X]/p(X)$. This ring is a field if and only if $p(X)$ is irreducible over $\mathbb{F}$.

In the case where $\mathbb{F} = \mathbb{Z}_p$ for a prime number p, and the irreducible polynomial is of degree $k > 1$, the resulting field has p^k elements and is isomorphic to the Galois field $GF(p^k)$.

There is always at least one element ρ of $GF(p^k)$ with the property that $\rho^{p^k-1} = 1$ and $\rho^m \neq 1$ for $m < p^k - 1$. These elements have a special name.

DEFINITION 8.3 Primitive Element ρ *is a primitive element of the field* $\mathbb{F}$ *with* n *elements if* $\rho^{n-1} = 1$ *and* $\rho^m \neq 1$ *for all* $m < n - 1$.

The following examples illustrate the construction of Galois fields.

EXAMPLE 8.12

The polynomial $(X^2 + X + 1)$ in $\mathbb{B}[X]$ is irreducible over $\mathbb{B}$. $\mathbb{B}[X]/(X^2 + X + 1)$ is a finite field with $2^2 = 4$ elements, isomorphic to $GF(4)$.

To construct the addition and multiplication tables of the field, we use α to denote the coset of X. Since taking the quotient modulo $(X^2 + X + 1)$ means that the coset of $(X^2 + X + 1)$ is the coset of 0, it follows that $\alpha^2 + \alpha + 1 = 0$ or $\alpha^2 = \alpha + 1$. The other two elements of the field are 0 and 1. This gives us the addition table.

$+$	0	1	α	α^2
0	0	1	α	α^2
1	1	0	α^2	α
α	α	α^2	0	1
α^2	α^2	α	1	0

Since $\alpha^2 = \alpha + 1$, multiplying by α gives $\alpha^3 = \alpha^2 + \alpha$. Substituting for α^2 gives $\alpha^3 = 1$, which enables us to construct the multiplication table.

$\times$	0	1	α	α^2
0	0	0	0	0
1	0	1	α	α^2
α	0	α	α^2	1
α^2	0	α^2	1	α

To see that this is isomorphic to $GF(4)$ as given in the example above, we use the mapping $0 \leftrightarrow 0$, $1 \leftrightarrow 1$, $\alpha \leftrightarrow 2$, $\alpha^2 \leftrightarrow 3$.

We can use the result $\alpha^2 = \alpha + 1$ to write out the addition and multiplication tables in terms of the elements $0, 1, \alpha, \alpha + 1$:

$+$	0	1	α	$\alpha + 1$
0	0	1	α	$\alpha + 1$
1	1	0	$\alpha + 1$	α
α	α	$\alpha + 1$	0	1
$\alpha + 1$	$\alpha + 1$	α	1	0

$\times$	0	1	α	$\alpha + 1$
0	0	0	0	0
1	0	1	α	$\alpha + 1$
α	0	α	$\alpha + 1$	1
$\alpha + 1$	0	$\alpha + 1$	1	α

EXAMPLE 8.13

$(X^3 + X^2 + 1) \in \mathbb{B}[X]$ is irreducible over $\mathbb{B}$. The quotient $\mathbb{B}[X]/(X^3 + X^2 + 1)$ is a finite field with $2^3 = 8$ elements, isomorphic to $GF(8)$.

As before, we will use α to stand for the coset of X. Then

$$\begin{aligned} \alpha^2 \quad &\text{is the coset of } X^2 \\ \alpha^3 \quad &\text{is the coset of } X^2 + 1 \\ \alpha^4 \quad &\text{is the coset of } X^2 + X + 1 \\ \alpha^5 \quad &\text{is the coset of } X + 1 \\ \alpha^6 \quad &\text{is the coset of } X^2 + X, \end{aligned}$$

where we have used that fact that the coset of $X^3 + X^2 + 1$ is the coset of 0.

These relationships allow us to construct the addition table.

+	0	1	α	α^2	α^3	α^4	α^5	α^6
0	0	1	α	α^2	α^3	α^4	α^5	α^6
1	1	0	α^5	α^3	α^2	α^6	α	α^4
α	α	α^5	0	α^6	α^4	α^3	1	α^2
α^2	α^2	α^3	α^6	0	1	α^5	α^4	α
α^3	α^3	α^2	α^4	1	0	α	α^6	α^5
α^4	α^4	α^6	α^3	α^5	α	0	α^2	1
α^5	α^5	α	1	α^4	α^6	α^2	0	α^3
α^6	α^6	α^4	α^2	α	α^5	1	α^3	0

Substituting for powers of α, we see that

$$\alpha^7 = \alpha^2 \alpha^5 = \alpha^2(\alpha + 1) = \alpha^3 + \alpha^2 = \alpha^2 + 1 + \alpha^2 = 1.$$

This gives us the multiplication table.

$\times$	0	1	α	α^2	α^3	α^4	α^5	α^6
0	0	0	0	0	0	0	0	0
1	0	1	α	α^2	α^3	α^4	α^5	α^6
α	0	α	α^2	α^3	α^4	α^5	α^6	1
α^2	0	α^2	α^3	α^4	α^5	α^6	1	α
α^3	0	α^3	α^4	α^5	α^6	1	α	α^2
α^4	0	α^4	α^5	α^6	1	α	α^2	α^3
α^5	0	α^5	α^6	1	α	α^2	α^3	α^4
α^6	0	α^6	1	α	α^2	α^3	α^4	α^5

To show that this field is isomorphic to $GF(8)$ in the example above, we use the mapping $0 \leftrightarrow 0$, $1 \leftrightarrow 1$, $\alpha \leftrightarrow 2$, $\alpha^2 \leftrightarrow 3$, $\alpha^3 \leftrightarrow 4$, $\alpha^4 \leftrightarrow 5$, $\alpha^5 \leftrightarrow 6$, $\alpha^6 \leftrightarrow 7$.

It is possible to write out the addition and multiplication tables for this field in terms of the elements $0, 1, \alpha, \alpha + 1, \alpha^2, \alpha^2 + 1, \alpha^2 + \alpha, \alpha^2 + \alpha + 1$; see Exercise 6. □

EXAMPLE 8.14

$(X^2 + 1) \in \mathbb{Z}_3[X]$ is irreducible over $\mathbb{Z}_3$. The quotient $\mathbb{Z}_3[X]/(X^2 + 1)$ is a finite field with $3^2 = 9$ elements, isomorphic to $GF(9)$.

To construct the addition and multiplication tables, we will use α to denote the coset of $(X + 1)$. Then:

$$\begin{array}{ll} \alpha^2 & \text{is the coset of } 2X \\ \alpha^3 & \text{is the coset of } 2X + 1 \\ \alpha^4 & \text{is the coset of } 2 \\ \alpha^5 & \text{is the coset of } 2X + 2 \\ \alpha^6 & \text{is the coset of } X \\ \alpha^7 & \text{is the coset of } X + 2, \end{array}$$

where we have used the fact that the coset of $(X^2 + 1)$ is the coset of 0. This gives us the addition table.

+	0	1	α	α^2	α^3	α^4	α^5	α^6	α^7
0	0	1	α	α^2	α^3	α^4	α^5	α^6	α^7
1	1	α^4	α^7	α^3	α^5	0	α^2	α	α^6
α	α	α^7	α^5	1	α^4	α^6	0	α^3	α^2
α^2	α^2	α^3	1	α^6	α	α^5	α^7	0	α^4
α^3	α^3	α^5	α^4	α	α^7	α^2	α^6	1	0
α^4	α^4	0	α^6	α^5	α^2	1	α^3	α^7	α
α^5	α^5	α^2	0	α^7	α^6	α^3	α	α^4	1
α^6	α^6	α	α^3	0	1	α^7	α^4	α^2	α^5
α^7	α^6	α^6	α^2	α^4	0	α	1	α^5	α^3

We also have

$$\alpha^8 = \alpha^4 \times \alpha^4 = 2 \times 2 = 1.$$

This gives us the multiplication table.

×	0	1	α	α^2	α^3	α^4	α^5	α^6	α^7
0	0	0	0	0	0	0	0	0	α^7
1	0	1	α	α^2	α^3	α^4	α^5	α^6	α^7
α	0	α	α^2	α^3	α^4	α^5	α^6	α^7	1
α^2	0	α^2	α^3	α^4	α^5	α^6	α^7	1	α
α^3	0	α^3	α^4	α^5	α^6	α^7	1	α	α^2
α^4	0	α^4	α^5	α^6	α^7	1	α	α^2	α^3
α^5	0	α^5	α^6	α^7	1	α	α^2	α^3	α^4
α^6	0	α^6	α^7	1	α	α^2	α^3	α^4	α^5
α^7	0	α^7	1	α	α^2	α^3	α^4	α^5	α^6

The isomorphism between this field and GF(9) in the example above is given by $0 \leftrightarrow 0, 1 \leftrightarrow 1, \alpha^k \leftrightarrow k + 1$, for $k = 1, 2, \dots, 7$.

It is also possible to write out the addition and multiplication tables of this field in terms of the elements $0, 1, 2, \alpha, \alpha + 1, \alpha + 2, 2\alpha, 2\alpha + 1, 2\alpha + 2$; see Exercise 8.

☐

EXAMPLE 8.15

We have shown that $(X^7 + X + 1)$ is irreducible over $\mathbb{B}$. We will now use this result to construct the field $GF(2^7) = GF(128)$ with 128 elements.

$GF(128)$ is the quotient $\mathbb{B}[X]/(X^7 + X + 1)$. If we denote the coset of X by α, the powers of α must satisfy the equation

$$\alpha^7 = \alpha + 1.$$

It follows that

$$\alpha^8 = \alpha^2 + \alpha,$$
$$\alpha^9 = \alpha^3 + \alpha^2,$$
$$\alpha^{10} = \alpha^4 + \alpha^3,$$
$$\alpha^{11} = \alpha^5 + \alpha^4,$$
$$\alpha^{12} = \alpha^6 + \alpha^5,$$
$$\alpha^{13} = \alpha^6 + \alpha + 1,$$
$$\alpha^{14} = \alpha^2 + 1,$$
$$\alpha^{15} = \alpha^3 + \alpha,$$

and so on. These equations can be used to construct the addition and multiplication tables of $GF(128)$. As these tables have over one hundred rows and columns, we will not show the entire tables here. We will show only the top left corners of the tables.

We can express the elements of $GF(128)$ as sums of the powers α, α^2, α^3, α^4, α^5 and α^6. If we do this, the corners of the addition and multiplication tables are:

$+$	0	1	α	α^2	α^3	α^4	α^5	α^6	$\ldots$
0	0	1	α	α^2	α^3	α^4	α^5	α^6	$\ldots$
1	1	0	$\alpha+1$	α^2+1	α^3+1	α^4+1	α^5+1	α^6+1	$\ldots$
α	α	$\alpha+1$	0	$\alpha^2+\alpha$	$\alpha^3+\alpha$	$\alpha^4+\alpha$	$\alpha^5+\alpha$	$\alpha^6+\alpha$	$\ldots$
α^2	α^2	α^2+1	$\alpha^2+\alpha$	0	$\alpha^3+\alpha^2$	$\alpha^4+\alpha^2$	$\alpha^5+\alpha^2$	$\alpha^6+\alpha^2$	$\ldots$
α^3	α^3	α^3+1	$\alpha^3+\alpha$	$\alpha^3+\alpha^2$	0	$\alpha^4+\alpha^3$	$\alpha^5+\alpha^3$	$\alpha^6+\alpha^3$	$\ldots$
α^4	α^4	α^4+1	$\alpha^4+\alpha$	$\alpha^4+\alpha^2$	$\alpha^4+\alpha^3$	0	$\alpha^5+\alpha^4$	$\alpha^6+\alpha^4$	$\ldots$
α^5	α^5	α^5+1	$\alpha^5+\alpha$	$\alpha^5+\alpha^2$	$\alpha^5+\alpha^3$	$\alpha^5+\alpha^4$	0	$\alpha^6+\alpha^5$	$\ldots$
α^6	α^6	α^6+1	$\alpha^6+\alpha$	$\alpha^6+\alpha^2$	$\alpha^6+\alpha^3$	$\alpha^6+\alpha^4$	$\alpha^6+\alpha^5$	0	$\ldots$
$\vdots$				$\vdots$					$\ddots$

and

$\times$	0	1	α	α^2	α^3	α^4	α^5	α^6	$\ldots$
0	0	0	0	0	0	0	0	0	$\ldots$
1	0	1	α	α^2	α^3	α^4	α^5	α^6	$\ldots$
α	0	α	α^2	α^3	α^4	α^5	α^6	$\alpha+1$	$\ldots$
α^2	0	α^2	α^3	α^4	α^5	α^6	$\alpha+1$	$\alpha^2+\alpha$	$\ldots$
α^3	0	α^3	α^4	α^5	α^6	$\alpha+1$	$\alpha^2+\alpha$	$\alpha^3+\alpha^2$	$\ldots$
α^4	0	α^4	α^5	α^6	$\alpha+1$	$\alpha^2+\alpha$	$\alpha^3+\alpha^2$	$\alpha^4+\alpha^3$	$\ldots$
α^5	0	α^5	α^6	$\alpha+1$	$\alpha^2+\alpha$	$\alpha^3+\alpha^2$	$\alpha^4+\alpha^3$	$\alpha^4+\alpha^5$	$\ldots$
α^6	0	α^6	$\alpha+1$	$\alpha^2+\alpha$	$\alpha^3+\alpha^2$	$\alpha^4+\alpha^3$	$\alpha^4+\alpha^5$	$\alpha^6+\alpha^5$	$\ldots$
$\vdots$				$\vdots$					$\ddots$

Starting from $\alpha^{15} = \alpha^3 + \alpha$, we can compute $\alpha^{30} = \alpha^6 + \alpha^2$, $\alpha^{60} = \alpha^6 + \alpha^5 + \alpha^4$, $\alpha^{120} = \alpha^6 + \alpha^5 + \alpha^4 + \alpha^3 + \alpha^2 + \alpha$, and finally $\alpha^{127} = 1$, showing that α is a primitive element of $GF(128)$. This means we can label the elements of $GF(128)$ by powers of α, so that

$$GF(128) = \{0, 1, \alpha, \alpha^2, \ldots, \alpha^{126}\}.$$

If we do this the corner of the addition table becomes

+	0	1	α	α^2	α^3	α^4	α^5	α^6	$\cdots$
0	0	1	α	α^2	α^3	α^4	α^5	α^6	$\cdots$
1	1	0	α^7	α^{14}	α^{63}	α^{28}	α^{54}	α^{126}	$\cdots$
α	α	α^7	0	α^8	α^{15}	α^{64}	α^{29}	α^{55}	$\cdots$
α^2	α^2	α^{14}	α^8	0	α^9	α^{16}	α^{65}	α^{30}	$\cdots$
α^3	α^3	α^{63}	α^{15}	α^9	0	α^{10}	α^{17}	α^{66}	$\cdots$
α^4	α^4	α^{28}	α^{64}	α^{16}	α^{10}	0	α^{11}	α^{18}	$\cdots$
α^5	α^5	α^{54}	α^{29}	α^{65}	α^{17}	α^{11}	0	α^{12}	$\cdots$
α^6	α^6	α^{126}	α^{55}	α^{30}	α^{66}	α^{18}	α^{12}	0	$\cdots$
$\vdots$					$\vdots$				$\ddots$

and the corner of the multiplication table is

$\times$	0	1	α	α^2	α^3	α^4	α^5	α^6	$\cdots$
0	0	0	0	0	0	0	0	0	$\cdots$
1	0	1	α	α^2	α^3	α^4	α^5	α^6	$\cdots$
α	0	α	α^2	α^3	α^4	α^5	α^6	α^7	$\cdots$
α^2	0	α^2	α^3	α^4	α^5	α^6	α^7	α^8	$\cdots$
α^3	0	α^3	α^4	α^5	α^6	α^7	α^8	α^9	$\cdots$
α^4	0	α^4	α^5	α^6	α^7	α^8	α^9	α^{10}	$\cdots$
α^5	0	α^5	α^6	α^7	α^8	α^9	α^{10}	α^{11}	$\cdots$
α^6	0	α^6	α^7	α^8	α^9	α^{10}	α^{11}	α^{12}	$\cdots$
$\vdots$					$\vdots$				$\ddots$

Finally, we can label the elements of $GF(128)$ with the integers from 0 to 127, using k to represent the $(k-1)$th power of α. When we do this, the corner of the addition table becomes

+	0	1	2	3	4	5	6	7	8	9	...
0	0	1	2	3	4	5	6	7	8	9	...
1	1	0	8	15	64	29	55	127	2	57	...
2	2	8	0	9	16	65	30	56	1	3	...
3	3	15	9	0	10	17	66	31	57	2	...
4	4	64	16	10	0	11	18	67	32	58	...
5	5	29	65	17	11	0	12	19	68	33	...
6	6	55	30	66	18	12	0	13	20	69	...
7	7	127	56	31	67	19	13	0	14	21	...
8	8	2	1	57	32	68	20	14	0	15	...
9	9	57	3	2	58	33	69	21	15	0	...
⋮											⋱

and the corner of the multiplication table becomes

×	0	1	2	3	4	5	6	7	8	9	...
0	0	0	0	0	0	0	0	0	0	0	...
1	0	1	2	3	4	5	6	7	8	9	...
2	0	2	3	4	5	6	7	8	9	10	...
3	0	3	4	5	6	7	8	9	10	11	...
4	0	4	5	6	7	8	9	10	11	12	...
5	0	5	6	7	8	9	10	11	12	13	...
6	0	6	7	8	9	10	11	12	13	14	...
7	0	7	8	9	10	11	12	13	14	15	...
8	0	8	9	10	11	12	13	14	15	16	...
9	0	9	10	11	12	13	14	15	16	17	...
⋮											⋱

The following table shows the relationships between the three labellings of the elements of $GF(128)$.

0	0	0	1	1	1
2	α^1	α	3	α^2	α^2
4	α^3	α^3	5	α^4	α^4
6	α^5	α^5	7	α^6	α^6
8	α^7	$\alpha+1$	9	α^8	$\alpha^2+\alpha$
10	α^9	$\alpha^3+\alpha^2$	11	α^{10}	$\alpha^4+\alpha^3$
12	α^{11}	$\alpha^5+\alpha^4$	13	α^{12}	$\alpha^6+\alpha^5$
14	α^{13}	$\alpha^6+\alpha+1$	15	α^{14}	α^2+1
16	α^{15}	$\alpha^3+\alpha$	17	α^{16}	$\alpha^4+\alpha^2$
18	α^{17}	$\alpha^5+\alpha^3$	19	α^{18}	$\alpha^6+\alpha^4$
20	α^{19}	$\alpha^5+\alpha+1$	21	α^{20}	$\alpha^6+\alpha^2+\alpha$
22	α^{21}	$\alpha^3+\alpha^2+\alpha+1$	23	α^{22}	$\alpha^4+\alpha^3+\alpha^2+\alpha$
24	α^{23}	$\alpha^5+\alpha^4+\alpha^3+\alpha^2$	25	α^{24}	$\alpha^6+\alpha^5+\alpha^4+\alpha^3$
26	α^{25}	$\alpha^6+\alpha^5+\alpha^4+\alpha+1$	27	α^{26}	$\alpha^6+\alpha^5+\alpha^2+1$
28	α^{27}	$\alpha^6+\alpha^3+1$	29	α^{28}	α^4+1
30	α^{29}	$\alpha^5+\alpha$	31	α^{30}	$\alpha^6+\alpha^2$
32	α^{31}	$\alpha^3+\alpha+1$	33	α^{32}	$\alpha^4+\alpha^2+\alpha$
34	α^{33}	$\alpha^5+\alpha^3+\alpha^2$	35	α^{34}	$\alpha^6+\alpha^4+\alpha^3$
36	α^{35}	$\alpha^5+\alpha^4+\alpha+1$	37	α^{36}	$\alpha^6+\alpha^5+\alpha^2+\alpha$
38	α^{37}	$\alpha^6+\alpha^3+\alpha^2+\alpha+1$	39	α^{38}	$\alpha^4+\alpha^3+\alpha^2+1$
40	α^{39}	$\alpha^5+\alpha^4+\alpha^3+\alpha$	41	α^{40}	$\alpha^6+\alpha^5+\alpha^4+\alpha^2$
42	α^{41}	$\alpha^6+\alpha^5+\alpha^3+\alpha+1$	43	α^{42}	$\alpha^6+\alpha^4+\alpha^2+1$
44	α^{43}	$\alpha^5+\alpha^3+1$	45	α^{44}	$\alpha^6+\alpha^4+\alpha$
46	α^{45}	$\alpha^5+\alpha^2+\alpha+1$	47	α^{46}	$\alpha^6+\alpha^3+\alpha^2+\alpha$
48	α^{47}	$\alpha^4+\alpha^3+\alpha^2+\alpha+1$	49	α^{48}	$\alpha^5+\alpha^4+\alpha^3+\alpha^2+\alpha$
50	α^{49}	$\alpha^6+\alpha^5+\alpha^4+\alpha^3+\alpha^2$	51	α^{50}	$\alpha^6+\alpha^5+\alpha^4+\alpha^3+\alpha+1$
52	α^{51}	$\alpha^6+\alpha^5+\alpha^4+\alpha^2+1$	53	α^{52}	$\alpha^6+\alpha^5+\alpha^3+1$
54	α^{53}	$\alpha^6+\alpha^4+1$	55	α^{54}	α^5+1
56	α^{55}	$\alpha^6+\alpha$	57	α^{56}	$\alpha^2+\alpha+1$
58	α^{57}	$\alpha^3+\alpha^2+\alpha$	59	α^{58}	$\alpha^4+\alpha^3+\alpha^2$
60	α^{59}	$\alpha^5+\alpha^4+\alpha^3$	61	α^{60}	$\alpha^6+\alpha^5+\alpha^4$
62	α^{61}	$\alpha^6+\alpha^5+\alpha+1$	63	α^{62}	$\alpha^6+\alpha^2+1$
64	α^{63}	α^3+1	65	α^{64}	$\alpha^4+\alpha$
66	α^{65}	$\alpha^5+\alpha^2$	67	α^{66}	$\alpha^6+\alpha^3$
68	α^{67}	$\alpha^4+\alpha+1$	69	α^{68}	$\alpha^5+\alpha^2+\alpha$
70	α^{69}	$\alpha^6+\alpha^3+\alpha^2$	71	α^{70}	$\alpha^4+\alpha^3+\alpha+1$
72	α^{71}	$\alpha^5+\alpha^4+\alpha^2+\alpha$	73	α^{72}	$\alpha^6+\alpha^5+\alpha^3+\alpha^2$
74	α^{73}	$\alpha^6+\alpha^4+\alpha^3+\alpha+1$	75	α^{74}	$\alpha^5+\alpha^4+\alpha^2+1$
76	α^{75}	$\alpha^6+\alpha^5+\alpha^3+\alpha$	77	α^{76}	$\alpha^6+\alpha^4+\alpha^2+\alpha+1$
78	α^{77}	$\alpha^5+\alpha^3+\alpha^2+1$	79	α^{78}	$\alpha^6+\alpha^4+\alpha^3+\alpha$
80	α^{79}	$\alpha^5+\alpha^4+\alpha^2+\alpha+1$	81	α^{80}	$\alpha^6+\alpha^5+\alpha^3+\alpha^2+\alpha$
82	α^{81}	$\alpha^6+\alpha^4+\alpha^3+\alpha^2+\alpha+1$	83	α^{82}	$\alpha^5+\alpha^4+\alpha^3+\alpha^2+1$
84	α^{83}	$\alpha^6+\alpha^5+\alpha^4+\alpha^3+\alpha$	85	α^{84}	$\alpha^6+\alpha^5+\alpha^4+\alpha^2+\alpha+1$
86	α^{85}	$\alpha^6+\alpha^5+\alpha^3+\alpha^2+1$	87	α^{86}	$\alpha^6+\alpha^4+\alpha^3+1$
88	α^{87}	$\alpha^5+\alpha^4+1$	89	α^{88}	$\alpha^6+\alpha^5+\alpha$
90	α^{89}	$\alpha^6+\alpha^2+\alpha+1$	91	α^{90}	$\alpha^3+\alpha^2+1$

92	α^{91}	$\alpha^4+\alpha^3+\alpha$	93	α^{92}	$\alpha^5+\alpha^4+\alpha^2$
94	α^{93}	$\alpha^6+\alpha^5+\alpha^3$	95	α^{94}	$\alpha^6+\alpha^4+\alpha+1$
96	α^{95}	$\alpha^5+\alpha^2+1$	97	α^{96}	$\alpha^6+\alpha^3+\alpha$
98	α^{97}	$\alpha^4+\alpha^2+\alpha+1$	99	α^{98}	$\alpha^5+\alpha^3+\alpha^2+\alpha$
100	α^{99}	$\alpha^6+\alpha^4+\alpha^3+\alpha^2$	101	α^{100}	$\alpha^5+\alpha^4+\alpha^3+\alpha+1$
102	α^{101}	$\alpha^6+\alpha^5+\alpha^4+\alpha^2+\alpha$	103	α^{102}	$\alpha^6+\alpha^5+\alpha^3+\alpha^2+\alpha+1$
104	α^{103}	$\alpha^6+\alpha^4+\alpha^3+\alpha^2+1$	105	α^{104}	$\alpha^5+\alpha^4+\alpha^3+1$
106	α^{105}	$\alpha^6+\alpha^5+\alpha^4+\alpha$	107	α^{106}	$\alpha^6+\alpha^5+\alpha^2+\alpha+1$
108	α^{107}	$\alpha^6+\alpha^3+\alpha^2+1$	109	α^{108}	$\alpha^4+\alpha^3+1$
110	α^{109}	$\alpha^5+\alpha^4+\alpha$	111	α^{110}	$\alpha^6+\alpha^5+\alpha^2$
112	α^{111}	$\alpha^6+\alpha^3+\alpha+1$	113	α^{112}	$\alpha^4+\alpha^2+1$
114	α^{113}	$\alpha^5+\alpha^3+\alpha$	115	α^{114}	$\alpha^6+\alpha^4+\alpha^2$
116	α^{115}	$\alpha^5+\alpha^3+\alpha+1$	117	α^{116}	$\alpha^6+\alpha^4+\alpha^2+\alpha$
118	α^{117}	$\alpha^5+\alpha^3+\alpha^2+\alpha+1$	119	α^{118}	$\alpha^6+\alpha^4+\alpha^3+\alpha^2+\alpha$
120	α^{119}	$\alpha^5+\alpha^4+\alpha^3+\alpha^2+\alpha+1$	121	α^{120}	$\alpha^6+\alpha^5+\alpha^4+\alpha^3+\alpha^2+\alpha$
122	α^{121}	$\alpha^6+\alpha^5+\alpha^4+\alpha^3+\alpha^2+\alpha+1$	123	α^{122}	$\alpha^6+\alpha^5+\alpha^4+\alpha^3+\alpha^2+1$
124	α^{123}	$\alpha^6+\alpha^5+\alpha^4+\alpha^3+1$	125	α^{124}	$\alpha^6+\alpha^5+\alpha^4+1$
126	α^{125}	$\alpha^6+\alpha^5+1$	127	α^{126}	α^6+1

The labelling that uses sums of powers of α is convenient to use when computations involving addition are involved. The labelling that uses all the powers of α is more convenient for computations involving multiplications.

$\square$

A Galois field that is constructed by taking the quotient $\mathbb{F}[X]/p(X)$ for some irreducible polynomial p also has the structure of a linear space over $\mathbb{F}$. In the examples above, the powers of α are the elements of a basis for the linear space. We will not use these facts, but they are important results in the theory of finite fields.

In the quotient ring $\mathbb{F}[X]/p(X)$, the coset of $p(X)$ is 0. If $p(X)$ is irreducible over $\mathbb{F}$, then $p(a) \neq 0$ for all $a \in \mathbb{F}$. If α denotes the coset of X, then $p(\alpha) = p(X) = 0$. So α is a root of $p(X)$ in $\mathbb{F}[X]/p(X)$, and $p(X)$ is not irreducible over $\mathbb{F}[X]/p(X)$.

DEFINITION 8.4 Order of the Root of a Polynomial   *If $p(X) \in \mathbb{F}[X]$ is an irreducible polynomial over $\mathbb{F}$, and α is a root of $p(X)$ in the quotient field $\mathbb{F}[X]/p(X)$, the* order *of α is the least positive integer p for which $\alpha^p = 1$.*

All the roots of $p(X)$ have the same order.

EXAMPLE 8.16

In the construction of $\mathbb{B}[X]/(X^2 + X + 1)$, we used α to denote the coset of X, which became a root of $(X^2 + X + 1)$. Since $\alpha^3 = 1$, α is a root of order 3.

If we consider

$$(\alpha^2)^2 + \alpha^2 + 1 = \alpha^4 + \alpha^2 + 1 = \alpha + \alpha^2 + 1 = 0,$$

we see that α^2 is also a root of $(X^2 + X + 1)$. Since $(\alpha^2)^3 = \alpha^6 = 1$, it is also a root of order 3.

□

EXAMPLE 8.17

In the construction of $\mathbb{B}[X]/(X^3 + X^2 + 1)$, we used α to denote the coset of X, which became a root of $(X^3 + X^2 + 1)$. Since $\alpha^7 = 1$, α is a root of order 7.

α^2 is also a root of $(X^3 + X^2 + 1)$, since

$$\begin{aligned}
(\alpha^2)^3 + (\alpha^2)^2 + 1 &= \alpha^6 + \alpha^4 + 1 \\
&= \alpha^2\alpha + \alpha^2\alpha + 1 + 1 \\
&= 0.
\end{aligned}$$

Since $(\alpha^2)^7 = (\alpha^7)^2 = 1$, it is also a root of order 7.

The third root of $(X^3 + X^2 + 1)$ is α^4, since

$$\begin{aligned}
(\alpha^4)^3 + (\alpha^4)^2 + 1 &= \alpha^{12} + \alpha^8 + 1 \\
&= \alpha^5 + \alpha + 1 \\
&= \alpha + 1 + \alpha + 1 \\
&= 0.
\end{aligned}$$

It is also a root of order 7.

□

EXAMPLE 8.18

In the construction of $\mathbb{Z}_3[X]/(X^2 + 1)$, α was used to denote the coset of $(X + 1)$. In this case, α is not a root of $(X^2 + 1)$. Instead, α^2 and α^6 are roots of $(X^2 + 1)$, since

$$(\alpha^2)^2 + 1 = \alpha^4 + 1 = 2 + 1 = 0,$$

and

$$(\alpha^6)^2 + 1 = (\alpha^4)^3 + 1 = 2 + 1 = 0.$$

They are both roots of order 4.

□

8.5   Bursts of Errors

Before we apply the concepts developed above to the design of codes, we will define bursts of errors and state some simple results about them.

DEFINITION 8.5 Burst   *A burst of length d is a vector whose only non-zero components belong to a set of d successive components, of which the first and last are not zero.*

A burst of length d will change a code word into another word whose Hamming distance from the original code word will be between 1 and d. This means that a code that can detect a burst of length d must have a minimum distance greater than d. Roughly speaking, adding a parity check symbol to a code increases the minimum distance of the code by 1. It follows that a linear code of length n can detect all bursts of errors of length d or less if and only if it has d parity-check symbols. Similar arguments show that, in order to correct all bursts of errors of length b or less, the code must have at least $2b$ parity-check symbols, and to correct all bursts of errors of length b or less and simultaneously detect all bursts of length $d \geq b$ or less, it must have at least $b + d$ parity-check symbols.

More detailed analyses of the burst detection and correction capabilities of linear and cyclic codes can be found in [2], Chapters 8 and 9; [3], Section 8.4 and [4], Chapter 10.

The following sections describe types of codes that are designed to correct bursts of errors.

8.6   Fire Codes

Fire codes correct single-burst errors in code vectors.

DEFINITION 8.6 Fire Code   *A Fire code is a cyclic code with a generator polynomial of the form*

$$g(X) = p(X)(X^c + 1), \qquad (8.1)$$

where $p(X)$ is an irreducible polynomial over $\mathbb{B}$ of degree m, whose roots have order r and c is not divisible by r.

The length n of the code words in a Fire Code is the least common multiple of c and r, the number of parity check bits is $c + m$ and the number of information bits is $n - c - m$. The code is capable of correcting a single burst of length b and simultaneously detecting a burst of length $d \geq b$ or less if $b \leq m$ and $b + d \leq c + 1$.

EXAMPLE 8.19

We have seen that $(X^2 + X + 1)$ is irreducible over $\mathbb{B}$ and that the order of its roots is 3. We can use this polynomial to construct a generator polynomial for a Fire code by multiplying it by $(X^4 + 1)$ to give

$$g(X) = (X^2 + X + 1)(X^4 + 1) = (X^6 + X^5 + X^4 + X^2 + X + 1).$$

In this case, we have $c = 4$, $m = 2$ and $r = 3$. The code has code words that are twelve bits long, with six information bits and six parity check bits. It can correct bursts up to two bits long.

Its generator matrix is

$$G = \begin{bmatrix} 1 & 1 & 1 & 0 & 1 & 1 & 1 & 0 & 0 & 0 & 0 & 0 \\ 0 & 1 & 1 & 1 & 0 & 1 & 1 & 1 & 0 & 0 & 0 & 0 \\ 0 & 0 & 1 & 1 & 1 & 0 & 1 & 1 & 1 & 0 & 0 & 0 \\ 0 & 0 & 0 & 1 & 1 & 1 & 0 & 1 & 1 & 1 & 0 & 0 \\ 0 & 0 & 0 & 0 & 1 & 1 & 1 & 0 & 1 & 1 & 1 & 0 \\ 0 & 0 & 0 & 0 & 0 & 1 & 1 & 1 & 0 & 1 & 1 & 1 \end{bmatrix}.$$

EXAMPLE 8.20

It is easy to construct Fire codes with long code words. Consider

$$g(X) = (X^7 + X^3 + 1)(X^8 + 1).$$

$(X^7 + X^3 + 1)$ is irreducible over $\mathbb{B}$, and the order of its roots is 127. We therefore have $m = 7$, $r = 127$ and $c = 8$. The least common multiple of c and r is $8 \times 127 = 1016$. The code has code words that are 1016 bits long, with 15 parity check bits and 1001 information bits. It can correct bursts up to seven bits long.

8.7   Minimum Polynomials

Let $\mathbb{F}$ be any field. For any $a \in \mathbb{F}$, the polynomial $(X - a)$ has a as a zero. It is irreducible over $\mathbb{F}$.

If $p(X) \in \mathbb{F}[X]$ is irreducible over $\mathbb{F}$, and the degree of $p(X)$ is greater than 1, then $\mathbb{F}[X]/p(X)$ is a field in which the coset of X is a zero of $p(X)$.

There may be many polynomials that have a given zero, some of which are irreducible and some of which are not. Of these, there is one special polynomial that has the smallest degree. To specify it, we need the following definitions.

DEFINITION 8.7 Extension Field   *Let $\mathbb{F}$ and $\mathbb{G}$ be two fields. $\mathbb{G}$ is an* extension field *of $\mathbb{F}$ if there exists a ring isomorphism from $\mathbb{F}$ onto a subset of $\mathbb{G}$.*

DEFINITION 8.8 Monic Polynomial   *A polynomial of degree $n \geq 1$ in which the coefficient of X^n is 1 is a* monic polynomial

DEFINITION 8.9 Minimum Polynomial   *Let $\mathbb{F}$ be a field, and let a belong either to $\mathbb{F}$ or an extension field of $\mathbb{F}$. If $p(X) \in \mathbb{F}[X]$ is an irreducible monic polynomial of which a is a zero, and there is no polynomial of lesser degree of which a is a zero, then $p(X)$ is the* minimum polynomial *of a over $\mathbb{F}$.*

Note that if $p(X)$ is the minimum polynomial of α over $\mathbb{F}$, then $p(X)$ will be a factor of any other polynomial of which α is a zero.

EXAMPLE 8.21

$GF(4)$ is an extension field of $\mathbb{B}$. If we let $GF(4) = \mathbb{B}[X]/(X^2 + X + 1)$, and let α denote the coset of X, then α is a zero of $(X^2 + X + 1)$, $(X^3 + 1)$, and $(X^5 + X^3 + X^2 + 1)$, and other polynomials. All these polynomials are monic, but only $(X^2 + X + 1)$ is irreducible over $\mathbb{B}$. It is the minimum polynomial of α, and is a factor of $(X^3 + 1)$ and $(X^5 + X^3 + X^2 + 1)$.   ☐

EXAMPLE 8.22

We can construct $GF(25)$ as the quotient $\mathbb{Z}_5[X]/(X^2 + 1)$. If α denotes the coset of X in $GF(25)$, it is a zero of $(X^2 + 1)$. It is also a zero of $(2X^2 + 2)$, $(3X^2 + 3)$, $(4X^2 + 4)$, $(X^3 + X^2 + X + 1)$, $(X^4 + 2X^2 + 1)$, and other polynomials. $(2X^2 + 2)$, $(3X^2 + 3)$ and $(4X^2 + 4)$ are not monic polynomials and $(X^3 + X^2 + X + 1)$ and $(X^4 + 2X^2 + 1)$ are not irreducible over $\mathbb{Z}_5$. $(X^2 + 1)$ is the minimum polynomial of α over $\mathbb{Z}_5$.   ☐

8.8   Bose-Chaudhuri-Hocquenghem Codes

Bose-Chaudhuri-Hocquenghem (BCH) codes are cyclic codes whose generator polynomial has been chosen to make the distance between code words large, and for which effective decoding procedures have been devised. The construction of BCH codes uses roots of unity.

DEFINITION 8.10 nth Root of Unity   Let $\mathbb{F}$ be a field. An nth root of unity *is a zero of the polynomial* $(X^n - 1) \in \mathbb{F}[X]$.

1 is obviously always an nth root of unity, but in most cases, the roots of unity will not belong to $\mathbb{F}$, but to some extension field of $\mathbb{F}$. For a Galois field $GF(p)$ there will be some m such that the nth roots of unity belong to $GF(p^m)$. In this case, n must divide $p^m - 1$. (This means that n and p cannot have any common factors.)

DEFINITION 8.11 Primitive Root of Unity   Let $GF(p^m)$ be the Galois field that *contains the nth roots of unity of* $GF(p)$. *Let* α *be one of these roots of unity. If* α, α^2, $\alpha^3, \ldots, \alpha^n$ *are all distinct roots of unity, then* α *is called a* primitive root of unity.

EXAMPLE 8.23

The 3rd roots of unity of $\mathbb{B}$ have $n = 3$, $p = 2$. Since $2^2 - 1 = 3$, we have $m = 2$. The roots of unity are the three non-zero elements of $GF(4)$. Since

$$(X^3 + 1) = (X + 1)(X^2 + X + 1),$$

the minimum polynomial of 1 is $(X + 1)$ and the minimum polynomial of the other roots of unity is $(X^2 + X + 1)$. The zeros of $(X^2 + X + 1)$ in $GF(4)$ are primitive roots of unity in $\mathbb{B}$. ▯

EXAMPLE 8.24

The fifth roots of unity in $GF(3)$ have $n = 5$, $p = 3$. Since $3^4 - 1 = 80$, $m = 4$. The roots belong to $GF(3^4)$. In $GF(3)[X]$, $(X^5 - 1) = (X^5 + 2)$, and

$$(X^5 + 2) = (X + 2)(X^4 + X^3 + X^2 + X + 1)$$

so 1 is a fifth root of unity with minimum polynomial $(X + 2)$, and the other four fifth roots of unity have minimum polynomial $(X^4 + X^3 + X^2 + X + 1)$, which is irreducible over $GF(3)$ (see Exercise 5). ▯

DEFINITION 8.12 Least Common Multiple   *The* least common multiple *of a set of polynomials is the polynomial of minimum degree that is divisible by all the polynomials in the set.*

EXAMPLE 8.25

In $\mathbb{B}[X]$, the least common multiple of X and $(X + 1)$ is $(X^2 + X)$.

The least common multiple of $(X + 1)$ and $(X^2 + 1)$ is $(X^2 + 1)$, since $(X + 1)$ divides $(X^2 + 1)$.

The least common multiple of $(X^2 + 1)$ and $(X^3 + 1)$ can be found by finding the factors of these polynomials and multiplying together those that appear in at least one of the polynomials. Since

$$(X^2 + 1) = (X + 1)(X + 1)$$

and

$$(X^3 + 1) = (X + 1)(X^2 + X + 1)$$

their least common multiple is given by

$$(X + 1)(X + 1)(X^2 + X + 1) = (X^4 + X^3 + X + 1).$$

$\square$

DEFINITION 8.13 Bose-Chaudhuri-Hocquenghem (BCH) Code   *A Bose-Chaudhuri-Hocquenghem (BCH) code is a cyclic code of length n whose generator polynomial is the least common multiple of the minimal polynomials of successive powers of a primitive nth root of unity in* $\mathbb{B}$.

From the above, there is some m such that $GF(2^m)$ contains a primitive nth root of unity in $\mathbb{B}$. If b and δ are positive integers, then $\alpha^b, \alpha^{b+1} \ldots \alpha^{b+\delta-2}$ are successive powers of α. Each of these powers will have a minimal polynomial in $\mathbb{B}[X]$. The least common multiple of these minimal polynomials will be the generator polynomial of a cyclic code whose minimum distance will be no less than δ. δ is the *designed distance* of the code.

The most important BCH codes are obtained by taking $b = 1$. It can be shown that for any positive integers m and t, there is a BCH binary code of length $n = 2^m - 1$ which corrects all combinations of t or fewer errors and has no more than mt parity-check bits. In particular, the code will correct bursts of length t or less.

EXAMPLE 8.26

$p(X) = X^3 + X^2 + 1$ is irreducible over $\mathbb{B}$. If we let α be the coset of X in $\mathbb{B}[X]/p(X)$, and take $m = 3$, $b = 1$ and $\delta = 3$, we get a BCH code whose code words are 7 bits long. The generator polynomial of this code is the polynomial in $\mathbb{B}[X]$ of minimal degree whose roots include α and α^2. The polynomial $X^3 + X^2 + 1$ has this property, since

$$(X^3 + X^2 + 1) = (X + \alpha)(X + \alpha^2)(X + \alpha^2 + \alpha + 1).$$

The generator matrix of the code is

$$G = \begin{bmatrix} 1 & 0 & 1 & 1 & 0 & 0 & 0 \\ 0 & 1 & 0 & 1 & 1 & 0 & 0 \\ 0 & 0 & 1 & 0 & 1 & 1 & 0 \\ 0 & 0 & 0 & 1 & 0 & 1 & 1 \end{bmatrix}.$$

The code has three information bits and four parity check bits.

⬜

There are error-correction procedures for BCH codes which identify the locations of errors, in a way similar to the procedure for Hamming codes.

8.9 Other Fields

The error-correcting codes that we have discussed so far have all been binary codes. The code words of these codes are strings consisting of the characters 0 and 1, and no others. The theory that underlies the construction and use of these codes is based on the fact that $\mathbb{Z}_2$, or $\mathbb{B}$, is a finite field. From this fact it follows that $\mathbb{B}^n$ is a linear space over $\mathbb{B}$ and that if $p(X)$ is a polynomial with coefficients in $\mathbb{B}$, $\mathbb{B}[X]/p(X)$ is a ring in general, and a field if $p(X)$ is irreducible over $\mathbb{B}$.

It is possible to develop the theory in exactly the same way for other finite fields, such as $\mathbb{Z}_p$ for prime numbers p or $GF(q)$, and devise error-correcting codes with alphabets other than $\{0, 1\}$. If $\mathbb{F}$ stands for $\mathbb{Z}_p$ or $GF(q)$, then just as in the case of $\mathbb{B}$, $\mathbb{F}^n$ is a linear space over $\mathbb{F}$, and if $p(X)$ is a polynomial with coefficients in $\mathbb{F}$, then $\mathbb{F}[X]/p(X)$ is ring in general and a field if $p(X)$ is irreducible over $\mathbb{F}$.

There are three important differences between the binary case and all the other cases, however. First, when reducing the generator matrix to canonical form, the row operations are performed by replacing a row with the result of multiplying the row by some element of the field, or by multiplying a row by some element of the field and then adding the result to another row. Second, when constructing the parity check matrix from the canonical form of the generator matrix, we use $-A^T$, not A^T. Third, for cyclic codes, we considered the rings $\mathbb{B}[X]/(X^n + 1)$, but in all other cases, the cyclic codes are derived from operations in the ring $\mathbb{F}[X]/(X^n - 1)$.

The following examples construct linear and cyclic codes based on various fields.

EXAMPLE 8.27

Consider $\mathbb{Z}_3$, the finite field with elements 0, 1 and 2. Let us construct a linear code with code words that are four ternary digits long, that is, code words that are members of $\mathbb{Z}_3^4$. For a linear code of dimension 2, we choose two basis elements, say 1001 and 0202. The code consists of 3^2 code words, obtained by multiplying the basis elements by 0, 1 and 2 and adding the results. This gives the subspace $\{0000, 1001, 2002, 0101, 0202, 1102, 2100, 1200, 2201\}$. The generator matrix is

$$G = \begin{bmatrix} 1 & 0 & 0 & 1 \\ 0 & 2 & 0 & 2 \end{bmatrix}.$$

To reduce it to canonical form, we multiply the second row by 2 to get

$$G_c = \begin{bmatrix} 1 & 0 & 0 & 1 \\ 0 & 1 & 0 & 1 \end{bmatrix},$$

and the canonical form of the parity check matrix is

$$H_c = \begin{bmatrix} 0 & 0 & 1 & 0 \\ 2 & 2 & 0 & 1 \end{bmatrix}.$$

As we did not permute the columns of G to find G_c, H_c is also the parity check matrix of G.

▯

EXAMPLE 8.28

$GF(4)$ is the finite field with the elements 0, 1, 2 and 3, in which $1+1 = 0, 2+2 = 0$ and $3+3 = 0$. We will construct a linear code of length 4 and dimension 3. The basis elements are 1001, 0102 and 0013 and there are $4^3 = 64$ code words in the code. The generator matrix is

$$G = \begin{bmatrix} 1 & 0 & 0 & 1 \\ 0 & 1 & 0 & 2 \\ 0 & 0 & 1 & 3 \end{bmatrix}.$$

This is in canonical form, so the parity check matrix is

$$H = \begin{bmatrix} 1 & 2 & 3 & 1 \end{bmatrix},$$

where we have used the fact that $-a = a$ in GF(4).

Using the parity check matrix, we see that 1110 belongs to the code, since $1110H^T = 0$, while 1213 does not belong to the code, as $1213H^T = 2$.   ⬚

EXAMPLE 8.29

To construct a cyclic code of length 3 using the finite field $\mathbb{Z}_5$, we have to use polynomials in the ring $\mathbb{Z}_5[X]/(X^3 - 1)$.

The generator polynomial will be $(X - 1)$. This must divide $(X^3 - 1)$, and in fact,

$$(X^3 - 1) = (X - 1)(X^2 + X + 1).$$

The generator matrix is

$$G = \begin{bmatrix} -1 & 1 & 0 \\ 0 & -1 & 1 \end{bmatrix},$$

or,

$$G = \begin{bmatrix} 4 & 1 & 0 \\ 0 & 4 & 1 \end{bmatrix},$$

since $-1 = 4$ in $\mathbb{Z}_5$.

The canonical form of G is

$$G_c = \begin{bmatrix} 1 & 0 & 4 \\ 0 & 1 & 4 \end{bmatrix},$$

after permuting the columns of G.

The canonical form of the parity check matrix is

$$H_c = \begin{bmatrix} -4 & -4 & 1 \end{bmatrix},$$

or

$$H_c = \begin{bmatrix} 1 & 1 & 1 \end{bmatrix}.$$

H_c is also the parity check matrix of G.

⬚

EXAMPLE 8.30

The *ternary Golay code* is the only known perfect code on a field other than $\mathbb{B}$. It has code words in $\mathbb{Z}_3[X]/(X^{11} - 1)$, which are 11 ternary digits long with 6 information digits and 5 check digits.

The code can be generated by either of the polynomials

$$g_1(X) = X^5 + X^4 - X^3 + X^2 - 1 = X^5 + X^4 + 2X^3 + X^2 + 2,$$

whose generator matrix is

$$G_1 = \begin{bmatrix} 2 & 0 & 1 & 2 & 1 & 1 & 0 & 0 & 0 & 0 & 0 \\ 0 & 2 & 0 & 1 & 2 & 1 & 1 & 0 & 0 & 0 & 0 \\ 0 & 0 & 2 & 0 & 1 & 2 & 1 & 1 & 0 & 0 & 0 \\ 0 & 0 & 0 & 2 & 0 & 1 & 2 & 1 & 1 & 0 & 0 \\ 0 & 0 & 0 & 0 & 2 & 0 & 1 & 2 & 1 & 1 & 0 \\ 0 & 0 & 0 & 0 & 0 & 2 & 0 & 1 & 2 & 1 & 1 \end{bmatrix},$$

or

$$g_2(X) = X^5 - X^3 + X^2 - X - 1 = X^5 + 2X^3 + X^2 + 2X + 2,$$

whose generator matrix is

$$G_2 = \begin{bmatrix} 2 & 2 & 1 & 2 & 0 & 1 & 0 & 0 & 0 & 0 & 0 \\ 0 & 2 & 2 & 1 & 2 & 0 & 1 & 0 & 0 & 0 & 0 \\ 0 & 0 & 2 & 2 & 1 & 2 & 0 & 1 & 0 & 0 & 0 \\ 0 & 0 & 0 & 2 & 2 & 1 & 2 & 0 & 1 & 0 & 0 \\ 0 & 0 & 0 & 0 & 2 & 2 & 1 & 2 & 0 & 1 & 0 \\ 0 & 0 & 0 & 0 & 0 & 2 & 2 & 1 & 2 & 0 & 1 \end{bmatrix}.$$

⬚

These codes can be decoded using syndromes.

8.10   Reed-Solomon Codes

> **DEFINITION 8.14 Reed-Solomon Code**   A Reed-Solomon code *is a BCH code with parameters* $m = 1$ *and* $b = 1$.

Reed-Solomon codes are an important subclass of BCH codes. They are constructed in the following manner. Let $\mathbb{F}$ be a finite field, and let n be the order of $\alpha \in \mathbb{F}$, that is, $\alpha^n = 1$. The polynomial

$$g(X) = (X - \alpha)(X - \alpha^2) \cdots (X - \alpha^{d-1}) \tag{8.2}$$

is the generator polynomial of a code whose words have n digits, $(d-1)$ parity check digits and minimum distance d.

EXAMPLE 8.31

In $\mathbb{Z}_7$ the powers of 3 are 3, 2, 6, 4, 5, 1, so $n = 6$. If we take $d = 4$, we get

$$g(X) = (X - 3)(X - 2)(X - 6) = (X + 4)(X + 5)(X + 1) = X^3 + 3X^2 + X + 6.$$

This gives us a code whose generator matrix is

$$G = \begin{bmatrix} 6 & 1 & 3 & 1 & 0 & 0 \\ 0 & 6 & 1 & 3 & 1 & 0 \\ 0 & 0 & 6 & 1 & 3 & 1 \end{bmatrix}.$$

⬚

If the number of elements in $\mathbb{F}$ is a power of two, we can construct a binary code from the Reed-Solomon code by renaming the elements, as shown in the following example.

EXAMPLE 8.32

In $\mathbb{B}[X]/(X^3 + X^2 + 1)$, if a denotes the coset of X, $a^7 = 1$. If we take $d = 4$, our generator polynomial is

$$g(X) = (X + a)(X + a^2)(X + a^3) = X^3 + a^5 X^2 + X + a^6.$$

If we rename the elements of $\mathbb{B}[X]/(X^3 + X^2 + 1)$ with the digits 0, 1, 2, 3, 4, 5, 6, 7, the generator polynomial is

$$g(X) = X^3 + 6X^2 + X + 7,$$

and we have a generator matrix

$$G = \begin{bmatrix} 7 & 1 & 6 & 1 & 0 & 0 & 0 \\ 0 & 7 & 1 & 6 & 1 & 0 & 0 \\ 0 & 0 & 7 & 1 & 6 & 1 & 0 \\ 0 & 0 & 0 & 7 & 1 & 6 & 1 \end{bmatrix}$$

for a code on $\{0, 1, \ldots, 7\}$.

If we express the digits in the generator matrix in binary notation, we get the generator matrix for a binary code with code words that are 21 bits long:

$$G = \begin{bmatrix} 111\ 001\ 110\ 001\ 000\ 000\ 000 \\ 000\ 111\ 001\ 110\ 001\ 000\ 000 \\ 000\ 000\ 111\ 001\ 110\ 001\ 000 \\ 000\ 000\ 000\ 111\ 001\ 110\ 001 \end{bmatrix}.$$

Note that if G is used to generate code words, the operations of $\mathbb{B}[X]/(X^3 + X^2 + 1)$, with the elements suitably renamed, must be used in the computations. ⬚

This procedure can be used generally to produce binary Reed-Solomon codes with code words that are $m(2^m - 1)$ bits long with $m(2^m - 1 - 2t)$ information bits. They are capable of correcting errors occurring in up to t $m-$bit blocks.

EXAMPLE 8.33

In previous examples, we have shown that $(X^7 + X + 1) \in \mathbb{B}[X]$ is irreducible over $\mathbb{B}$ and used this to study the structure of $GF(128) = GF(2^7) = \mathbb{B}[X]/(X^7 + X + 1)$. If α denotes the coset of X in $GF(128)$, then $\alpha^{127} = 1$, and we can use the polynomial

$$g(X) = (X + \alpha)(X + \alpha^2)(X + \alpha^3)(X + \alpha^4)$$

as the generator polynomial of a Reed-Solomon code.

If we expand $g(X)$ and simplify the result, we get

$$g(X) = X^4 + (\alpha^4 + \alpha^3 + \alpha^2 + \alpha)X^3 + (\alpha^6 + \alpha^4 + \alpha^3 + \alpha + 1)X^2$$
$$+(\alpha^6 + \alpha^3 + 1)X + (\alpha^4 + \alpha^3).$$

Labelling the elements of $GF(128)$ with the integers $0, 1, \ldots, 127$, we can write this as

$$g(X) = X^4 + 23X^3 + 74X^2 + 28X + 11.$$

The generator matrix for the code has 123 rows and 127 columns:

$$G = \begin{bmatrix} 11 & 28 & 74 & 23 & 1 & 0 & 0 \ldots 0 \\ 0 & 11 & 28 & 74 & 23 & 1 & 0 \ldots 0 \\ \vdots & & & \vdots & & \ddots & \vdots \\ 0 & 0 \ldots & & 0 & 11 & 28 & 74 & 23 & 1 \end{bmatrix}.$$

We can construct binary code words that are 889 bits long by replacing the integer labels with their decimal equivalents. If we do this, the code word corresponding to the generator polynomial is

0001011 0011100 1001010 0010111 0000001 0000000 . . . 0000000,

where the spaces have been introduced for clarity. The code words have 861 information bits and are capable of correcting burst errors in up to two seven-bit blocks.

☐

8.11 Exercises

1. The first five Galois fields are $GF(2)$, $GF(3)$, $GF(4)$, $GF(5)$ and $GF(7)$. What are the next ten Galois fields?

2. Find the factors of $(X^5 + X + 1)$ in $\mathbb{B}[X]$.

3. Show that $(X^5 + X^2 + 1)$ is irreducible over $\mathbb{B}$.

4. Is $(X^6 + X^5 + X^4 + X^3 + X^2 + X + 1)$ irreducible over $\mathbb{B}$?

5. Show that $(X^4 + X^3 + X^2 + X + 1)$ is irreducible over $GF(3)$.

6. Write out the addition and multiplication tables of the field

$$GF(8) = \mathbb{B}[X]/(X^3 + X^2 + 1)$$

in terms of the elements $0, 1, \alpha, \alpha + 1, \alpha^2, \alpha^2 + 1, \alpha^2 + \alpha, \alpha^2 + \alpha + 1$.

7. Write out the addition and multiplication tables of the field

$$\mathbb{B}[X]/(X^3 + X + 1)$$

and show that it is isomorphic to $GF(8)$.

8. Write out the addition and multiplication tables of the field

$$GF(9) = \mathbb{Z}_3[X]/(X^2 + 1)$$

in terms of the elements $0, 1, 2, \alpha, \alpha + 1, \alpha + 2, 2\alpha, 2\alpha + 1, 2\alpha + 2$.

9. Show that $(X^2 + X + 1)$ is irreducible over $\mathbb{Z}_5$. Construct the Galois Field $GF(25) = \mathbb{Z}_5[X]/(X^2 + X + 1)$. Let α denote the coset of $X + 2$ in $GF(25)$. Draw up the addition and multiplication tables in terms of:

   (a) the elements $\{0, 1, \alpha, \alpha^2, \ldots, \alpha^{23}\}$;

   (b) terms of the form $(i\alpha + j)$, for $i, j \in \mathbb{Z}_5$;

   (c) the integers $\{0, 1, \ldots, 24\}$, where k represents the $(k-1)$th power of α.

10. Show that the polynomial $(X^2 + 1)$ is irreducible over $\mathbb{R}$. Find its roots in $\mathbb{C} = \mathbb{R}[X]/(X^2 + 1)$. What is their order?

11. Find the roots of $(X^2 + X + 1)$ in $\mathbb{Z}_5[X]/(X^2 + X + 1)$. What is their order?

12. $(X^3 + X^2 + 1)$ is irreducible over $\mathbb{B}$ and the order of its roots is 7. Use this fact to construct a Fire code whose code words have length 14. Write down the generator polynomial and generator matrix for this code. How many information bits and how many parity check bits does the code have?

13. The real numbers, $\mathbb{R}$, form an extension field of the rational numbers, $\mathbb{Q}$. What is the minimum polynomial in $\mathbb{Q}[X]$ of $5 \in \mathbb{Q}$? What is the minimum polynomial in $\mathbb{Q}[X]$ of $\sqrt[3]{5} \in \mathbb{R}$? What is the minimum polynomial in $\mathbb{Q}[X]$ of $(1 + \sqrt{5}) \in \mathbb{R}$?

*14. Show that $GF(9)$ is an extension field of $\mathbb{Z}_3$ by finding a mapping from $\mathbb{Z}_3$ into a subset of $GF(9)$ that is a ring isomorphism. Find the minimum polynomial in $\mathbb{Z}_3[X]$ of $2 \in \mathbb{Z}_3$ and the minimum polynomial in $\mathbb{Z}_3[X]$ of $5 \in GF(9)$. (Use the addition and multiplication tables for $GF(9)$ given in Example 8.4.)

15. Find the least common multiple of the following sets of polynomials:

    (a) $(X + 1)$ and $(X^2 + 1)$ in $\mathbb{B}[X]$;

    (b) $(X^2 + 1)$ and $(X^3 + 1)$ in $\mathbb{B}[X]$;

    (c) $(X^2 + 2)$ and $(X^2 + X)$ in $\mathbb{Z}_3[X]$;

    (d) $(X^2 + X + 2)$ and $(X^2 + 3X + 2)$ in $\mathbb{Z}_4[X]$;

    (e) $(X^2 + X + 3)$ and $(X^2 + 4X + 3)$ in $\mathbb{Z}_5[X]$.

16. Construct a linear code whose code words are five ternary digits long and whose dimension is 2. List all its code words and find its parity check matrix.

17. Construct a linear code whose code words belong to $\mathbb{Z}_5^4$ and whose dimension is 2. List all its code words and find its parity check matrix.

18. Construct a cyclic code whose code words are six ternary digits long and whose dimension is 3. List all its code words and find its parity check matrix.

19. Find the powers of 3 in $\mathbb{Z}_5$. Use them to construct a Reed-Solomon code with $n = 4$ and $d = 3$.

20. Find the powers of 2 in $\mathbb{Z}_{11}$. Use them to construct a Reed-Solomon code with $n = 10$ and $d = 5$.

8.12 References

[1] J. B. Fraleigh, *A First Course in Abstract Algebra,* 5th ed., Addison-Wesley, Reading, MA, 1994.

[2] S. Lin, *An Introduction of Error-Correcting Codes,* Prentice-Hall, Englewood Cliffs, NJ, 1970.

[3] R. J. McEliece, *The Theory of Information and Coding,* 2nd ed., Cambridge University Press, Cambridge, 2002.

[4] W. W. Peterson, *Error-correcting Codes,* MIT Press, Cambridge, MA, 1961.

Chapter 9

Convolutional Codes

9.1 Introduction

In this chapter we will study *convolutional codes*. These are a generalisation of the block codes that have been described in the previous chapters. They introduce memory into the coding process to improve the error-correcting capabilities of the codes. However, they require more complicated decoding procedures. After looking at convolutional codes in general, we will give a brief description of the special case of trellis modulation, which is a very important practical application of convolutional codes in telecommunications. We will also give a brief introduction to parallel concatenated coding schemes, also known as Turbo codes.

9.2 A Simple Example

The coding and decoding processes that are applied to error-correcting block codes are memoryless. The encoding and decoding of a block depends only on that block and is independent of any other block. Convolutional codes introduce memory into the error-correcting process. They do this by making the parity checking bits dependent on the bit values in several consecutive blocks. How this works is illustrated by the following simple example.

EXAMPLE 9.1

Figure 9.1 shows a simple convolutional encoder. The input to the encoder is a stream of single bits. The encoder consists of three shift registers, each of which holds one bit. Before the coding begins, all the registers hold the value 0. As each input bit is received, the contents of the first register are shifted into the second register, the contents of the second register are shifted into the third register and the contents of the third register are discarded. The input bit is shifted into the first register. The output of the encoder is a pair of bits for each input bit. The output bits are the input

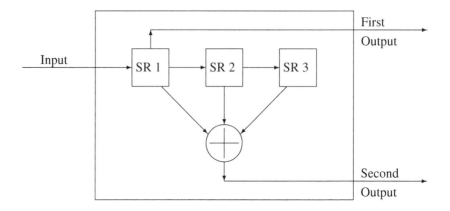

FIGURE 9.1
A simple convolutional encoder.

bit and the sum modulo 2 of the contents of the three shift registers, that is, the sum of three consecutive input bits.

The table shows the inputs and outputs for a particular input sequence. Note that the construction of the encoder is such that the input sequence is a subsequence of the output sequence.

Time	1	2	3	4	5	6	7	8	9	10	11	12	13	14	15	16	17	18
Input	1	1	0	0	0	0	1	1	1	0	0	1	0	1	0	0	1	0
First output bit	1	1	0	0	0	0	1	1	1	0	0	1	0	1	0	0	1	0
Second output bit	1	0	0	1	0	0	1	0	1	0	1	1	1	0	1	1	1	1

Suppose that the first output bit at time 8 gets corrupted. The inputs and outputs will then be

Time	1	2	3	4	5	6	7	8	9	10	11	12	13	14	15	16	17	18
Input	1	1	0	0	0	0	1	1	1	0	0	1	0	1	0	0	1	0
First output bit	1	1	0	0	0	0	1	0	1	0	0	1	0	1	0	0	1	0
Second output bit	1	0	0	1	0	0	1	0	1	0	1	1	1	0	1	1	1	1

This change will make the parity check bits at times 8, 9 and 10 incorrect. In general, changing the value of any one of the first output bits will cause the parity bits at three successive times to be incorrect.

Suppose instead that the second output bit at time 8 is changed. The inputs and outputs will be

Time	1	2	3	4	5	6	7	8	9	10	11	12	13	14	15	16	17	18
Input	1	1	0	0	0	0	1	1	1	0	0	1	0	1	0	0	1	0
First output bit	1	1	0	0	0	0	1	1	1	0	0	1	0	1	0	0	1	0
Second output bit	1	0	0	1	0	0	1	1	1	0	1	1	1	0	1	1	1	1

In this case, only the parity check at time 8 will fail.

As a final example, consider what happens if the first output bits at times 8 and 9 are corrupted. The inputs and outputs will be

Time	1	2	3	4	5	6	7	8	9	10	11	12	13	14	15	16	17	18
Input	1	1	0	0	0	0	1	1	1	0	0	1	0	1	0	0	1	0
First output bit	1	1	0	0	0	0	1	0	0	0	0	1	0	1	0	0	1	0
Second output bit	1	0	0	1	0	0	1	0	1	0	1	1	1	0	1	1	1	1

The parity checks at times 8 and 11 will fail. This cannot be distinguished from the case where the second output bits at those times are corrupted.

There are different patterns of parity check failures which indicate different patterns of corruption.

☐

The operation of a convolutional encoder can be described in a number of ways. A simple way is in terms of a *Finite State Machine (FSM)*. The FSM is an extension of the Markov information sources that were discussed in Chapter 1. It has a number of internal states. When it receives an input, it changes its state and produces an output.

EXAMPLE 9.2

The encoder in the previous example can be described by a FSM with four states. The states can be represented by the values of the first two shift registers. The table below summarises the way the machine works.

Original state	Input	Output	Final State
00	0	00	00
	1	11	10
01	0	01	00
	1	10	10
10	0	01	01
	1	10	11
11	0	00	01
	1	11	11

Figure 9.2 shows the operation of the FSM in diagrammatic form. In the figure the label on each transition between states shows the input and the output associated with that transition. For example, the label $0 \rightarrow 01$ means that the input is 0 and the output is 01.

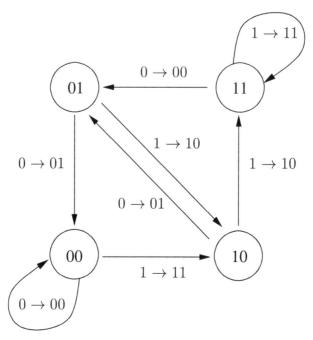

FIGURE 9.2
Diagrammatic representation of the operation of the FSM.

☐

For the purpose of studying the behaviour of the FSM in response to various input sequences, the most useful representation is the *trellis diagram*, which plots the succession of states of the FSM over time.

EXAMPLE 9.3

In a trellis diagram, time is represented horizontally from left to right, and the states of the FSM are located at various vertical positions. Transitions are represented by diagonal line segments. Figure 9.3 shows the operation of the FSM of Figure 9.2 when the input sequence is 10110100.

☐

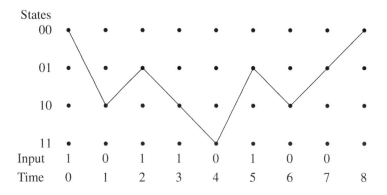

States

00									
01									
10									
11									
Input	1	0	1	1	0	1	0	0	
Time	0	1	2	3	4	5	6	7	8

FIGURE 9.3
Trellis diagram for input sequence 10110100.

Because of the internal structure of the FSM, there are sequences of states that can never be generated by any input sequence. This means that there are paths that can never appear on a trellis diagram. There are also output sequences that can never be generated by the FSM; so if such a sequence is received, it is clear that some corruption has occurred and that error-correction is necessary. The decoding and error-correction procedures that have been devised for convolutional codes are based on the comparison between paths in the trellis diagram.

The encoding process can continue indefinitely, and the path through the trellis can grow accordingly. In practice, it is necessary to limit the length of the paths to which the decoding process is applied. To do this, a sequence of zeros can be sent at regular intervals to drive the FSM into the initial state where all the shift registers hold the value zero and terminate the path at that state. The encoding then starts again with the FSM in the initial state.

In the example we have considered in this section, the input to the encoder is a single bit and the output consists of two bits. In the following sections we will consider cases where the input and output may be blocks of elements of $\mathbb{B}$ or some other finite field.

9.3   Binary Convolutional Codes

In this section we will consider the case where the input to the encoder will be a block of k bits and the output will be a block of n bits. The encoder has a memory that holds $(M + 1)$ k-bit blocks. There is a linear mapping that maps the contents of the memory to a n-bit output block. It is a linear mapping from $(\mathbb{B}^k)^{M+1}$ to $\mathbb{B}^n$. The input to the encoder consists of sequences of L k-bit blocks representing the information to be encoded, followed by M blocks of k 0s which drive the memory back to the initial state. (Note that the case $M = 0$, $L = 1$ is that of a linear block code.)

EXAMPLE 9.4

If we represent the contents of the three shift registers in the encoder illustrated in Figure 9.1 by a column vector, with the jth component representing the contents of the jth shift register, then the operation of the encoder is given by the linear transformation

$$\begin{bmatrix} c_1 \\ c_2 \end{bmatrix} = \begin{bmatrix} 1 & 0 & 0 \\ 1 & 1 & 1 \end{bmatrix} \begin{bmatrix} r_1 \\ r_2 \\ r_3 \end{bmatrix},$$

where c_1 and c_2 represent the first and second output bits, respectively, and r_1, r_2 and r_3 represent the contents of the shift registers.

If we have a sequence of input bits b_i and a sequence of output bits c_j, then the equation above means that

$$c_j = b_i$$

with $j = 2i$, and

$$c_{j+1} = b_i + b_{i-1} + b_{i-2}.$$

If we take $L = 6$, then we can represent the encoding of a sequence of six bits $b_1 b_2 b_3 b_4 b_5 b_6$ followed by 2 zeros into a sequence of 16 bits $c_1 \ldots c_{16}$ by the matrix equation

$$
\begin{bmatrix} c_1 \\ c_2 \\ c_3 \\ c_4 \\ c_5 \\ c_6 \\ c_7 \\ c_8 \\ c_9 \\ c_{10} \\ c_{11} \\ c_{12} \\ c_{13} \\ c_{14} \\ c_{15} \\ c_{16} \end{bmatrix} = \begin{bmatrix} 1\,0\,0\,0\,0\,0\,0\,0 \\ 1\,0\,0\,0\,0\,0\,0\,0 \\ 0\,1\,0\,0\,0\,0\,0\,0 \\ 1\,1\,0\,0\,0\,0\,0\,0 \\ 0\,0\,1\,0\,0\,0\,0\,0 \\ 1\,1\,1\,0\,0\,0\,0\,0 \\ 0\,0\,0\,1\,0\,0\,0\,0 \\ 0\,1\,1\,1\,0\,0\,0\,0 \\ 0\,0\,0\,0\,1\,0\,0\,0 \\ 0\,0\,1\,1\,1\,0\,0\,0 \\ 0\,0\,0\,0\,0\,1\,0\,0 \\ 0\,0\,0\,1\,1\,1\,0\,0 \\ 0\,0\,0\,0\,0\,0\,1\,0 \\ 0\,0\,0\,0\,1\,1\,1\,0 \\ 0\,0\,0\,0\,0\,0\,0\,1 \\ 0\,0\,0\,0\,0\,1\,1\,1 \end{bmatrix} \begin{bmatrix} b_1 \\ b_2 \\ b_3 \\ b_4 \\ b_5 \\ b_6 \\ 0 \\ 0 \end{bmatrix}.
$$

Note that the order in which the input bits are numbered is the reverse of the order in which the shift registers are numbered.

After the eight bits $b_1 b_2 b_3 b_4 b_5 b_6 00$ are encoded, the FSM is returned to its initial state and the encoding proceeds with the next six bits.

□

EXAMPLE 9.5

An encoder with four shift registers and parameters $k = 2$, $M = 1$ and $n = 4$ is given by the equation

$$
\begin{bmatrix} c_1 \\ c_2 \\ c_3 \\ c_4 \end{bmatrix} = \begin{bmatrix} 1\,0\,0\,0 \\ 0\,1\,0\,0 \\ 1\,0\,0\,1 \\ 0\,1\,1\,0 \end{bmatrix} \begin{bmatrix} r_1 \\ s_1 \\ r_2 \\ s_2 \end{bmatrix},
$$

where the c_j represent the four output bits, r_1 and r_2 represent the contents of the one pair of shift registers and s_1 and s_2 represent the contents of the other pair of shift registers.

A diagrammatic representation of the FSM of this encoder is given in Figure 9.4

If we take $L = 3$, then we can represent the encoding of a sequence of three pairs of bits $a_1 b_1$, $a_2 b_2$, $a_3 b_3$ followed by 2 zeros into a sequence of 16 bits $c_1 \ldots c_{16}$ by the

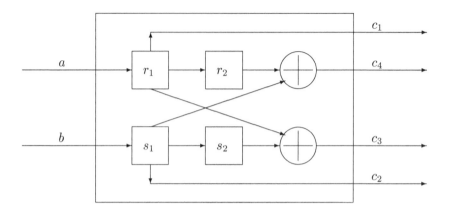

FIGURE 9.4
A convolutional encoder with $k = 2$, $M = 1$ and $n = 4$.

matrix equation

$$
\begin{bmatrix} c_1 \\ c_2 \\ c_3 \\ c_4 \\ c_5 \\ c_6 \\ c_7 \\ c_8 \\ c_9 \\ c_{10} \\ c_{11} \\ c_{12} \\ c_{13} \\ c_{14} \\ c_{15} \\ c_{16} \end{bmatrix}
=
\begin{bmatrix}
1\,0\,0\,0\,0\,0\,0\,0 \\
0\,1\,0\,0\,0\,0\,0\,0 \\
1\,0\,0\,0\,0\,0\,0\,0 \\
0\,1\,0\,0\,0\,0\,0\,0 \\
0\,0\,1\,0\,0\,0\,0\,0 \\
0\,0\,0\,1\,0\,0\,0\,0 \\
0\,1\,1\,0\,0\,0\,0\,0 \\
1\,0\,0\,1\,0\,0\,0\,0 \\
0\,0\,0\,0\,1\,0\,0\,0 \\
0\,0\,0\,0\,0\,1\,0\,0 \\
0\,0\,0\,1\,1\,0\,0\,0 \\
0\,0\,1\,0\,0\,1\,0\,0 \\
0\,0\,0\,0\,0\,0\,1\,0 \\
0\,0\,0\,0\,0\,0\,0\,1 \\
0\,0\,0\,0\,0\,1\,1\,0 \\
0\,0\,0\,0\,1\,0\,0\,1
\end{bmatrix}
\begin{bmatrix} a_1 \\ b_1 \\ a_2 \\ b_2 \\ a_3 \\ b_3 \\ 0 \\ 0 \end{bmatrix} .
$$

(As before, the order in which the input bits are numbered is the reverse of the order in which the shift registers are numbered.)

☐

There is an alternative representation of the operation of the encoder which facilitates the description of the coding process for sequences of blocks of bits. The representation uses *matrices of polynomials* to represent the structure of the encoder and matrix operations involving polynomial multiplication to describe the encoding process.

EXAMPLE 9.6

The generator matrix of the encoder shown in Figure 9.1 is

$$G = \begin{bmatrix} 1 & 1 + X + X^2 \end{bmatrix}.$$

We represent a sequence of bits as a polynomial by using the bits as the coefficients of the polynomial. For example, the sequence $b_0 b_1 b_2 b_3$ becomes the polynomial $b_0 + b_1 X + b_2 X^2 + b_3 X^3$. In this context, X can be considered to be like a *delay operator*.

As with linear block codes, the output of the encoder when the sequence is input is given by multiplying the generator matrix on the left by a vector representing the input sequence. In this case, we have

$$\begin{bmatrix} b_0 + b_1 X + b_2 X^2 + b_3 X^3 \end{bmatrix} \begin{bmatrix} 1 & 1 + X + X^2 \end{bmatrix}.$$

When we perform the multiplication, we obtain a matrix with two elements,

$$b_0 + b_1 X + b_2 X^2 + b_3 X^3$$

and

$$b_0 + (b_1 + b_0)X + (b_2 + b_1 + b_0)X^2 + (b_3 + b_2 + b_1)X^3 + (b_3 + b_2)X^4 + b_3 X^5.$$

The first element of the output simply repeats the input. The second element represents a sequence consisting of b_0 at time 0, $(b_1 + b_0)$ at time 1, $(b_2 + b_1 + b_0)$ at time 2, and so on.

We can use polynomials of arbitrarily high degree to compute the output generated by long input sequences.

$\Box$

EXAMPLE 9.7

The generator matrix of the encoder shown in Figure 9.4 is

$$G = \begin{bmatrix} 1 & 0 & 1 & X \\ 0 & 1 & X & 1 \end{bmatrix}.$$

The input of a_1 and b_1 followed by the input of a_2 and b_2 is represented by the row vector

$$I = \begin{bmatrix} a_1 + a_2 X & b_1 + b_2 X \end{bmatrix}.$$

The resulting output is

$$\begin{bmatrix} a_1 + a_2 X & b_1 + b_2 X \end{bmatrix} \begin{bmatrix} 1 & 0 & 1 & X \\ 0 & 1 & X & 1 \end{bmatrix} = \begin{bmatrix} a_1 + a_2 X \\ b_1 + b_2 X \\ a_1 + (a_2 + b_1)X + b_2 X^2 \\ b_1 + (a_1 + b_2)X + a_2 X^2 \end{bmatrix}^T .$$

$\square$

As matrices of numbers may be considered to be matrices of polynomials of degree 0, the linear block codes are again seen to be a special case of convolutional codes.

There is an extensive algebraic theory relating to the codes generated by matrices of polynomials, which is beyond the scope of this book. Details may be found in the paper by Forney [2] and in Chapter 4 of the book by Schlegel [4].

9.4 Decoding Convolutional Codes

Convolutional encoders are used to encode sequences of L k-bit blocks into sequences of $(L + M)$ n-bit blocks. The encoding process ensures that only certain sequences are valid output sequences. The decoding process has to convert the received sequence of $(L + M)$ blocks back to the original sequence of L blocks. The decoding process has to allow for the possibility that some of the bits in the received sequence may have been corrupted.

There are a number of algorithms that have been devised for the purpose of decoding convolutional codes. They attempt to reconstruct the sequence of states that the FSM in the encoder occupied during the encoding process from the received sequence, making allowance for the possibility that some of the received bits may be corrupted.

The *Viterbi algorithm* was proposed by Andrew Viterbi [5] for the purpose of decoding convolutional codes. It is in fact a very general algorithm that can be used to deduce the sequence of states of a FSM that generated a given output sequence. We will first describe it in general terms and then see how it is applied to decoding convolutional codes.

9.5   The Viterbi Algorithm

Let us suppose that we have a FSM with M states labelled $0, 1, \ldots, (M-1)$, and that each transition between states is accompanied by the output of a symbol from the alphabet $A = \{0, 1, \ldots, (N-1)\}$. If the output sequence is not corrupted, it is easy to recover the sequence of states that produced it by matching the symbols in the output string to a sequence of transitions between states.

EXAMPLE 9.8

Suppose our FSM has four states, and our alphabet is $\{0, 1\}$. The behaviour of the FSM is summarised by the following table and Figure 9.5.

Original state	Final state	Output
0	1	0
	3	1
1	2	0
	0	1
2	3	0
	1	1
3	0	0
	2	1

Suppose the FSM has 0 as its initial state and the output sequence is 00100100. What is the sequence of states that produced this output?

If we start at state 0, an output of 0 means that the second state is 1. From state 1, an output of 0 means that the following state is 2. From state 2, an output of 1 indicates that the next state is 1. If we go through the output sequence in this way, we find that the sequence of states was 012123230.

We can also represent this by the trellis diagram in Figure 9.6.

There is only one sequence of states that could have produced the output sequence 00100100 in the situation where none of the output symbols have been corrupted. In contrast, there are $2^8 = 256$ possible sequences of states that could have produced that sequence of symbols if it is possible that the output symbols may have been corrupted.

⬚

If each of the symbols of the alphabet is emitted by no more than one transition from any state, then the sequence of states can be inferred from the sequence of symbols in the absence of noise. Ambiguity in the output symbols and the presence of noise both complicate the process of recovering the sequence of states.

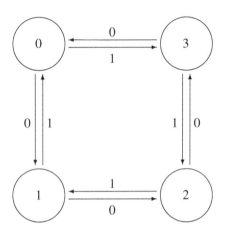

FIGURE 9.5
Diagrammatic representation of the operation of the FSM.

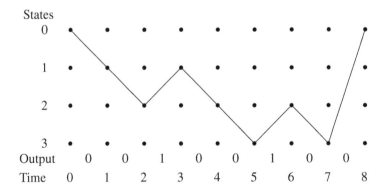

FIGURE 9.6
Trellis diagram for output sequence 00100100.

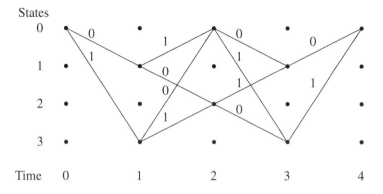

FIGURE 9.7
Trellis diagram showing output symbols.

For encoding purposes, we only consider sequences of states which have a fixed length and start and end at the same state. The constraints imposed by the transitions between states restrict the possible output sequences that can be produced by the encoder. The constraints also limit the paths through the trellis diagram that are possible.

EXAMPLE 9.9

The FSM of the previous example is such that there are only eight possible sequences of five states that begin and end at state 0. They are: 01010, with output sequence 0101; 03030, with output sequence 1010; 01030, with output sequence 0110; 03010, with output sequence 1001; 01210, with output sequence 0011; 03230, with output sequence 1100; 01230, with output sequence 0000; and 03210, with output sequence 1111.

The other eight four-bit sequences, such as 0001, can never be generated by this FSM if the initial and final states are 0.

Figure 9.7 shows the paths that can be taken in the trellis diagram. Each transition is labelled with the symbol that is output during the transition.

Note that some of the paths between states belong to more than one path from the initial state to the final state. ⬚

The Viterbi algorithm can be used to infer a sequence of states from a sequence of output symbols. Given a *distance function* or *metric* that quantifies the difference between two blocks of symbols, the algorithm will find the sequence of states that produces a block of symbols that is closer to a given block than the block of output symbols produced by any other sequence of states. It does this by assigning a length to each transition between states in such a way that the sum of the lengths of the transitions in a sequence of states is equal to the distance between the output block and the given block and then finding the sequence of states for which this length is a minimum.

The Viterbi algorithm is based on the observation that if we have a path between two points that is shorter than any other path between those points, then the section of the path between any intermediate points on it is also shorter than any other path joining the intermediate points. To see this, suppose that a shorter path joining the intermediate points did exist. Then the path consisting of the original path with the section between the intermediate points replaced by the shorter path would be shorter than the original path, contradicting the assumption that the original path was the shortest path joining its endpoints.

To use the Viterbi algorithm, we need to define a distance function between sequences of output symbols. A convenient distance function is the Hamming distance which is equal to the number of places where the sequences differ.

EXAMPLE 9.10

Returning to the FSM in the previous example, we saw that the only possible four-bit output sequences are 0000, 0011, 0101, 0110, 1001, 1010, 1100 and 1111. Suppose the string 0010 is received. How is the error to be corrected?

In this simple case, we can compute the Hamming distance between the valid output sequences and the received sequence and find which valid output sequence is closest to the received sequence. If we do this, we find that the Hamming distances are 1, 1, 3, 1, 3, 1, 3 and 3, respectively. We have a choice of four possible sequences to which to match the received sequence.

As an alternative, we can apply the Viterbi algorithm to find a matching sequence. To do this we need to define a length for each transition in the FSM such that the total of these lengths will add up to the Hamming distance between the output sequence generated by the transition and the received sequence. We can do this be assigning a distance of 0 to any transition where the output symbol matches the corresponding symbol in the received sequence and a distance of 1 to each of the other transitions.

We can now work our way through the trellis diagram of Figure 9.7. At time 0, the FSM is in state 0. There are two transitions from this state, one to state 1 with output symbol 0 and one to state 3 with output symbol 1. The first symbol of the received sequence is 0; so these transitions have lengths 0 and 1, respectively. So we have the beginnings of two paths through the trellis, 01 with length 0 and 03 with length 1. There are no transitions from state 0 to the states 0 and 2 at time 1.

At time 2, the path 01 can be extended to 010, and the transition from 0 to 1 outputs the symbol 1. The second bit of the received sequence is 0; so the length of this transition is 1. The length of the path 01 is 0; so the length of 010 is 1. The path 03 can be extended to 030. The transition from 3 to 0 outputs symbol 0; so its length is 0, making the length of the path 030 also 1. We can choose either of these paths as the shortest path from state 0 at time 0 to state 0 at time 2. We will choose 010.

Similarly, there are two paths from state 0 at time 0 to state 2 at time 2: 012 with length 0, and 032 with length 2. 012 is the shorter.

The path 010 can be extended to either state 1 or state 3 at time 3. The transitions have lengths 1 and 0, respectively, making the length of the path 0101 equal to 2 and the length of 0103 equal to 1. Extending the path 012 gives us paths 0121 with length 0 and path 0123 with length 1. So, at time 3, there are no paths ending at the states 0 and 2, the shortest path ending at state 1 is 0121 with length 0 and the two paths ending at state 3 have length 1. We choose 0123.

The final step gives us paths 01210 and 01230 corresponding to the output sequences 0011 and 0000, both of which have distance 1 from the received sequence 0010. The other two possibilities were eliminated when we chose one of two paths of equal length as we traversed the trellis.

□

To correct errors in the output of a convolutional encoder, the Viterbi algorithm can be used to infer a sequence of states from the received sequence of bits. This can then be used to generate a corrected output sequence. If the FSM is designed so that the input sequence is a subsequence of the output sequence, the input sequence can be recovered simply by discarding the parity checking bits.

EXAMPLE 9.11

Let us return to the convolutional encoder of Figure 9.1, whose FSM is described by the following table.

Original state	Input	Output	Final state
00	0	00	00
	1	11	10
01	0	01	00
	1	10	10
10	0	01	01
	1	10	11
11	0	00	01
	1	11	11

Referring to the table above, we see that the input sequence 11100 will drive the FSM through the sequence of states

$$00 \to 10 \to 11 \to 11 \to 01 \to 00$$

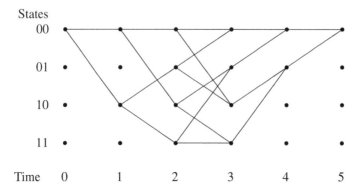

FIGURE 9.8
Trellis diagram for Viterbi decoding of 1110010000.

and generate the output sequence 1110110000. Suppose that this sequence is corrupted and the sequence 1110010000 is received.

To apply the Viterbi algorithm to this sequence, we need to consider the trellis diagram
for sequences of six states beginning and ending with the state 00. This is shown in
Figure 9.8.

We can trace the Viterbi algorithm through this trellis. Starting from the state 00 at
time 0, there are two paths, one to 00 with output 00 and one to 10 with output 11.
Since the first two bits of the received string are 11, these paths have distances 2 and
0, respectively. The following table shows the shortest paths to each state at time 1.

State at time 1	Shortest path to state	Length
00	00 → 00	2
01	No path	
10	00 → 10	0
11	No path	

We proceed to extend these paths. The following tables summarise the results at each
time step.

State at time 2	Shortest path to state	Length
00	00 → 00 → 00	3
01	00 → 10 → 01	2
10	00 → 00 → 10	3
11	00 → 10 → 11	0

State at time 3	Shortest path to state	Length
00	$00 \rightarrow 10 \rightarrow 01 \rightarrow 00$	2
01	$00 \rightarrow 10 \rightarrow 11 \rightarrow 01$	1
10	$00 \rightarrow 00 \rightarrow 00 \rightarrow 10$	4
11	$00 \rightarrow 10 \rightarrow 11 \rightarrow 11$	1

State at time 4	Shortest path to state	Length
00	$00 \rightarrow 10 \rightarrow 11 \rightarrow 01 \rightarrow 00$	2
01	$00 \rightarrow 10 \rightarrow 11 \rightarrow 11 \rightarrow 01$	1
10	No path	
11	No path	

Extending the two paths in the last table above to state 00 gives

$$00 \rightarrow 10 \rightarrow 11 \rightarrow 01 \rightarrow 00 \rightarrow 00$$

with length 3 and

$$00 \rightarrow 10 \rightarrow 11 \rightarrow 11 \rightarrow 01 \rightarrow 00$$

with length 1. We take the latter as the shortest path, with output sequence 1110110000, which is the correct output sequence. Deleting every second bit gives us 11100, which is the correct input sequence.

$\square$

When the Viterbi algorithm is applied to the output of a binary convolutional encoder with a memory holding $(M + 1)$ k-bit blocks, the decoding process will involve a FSM with 2^M states, and the complexity of the algorithm will grow exponentially with M. In contrast, the complexity of the algorithm only grows linearly with L, the number of blocks that are encoded before the FSM is driven back to the initial state. This means that it is possible to construct Viterbi decoders for codes with values of L in the thousands, but it is not practical to construct Viterbi decoders for M equal to 10 or more. Alternative methods are used to decode convolutional codes with large values of M.

9.6   Sequential Decoding

Sequential decoding algorithms are an alternative to the Viterbi algorithm for decoding convolutional codes with large values of M. They are based on representations of the encoding process using *tree diagrams*, instead of trellis diagrams.

A tree diagram is constructed by establishing a starting node corresponding to the initial state. The transitions from one state to another in response to an input are represented by branches between nodes. By convention, an upward branch represents a transition following an input of 0 and a downward branch represents a transition following an input of 1. If the branches are labelled with the outputs, tracing through the tree will give a list of all possible output sequences.

EXAMPLE 9.12

Consider the convolutional encoder of Figure 9.1, for which $M = 2$. If we take $L = 3$, so that the FSM returns to the initial state 00 at time 5, the possible sequences of states and the associated output sequences are shown in Figure 9.9. Transitions resulting from an input of 0 are shown as upwards branches, and those resulting from an input of 1 are shown as downwards branches. The nodes of the tree are labelled with the states, starting with 00, and the branches of the tree are labelled with the output associated with the corresponding transition between states.

If we trace through the diagram, we see that there are only eight possible sequences of states, corresponding to the eight possible input sequences that are three bits long. The following table shows the input sequences, the resulting sequences of states and the associated output sequences.

Input sequence	Sequence of states	Output sequence
00000	$00 \rightarrow 00 \rightarrow 00 \rightarrow 00 \rightarrow 00 \rightarrow 00$	0000000000
00100	$00 \rightarrow 00 \rightarrow 00 \rightarrow 10 \rightarrow 01 \rightarrow 00$	0000110101
01000	$00 \rightarrow 00 \rightarrow 10 \rightarrow 01 \rightarrow 00 \rightarrow 00$	0011010100
01100	$00 \rightarrow 00 \rightarrow 10 \rightarrow 11 \rightarrow 01 \rightarrow 00$	0011100001
10000	$00 \rightarrow 10 \rightarrow 01 \rightarrow 00 \rightarrow 00 \rightarrow 00$	1101010000
10100	$00 \rightarrow 10 \rightarrow 01 \rightarrow 10 \rightarrow 01 \rightarrow 00$	1101100101
11000	$00 \rightarrow 10 \rightarrow 11 \rightarrow 01 \rightarrow 00 \rightarrow 00$	1110000100
11100	$00 \rightarrow 10 \rightarrow 11 \rightarrow 11 \rightarrow 01 \rightarrow 00$	1110110001

Figure 9.9 is equivalent to the trellis diagram of Figure 9.8 and the information in the table above can also be derived from that diagram.

⬜

Sequential decoding algorithms, like the Viterbi algorithm, attempt to find the code word that best approximates the received sequence of bits. To do this, it requires a means of comparing the sequences generated by traversing the tree with the received word. Because the best path through the tree is built up in stages, this comparison has to work for paths that terminate at any node, and not just the paths that terminate at the terminal nodes.

A path through the tree will generate a sequence of k-bit blocks. A path $\mathbf{p}$ that includes $(n + 1)$ nodes will generate n blocks, $p_1, p_2, \ldots, p_n$. This has to be compared with a received sequence, $\mathbf{q}$, which consists of $(L + M)$ k-bits blocks, q_1,

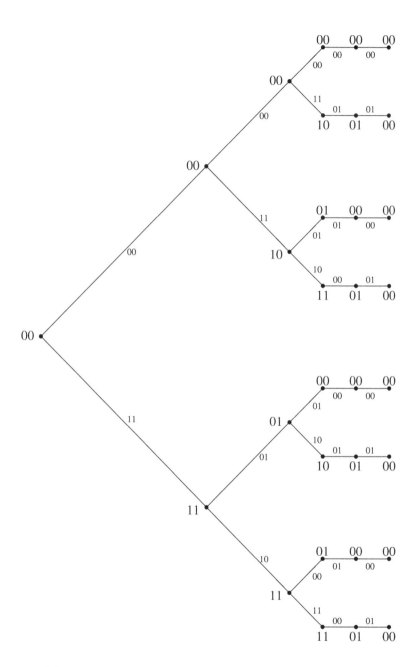

FIGURE 9.9
Tree diagram showing states and output sequences.

$q_2, \ldots, q_{L+M}$. The comparison cannot be carried out in general; some assumptions about the characteristics of the process that may have corrupted $\mathbf{q}$ (that is, the channel) must be made.

If we assume that the corrupting process is a discrete memoryless channel, then we can compute $P(q_j|p_j)$, the probability that the block q_j is received, given that p_j is transmitted. We can also compute the probability that a given q_j is received. This is just the sum of the $P(q_j|p_j)P(p_j)$ for all possible values of p_j.

The appropriate measure for comparing sequences generated by a path $\mathbf{p}$ of length n through the tree with a received string $\mathbf{q}$ is the *Fano metric* ([3], pg. 317)

$$m(\mathbf{p}) = \sum_{j=1}^{n} \left(\log \frac{P(q_j|p_j)}{P(q_j)} - R \right), \tag{9.1}$$

where $R = k/n$ is the rate of the encoder.

EXAMPLE 9.13

In the tree diagram of Figure 9.9, we have $k = 2$ and $R = \frac{1}{2}$. The conditional probabilities $P(q_j|p_j)$ for a binary symmetric channel with the probability of error equal to $\frac{1}{4}$ are given in the table below.

	$p_j = 00$	$p_j = 01$	$p_j = 10$	$p_j = 11$
$q_j = 00$	9/16	3/16	3/16	1/16
$q_j = 01$	3/16	9/16	1/16	3/16
$q_j = 10$	3/16	1/16	9/16	3/16
$q_j = 11$	1/16	3/16	3/16	9/16

If we assume that $P(p_j) = \frac{1}{4}$ for all the p_j, then $P(q_j) = \frac{1}{4}$ for all the q_j.

If $\mathbf{q} = 01001000111$, we can calculate the Fano metric for various paths through the tree as follows.

The path $\mathbf{p} = 00 \rightarrow 00 \rightarrow 00$ generates the output sequence 0000. We have $p_1 = 00$, $p_2 = 00$. Since $q_1 = 01$ and $q_2 = 00$,

$$m(\mathbf{p}) = \left[\log \left(\frac{3/16}{1/4} \right) - 0.25 \right] + \left[\log \left(\frac{9/16}{1/4} \right) - 0.25 \right]$$

$$= \log \left(\frac{3}{4} \right) + \log \left(\frac{9}{4} \right) - 0.5$$

$$= \log \left(\frac{27}{16} \right) - 0.5. \tag{9.2}$$

The path $\mathbf{p} = 00 \rightarrow 10 \rightarrow 11 \rightarrow 01$ generates the output sequence 111000. We have $p_1 = 11$, $p_2 = 10$, $p_3 = 00$. Since $q_1 = 01$, $q_2 = 00$, and $q_3 = 10$,

$$m(\mathbf{p}) = \left[\log \left(\frac{3/16}{1/4} \right) - 0.25 \right] + \left[\log \left(\frac{3/16}{1/4} \right) - 0.25 \right] + \left[\log \left(\frac{3/16}{1/4} \right) - 0.25 \right]$$

$$= \log\left(\frac{3}{4}\right) + \log\left(\frac{3}{4}\right) + \log\left(\frac{3}{4}\right) - 0.75$$

$$= \log\left(\frac{27}{64}\right) - 0.75. \tag{9.3}$$

$\square$

The sequential decoding algorithms are tree search algorithms which attempt to find the path through the tree for which the value of the Fano metric is the greatest. The simplest of these algorithms is the *stack algorithm*. It constructs a stack of paths through the tree in an iterative fashion.

The stack algorithm is initialised with a single path in the stack, the trivial path of length 0 that starts and ends at the root node of the tree. The metric of this path is defined to be zero.

At each step, the path at the top of the stack is selected and all the possible extensions of this path by a single node are constructed. The path at the top of the stack is deleted and its extensions are added to the stack. The insertions are made at points in the stack determined by the values of the Fano metric of the extensions. At all times the stack is arranged in decreasing order of the value of the Fano metric.

The extension process continues until a path that reaches a terminal node comes to the top of the stack. This path gives the best approximation of the received sequence.

In implementations of the stack algorithms, the Fano metric has to be computed for each extension of the path at the top of the stack at each stage. The complete list of all the extensions that have been constructed must also be retained throughout the search. This means that the size of this list can grow exponentially.

The *Fano algorithm* is an alternative to the stack algorithm. It prunes the search tree so as to ensure that the list of paths that has to be maintained does not grow unmanageably large.

Convolutional encoders and decoders can be constructed using any finite field not just the binary field. The details of the construction in terms of Finite State Machines and the algebraic theory have their counterparts in these cases. The Viterbi algorithm and the sequential decoding algorithms can also be implemented, though the complexity of the implementation increases.

9.7 Trellis Modulation

Convolutional encoders have been adopted very widely in telecommunications applications, such as the construction of data communications modems. The modulation schemes that have been developed for telecommunications provide natural alphabets which can be mapped to finite fields. These applications are quite specialised. A brief description of how these schemes work will be given here. A more detailed treatment may be found in the book by Schlegel [4].

Telecommunications systems use electromagnetic waves to transmit information from place to place. A high-frequency *carrier wave* is used to carry the information. The carrier wave is a continuous sinusoid that may be generated by an oscillator circuit, a laser or some other device. The carrier wave itself conveys no information. To transmit information, the carrier wave must be *modulated* in some way which reflects the desired information content. There are many modulation schemes that have been devised for the transmission of digital information.

The simplest modulation scheme is an *amplitude modulation* scheme which simply turns the carrier wave off and on for specified periods of time to represent 0s and 1s. From a mathematical point of view, this is equivalent to multiplying the carrier wave at each instant by a wave that consists of a series of rectangular pulses. This is not an efficient way of modulating the carrier wave. It tends to require large amounts of bandwidth relative to the information content of the signal being transmitted.

In addition to their amplitudes, carrier waves have two other parameters that can be used as the basis of modulation schemes. These are the *frequency* and the *phase*. Telecommunications systems may use modulation schemes that involve all three parameters.

The common modulation schemes include *binary phase-shift keying* (BPSK), *quadrature phase-shift keying* (QPSK), and *quadrature-amplitude modulation* (QAM).

BPSK, like simple amplitude modulation, encodes binary values. QPSK has four alternative states, and so can encode two bits of information in a modulated waveform. 8-PSK and 8-AMPM have eight alternative states and can encode three bits of information in a modulated waveform. 16-QAM and 64-QAM can encode four and six bits of information, respectively, in a modulated waveform.

Constellation diagrams are often used to describe modulation schemes. Each point in a constellation diagram represents a combination of two sinusoidal waveforms. The waveforms have the same frequency but differ in phase by $\pi/2$. The coordinates of the point represent the relative amplitudes of the two component waveforms. Figures 9.10 to 9.15 show the constellation diagrams of the six modulation schemes mentioned in the previous paragraph.

For error-correcting purposes, these modulation schemes are used in conjunction with a convolutional encoder. An encoder whose output consists of k-bit blocks is used with a modulation scheme that can encode k bits of information. Each block that is output by the encoder is translated to one of the states of the modulation scheme, which then determines the waveform of the carrier wave for a specified period. At the receiving end, the waveform is translated back to a block of bits and the sequence of blocks is decoded using a Viterbi decoder or a sequential decoder.

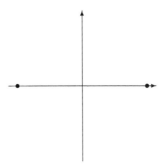

FIGURE 9.10
Constellation diagram for BPSK.

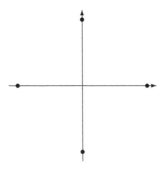

FIGURE 9.11
Constellation diagram for QPSK.

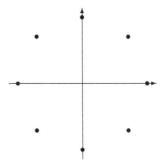

FIGURE 9.12
Constellation diagram for 8-PSK.

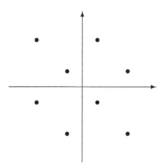

FIGURE 9.13
Constellation diagram for 8-AMPM.

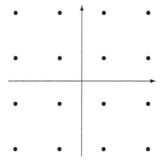

FIGURE 9.14
Constellation diagram for 16-QAM.

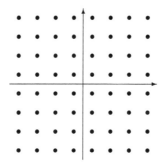

FIGURE 9.15
Constellation diagram for 64-QAM.

This combination of convolutional coding and complex modulation schemes makes it possible to transmit information over telecommunications channels at high bit rates with low error rates. Most of the improvement in the performance of modems in the past decade has been the result of the adoption of more complex error correction schemes which have made it possible to increase transmission rates without replacing the existing telecommunications infrastructure.

9.8   Turbo Codes

A recent development in the field of error-correcting codes has been the development of *Parallel Concatenated Coding Schemes*, also known as *Turbo codes*. These use two or more convolutional encoders operating in parallel to generate codes whose performance approaches the theoretical limits imposed by Shannon's Theorem.

The enhanced performance of Turbo codes arises from the introduction of an *interleaver* which changes the order of the input bits so that the two encoders operate on different input sequences, one of which is a permutation of the other. One consequence of the use of the interleaver is the improved error-correcting performance of the Turbo codes; another consequence is that the Viterbi and sequential decoding algorithms cannot be used and new decoding algorithms have had to be devised.

The input to the encoder is a sequence of blocks of bits, with N bits in each block. The bits in each block are input unchanged to the first convolutional encoder. They are also input to the interleaver, which permutes them according to some predetermined scheme and sends the permuted sequence to the input of the second convolutional encoder. The input bits are also sent to the output of the encoder and combined with the outputs of the two convolutional encoders. This results in a code with rate $1/3$. The rate can be reduced to $1/2$ by *puncturing* the output sequence, that is, by taking output bits from each of the convolutional encoders in turn. The output sequence therefore consists of blocks of two or three bits, made up of a *systematic bit*, which is the same as the input to the encoder, and one or two parity check bits. These are concatenated into the received sequence consisting of $2N$ or $3N$ bits.

EXAMPLE 9.14

A Turbo encoder is constructed from two convolutional encoders and an interleaver that operates on blocks of five bits. The FSM of the first encoder is described by the following table.

Original state	Input	Output	Final state
00	0	0	00
	1	1	10
01	0	1	00
	1	0	10
10	0	1	01
	1	0	11
11	0	0	01
	1	1	11

The FSM of the second encoder is described by the table below.

Original state	Input	Output	Final state
0	0	0	0
	1	1	1
1	0	1	1
	1	0	0

The interleaver permutes blocks of five bits according to the mapping:

$$(b_0, b_1, b_2, b_3, b_4) \mapsto (b_1, b_3, b_0, b_4, b_2).$$

The input sequence 11100 will drive the first encoder back to the initial state. The interleaver will permute this to 10101. Note that this will not drive the second encoder back to the initial state. It is possible to choose the permutation operation in such a way as to drive both encoders to initial state at the end of the block, but this restricts the choice of interleaver and does not produce any benefit.

The following table summarises the operation of the encoder when the block 11100 is input to it.

State of first FSM	Input to first FSM	Output of first FSM	State of second FSM	Input to second FSM	Output of second FSM	Concatenated output
00	1	0	0	1	1	101
10	1	0	1	0	1	101
11	1	1	1	1	1	111
11	0	0	0	0	0	000
01	0	1	0	1	1	011
00			1			

The full output sequence is therefore 101101111000011. If it is punctured, the resulting output sequence is 1011110001.

▯

The originators of Turbo codes proposed an iterative decoding scheme for them. Details can be found in the original paper [1] or in the book by Schlegel [4].

9.9 Exercises

1. Write down the output produced by the convolutional encoder of Figure 9.1 when the input is the sequence 0000111100110011001101010101.

2. Write down the output produced by the convolutional encoder of Figure 9.4 when the input is the sequence 00 00 11 11 00 11 00 11 01 01 01 01.

3. Write down the output produced by the convolutional encoder of Figure 9.16 when the input is the sequence 1100110001011101.

4. One bit of the output from the convolutional encoder of Figure 9.16 was corrupted to produce the following sequence:

    101 100 011 000 011 001 111 011.

   Find the bit that was corrupted and write down the input sequence.

5. Give a tabular description of the operation of the FSM that underlies the convolutional encoder of Figure 9.16.

6. Give a diagrammatic description of the operation of the FSM that underlies the convolutional encoder of Figure 9.16.

7. Draw trellis diagrams to show the sequences of states of the FSM of the convolutional encoder of Figure 9.1 that result from the following input sequences:

   (a) 11001100;
   (b) 11110101;
   (c) 10101010;
   (d) 01010101;
   (e) 11111000.

8. Draw trellis diagrams to show the sequences of states of the FSM of the convolutional encoder of Figure 9.4 that result from the following input sequences:

   (a) 11 00 11 00;
   (b) 11 11 01 01;
   (c) 10 10 10 10;
   (d) 01 01 01 01;
   (e) 11 11 10 00.

9. Draw trellis diagrams to show the sequences of states of the FSM of the convolutional encoder of Figure 9.16 that result from the following input sequences:

(a) 11001100;

(b) 11110101;

(c) 10101010;

(d) 01010101;

(e) 11111000.

10. Find the generator matrix of the convolutional encoder of Figure 9.16.

11. The following sequences are output sequences from the convolutional encoder of Figure 9.1 in which one bit has been corrupted:

(a) 11 01 11 11 01 01;

(b) 11 01 10 11 00 01;

(c) 11 00 00 10 01 01;

(d) 11 10 11 10 00 01.

Use the Viterbi algorithm to correct the sequences and recover the input sequences.

12. The convolutional encoder of Figure 9.16 has $M = 3$. Draw the tree diagram for $L = 2$, that is, for all sequences of six states starting and ending with 000.

13. A Turbo encoder is constructed from two convolutional encoders whose FSMs are described by the tables below.

Original state	Input	Output	Final state
00	0	0	00
	1	1	10
01	0	1	00
	1	0	10
10	0	1	01
	1	0	11
11	0	0	01
	1	1	11

Original state	Input	Output	Final state
00	0	0	00
	1	0	10
01	0	1	00
	1	1	10
10	0	1	01
	1	1	11
11	0	0	01
	1	0	11

The interleaver of the encoder operates on blocks of seven bits according to the mapping

$$(b_0, b_1, b_2, b_3, b_4, b_5, b_6) \mapsto (b_2, b_4, b_6, b_0, b_5, b_1, b_3).$$

The output of the Turbo encoder at each instant is the concatenation of the input bit and the two output bits generated by the two convolutional encoders. Find the output sequences that are produced by the Turbo encoder in response to the following input sequences:

(a) 1101100;

(b) 1010100;

(c) 0101000;

(d) 1111100.

*14. Consider the convolutional encoder whose generator matrix is

$$G = \begin{bmatrix} 1 + X & 1 + X^2 \end{bmatrix}.$$

Show that this encoder can be realised with three shift registers and describe the operation of the underlying FSM in tabular and diagrammatic form. What is the output sequence of the encoder when the input sequence is an infinite sequence of 1s?

*15. Encoders with the property that an input sequence with an infinite number of 1s generates an output sequence with a finite number of 1s are called *catastrophic encoders*, as the corruption of a finite number of bits in the received code word can cause an infinite number of decoding errors. Show that an encoder is catastrophic if there is a finite cycle of states which includes states other than the initial state with the property that the output from all of the states consists only of 0s. Find such a cycle of states for the encoder of the previous Exercise. Show that the encoder whose generator matrix is

$$G = \begin{bmatrix} 1 + X^2 & X + X^3 \end{bmatrix}.$$

is catastrophic by finding an appropriate cycle of states and find two input sequences with an infinite number of 1s that generate output sequences with a finite number of 1s.

9.10   References

[1] C. Berrou, A. Glavieux, and P. Thitimajshima, Near Shannon Limit Error-Correcting Coding and Decoding: Turbo-Codes (1), Proceedings of the 1993 IEEE International Conference on Communications, Geneva, 1993, 1064–1070.

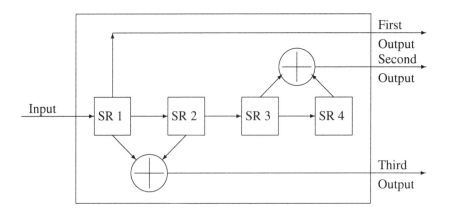

FIGURE 9.16
A convolutional encoder with one input and three output bits.

[2] G. D. Forney, Jr., Convolutional codes I: Algebraic structure, *IEEE Transactions on Information Theory*, IT-16(6), 720–738, November 1970,

[3] R. J. McEliece, *The Theory of Information and Coding,* 2nd ed., Cambridge University Press, Cambridge, 2002.

[4] C. Schlegel, *Trellis Coding,* IEEE Press, New York, 1997

[5] A. J. Viterbi, Error Bounds for Convolutional Codes and an Asymptotically Optimum Decoding Algorithm, *IEEE Transactions on Information Theory*, IT-13, (2), 260–269, April 1967.

Index

Milton Keynes UK
Ingram Content Group UK Ltd.
UKHW031126141024
449569UK00006B/415